CHEMICAL ENGINEERING METHODS AND TECHNOLOGY

TITANIUM DIOXIDE

CHEMICAL PROPERTIES, APPLICATIONS AND ENVIRONMENTAL EFFECTS

Chemical Engineering Methods and Technology

Additional books in this series can be found on Nova's website under the Series tab.

Additional e-books in this series can be found on Nova's website under the e-book tab.

CHEMICAL ENGINEERING METHODS AND TECHNOLOGY

TITANIUM DIOXIDE

CHEMICAL PROPERTIES, APPLICATIONS AND ENVIRONMENTAL EFFECTS

JERRI BROWN
EDITOR

nova publishers
New York

For permission to use material from this book please contact us:
Telephone 631-231-7269; Fax 631-231-8175
Web Site: http://www.novapublishers.com

Library of Congress Cataloging-in-Publication Data

Titanium dioxide : chemical properties, applications, and environmental effects/editors, Jerri Brown.
pages cm. -- (Chemical engineering methods and technology)
Includes bibliographical references and index.
ISBN 978-1-63321-391-3 (hardcover)
1. Titanium dioxide--Industrial applications. 2. Titanium dioxide--Environmental aspects. I. Brown, Jerri.
TP245.T85T58 2014
661'.0512--dc23

2014023754

Published by Nova Science Publishers, Inc. † New York

CONTENTS

PREFACE

Engineered nanomaterials (ENMs) have greatly impacted human society as well as natural environments over the past decade, in consequence of their production on a massive scale to support the emerging market for nanotechnology. Among these ENMs, titanium dioxide (TiO_2) nanoparticles have been extensively employed in the development of novel solutions for photocatalytic applications in environmental remediation. This book discusses the fate of TiO_2 and the effects in natural environments. Some other topics discussed include indoor environment control with functional mortars; the production, properties and applications of TiO_2; dye-sensitized TiO_2 for gas detection; TiO_2 induced photocatalytic reduction for organic synthesis; and the effects of TiO_2 nanoparticles on the inflammatory process.

Chapter 1 - Engineered nanomaterials (ENMs) have greatly impacted human society as well as natural environments over the past decade, in consequence of their production on a massive scale to support the emerging market for nanotechnology. Among these ENMs, TiO_2 nanoparticles have been extensively employed in the development of novel solutions for photocatalytic applications in environmental remediation. Inevitably, TiO_2 nanoparticles will be released into natural aquatic environments via discharges from sewage treatment, factory releases, and leaks or spills during transportation or use. Thus, they pose potential risks to living organisms in aquatic ecosystems, creating the urgent need for assessing and understanding the consequent environmental hazards. In this chapter, the authors critically reviewed the physicochemical properties and the fate of TiO_2 nanoparticles in aquatic environments and the subsequent impact on aquatic biota. Their intimate interactions with biotic and abiotic compartments in aquatic ecosystems, also known as nano-bio-eco interfaces, are noted and discussed. On the basis of literature reviewed, the proper assessment of their effects on specific model organisms could provide precise predictions of their adverse effects, as well as concepts for the design of sustainable-use, green nanoparticles.

Chapter 2 - In developed countries, people spend most of the time inside buildings, either their offices or houses. Indoor air quality has a strong impact on wellbeing since unhealthy or uncomfortable environments can cause illness and low productivity. The selection of materials that contribute to improve indoor air quality improves significantly the building quality and sustainability. Mortars, because they are one of the most commonly used materials, represent a strong environmental impact in construction. Preparing mortars with new and innovative functionalities, that promote energy savings and increase air quality, help to develop buildings with better environmental performance. However, adding a new functionality comprises the introduction of new additives into the mortar matrix, causing microstructural modifications that influence fresh and hardened state properties. Adding new additives always imply a compromise between functionality and durability. In this chapter, was investigated the introduction of a photocatalytic additive and a phase change material, in single and binary formulations. Latent heat storage, self-cleaning and depolluting properties were studied and the impact of the additives in the mortar final properties was assessed. This chapter proved that functional mortars maintain good fresh and hardened state properties improving sustainability in new and historical buildings.

Chapter 3 - Heterogeneous photocatalytic processes have been gaining increasing commercial interest worldwide mostly in the fields of environmental cleanup and energy. In particular, photocatalytic water treatment is considered a cost-effective and environmentally benign technology enabling the destruction of a great variety of hazardous emerging pollutants. One of the most challenging topics in heterogeneous photocatalysis is the development of efficient catalysts working under solar light irradiation. Titanium dioxide (TiO_2), the benchmark catalyst for this type of applications, requires UV-light excitation, which only accounts for 5% of the incoming solar energy. Loading TiO_2 with carbon materials and/or noble metal nanoparticles are effective techniques used to increase the photonic efficiency and to extend light absorption of TiO_2 towards the visible spectral range. In this chapter, the authors will describe some of their recent results on the development of TiO_2-based photocatalysts, namely TiO_2-activated carbon fiber composites and gold-loaded TiO_2, for the degradation of the antibiotic ciprofloxacin in water.

Chapter 4 - This study examines the photocatalytic activity (PCA) of coatings produced by atmospheric plasma spraying. The applied spraying tool was water-stabilized plasma gun. The TiO_2–Fe_2O_3 powders with and without Na_2SiO_3 additive were produced as agglomerates suitable for feeding into the plasma jet. The TiO_2–Fe_2O_3–Na_2SiO_3 coating was also post-treated in a furnace at 750°C for 2 hours. The composition of coatings was analyzed by X-ray diffraction. Photocatalytic degradation of butane under UV and under visible light (VIS) was quantified for all coatings. The coatings show a lamellar structure, as is typical for this process. Anatase titania from the feedstock powder was converted to rutile phase in all coatings but the doping with Na_2SiO_3 decreased the extent of this transformatuion. The coating FeT-VS was photocatalytically active especially under UV, whereas under visible light also FeT and FeT-VS-ann coatings were sufficiently active.

Chapter 5 - The article is an overview on titanium dioxide that describes the various production processes with beneficiation and upgrading of source materials and the important properties that led to potential applications. The article discusses on significant industrial production processes through several subsections. Following through the description of different natural and synthetic source materials, a discussion is given on the beneficiation and upgrading of ores. In that, several processes of upgrading the titanium slags have been given in detail through respective flowsheets. Existing commercial processes of production of TiO_2 for example, the sulphate and the chloride processes are discussed in terms of their advantages and disadvantages.

The second part of the article discusses on various properties of TiO_2 for example, structural, electronic and photocatalytic properties and their important applications. The artificial phases of the material initiated through structural phase transitions by the application of high pressure and temperature are discussed because of their implications in creating the doped materials for the possibility of having an altered semiconducting band configuration.

The fundamentals of the photocatalytic property and the advantage of using the nanoparticles have been discussed and elucidated through several applications. The synthesis of TiO_2 nanotube and the self-organized system of nanotube arrays are described comprehensively. Some properties of this amorphous material and its possible applications are indicated.

Chapter 6 - Porous Titanium Dioxide (TiO_2) electrodes having high dye adsorption characteristics is of interest biomolecule sensors and dye-sensitized solar cells. Dye adsorption characteristics directly affect the sensitivity and efficiency of devices. In recent years, the demand for gas sensors for safety control requirements and environmental monitoring has expanded enormously. There is an increasing need to detect low concentrations of air pollutants such as ammonia, amines, which are frequently encountered in everyday life. Films based on the semiconductor, titanium dioxide (TiO_2), have been investigated in this chapter. At low humidity, ammonia and amines can be ionized by the Cyanidin on the TiO_2 film and thereby decrease in the proton concentration at the surface. Consequently, more positively charged holes at the surface of the TiO_2 have to be extracted to neutralize the adsorbed cyanidin and water film. The resistance response to ammonia of the sensors was nearly independent on temperature from 10 to 50°C.

Chapter 7 - Titanium dioxide, as a chemically stable, nontoxic, highly efficient, and relatively inexpensive photocatalyst, has been attached much attention in recent years. However, practical applications of TiO_2 are limited because of the problems of low recycling efficiency and visible-light photocatalytic activity of TiO_2 fine particles. Natural zeolite supported ions doped nano TiO_2 photocatalyst, a photocatalyst loading nano TiO_2 on the surface of zeolite and doping different ions into the crystallite of TiO_2, which shows higher recycling efficiency, adsorption and photocatalytic activities than that of pure TiO_2, seems to be a promising photocatalyst for practical application. In this chapter, the authors introduced and concluded the preparation, characterization, photocatalytic activities, and the related mechanisms of natural zeolite supported ions doped TiO_2 photocatalysts based on their previous works. It shows that i) the photocatalysts can be successfully prepared by $TiCl_4$ hydrolysis method; ii) the zeolite substrate facilitates the formation of anatase, restrains the transformation of anatase to rutile and the growth of TiO_2 crystallite, and also improves the adsorption activity of the as-prepared photocatalysts; iii) the doped ions can diffuse into TiO_2 lattice and give TiO_2 the capacity to absorb light at lower energy levels, and certain ions with proper doping concentration can evidently improve photocatalytic activities of the photocatalysts; iv) Cr doped sample shows the highest photocatalytic activity than that of the other ions doped samples due to its lower energy gap and recombination probability of photo-generated electron-hole pair; v) the photocatalysts show photocatalysis-adsorption co-effect on the organic molecules, and this co-effect is helpful for photocatalytic degradation of the organic contaminant. The current problems and future developments of this photocatalyst are also prospected.

Chapter 8 - This chapter summarizes the synthetically useful photocatalytic reduction on a semiconductor particle titanium dioxide (TiO_2). TiO_2 has the potential to induce the reductive chemical transformation in the absence of oxygen (O_2), especially by choosing the suitable sacrificial hole scavenger such as alcohols or amines, since electrons photo-generated in the conduction band (CB) or those trapped at surface defect sites can transfer into an organic substrate adsorbed on a photocatalyst. The TiO_2-induced reduction of aldehydes or ketones gives the corresponding alcohols in the presence of alcohols as hole scavengers. The photocatalytic reduction of nitro compounds on TiO_2 gives the corresponding amines *via* the formation of hydroxylamine intermediates. Additionally, the reduction of nitro compounds was successfully applied into the synthesis of benzimidazoles or tetrahydroquinolines. Pt/TiO_2 is an effective catalyst for the reduction of imines. The reduction of imines was applied into several unique transformations involving both reductive and oxidative processes. The reduction of aromatic cyanides in ethanol gives toluene derivative and triethylamine *via* redox processes.

Chapter 9 - Nanosized titanium dioxide modifications are used in catalysis (for extraction of uranium compounds from sea water and producing hydrogen), sorption (for purification of aqueous media from a wide range of toxic ions), medicine (excellent cleaner of water and air environments from organic pollutants and bacteria), microelectronics (promising for solar cells) and so on. In comparison with other catalysts the semiconductor TiO_2 is chemically and biologically inert and stable to photodegradation, and unlike other disinfection technologies do not form dangerous (mutagenic and carcinogenic) compounds. Among other modifications of titania the most promising for application is anatase.

The known methods to obtain samples with nano-anatase were reviewed; their advantages and disadvantages were noted. The sulfate technique was selected as the easiest to performance and perspective. The samples obtained by this technique were characterized by a complex of methods: X-ray diffraction using X-ray (wide and small-angle scattering) and synchrotron radiation (distributing character of nanoparticles on sizes and the sizes of nano-objects, the composition of the samples, the sizes of coherent scattering regions and the quantity of nano-anatase microstressing), electron diffraction, scanning and transmission electron microscopy (the size of crystallite, nanoparticles, agglomerates and aggregates), quantitative electron probe microanalysis and X-ray fluorescence electron spectroscopy (elemental composition and charge state of components), low-temperature nitrogen adsorption (specific surface area, size of nano-objects, the volume of nano- and ultrananoporos), infrared spectroscopy and Raman spectroscopy.

It was established that the conditions of samples synthesis effect their composition, microstructure and substructure, electronic and surface properties. The possibility of directed synthesis of nano-anatase with necessary physical-chemical properties has been shown.

It was determined that the photocatalytic activity of the obtained samples is comparable to that of the commercial photocatalysts based on titanium dioxide, however, they can be multiple reused in a photocatalytic decomposition reaction of organic dyes without regenerating.

Chapter 10 - Titanium dioxide nanoparticles (TiO_2 NPs) are manufactured in large quantities worldwide for use in a wide range of applications. Recently, the IARC included TiO_2 as a type-2 carcinogenic compound. Additionally, at the nanoscale size, several toxic effects have been reported. Humans may be exposed to TiO_2 NPs through three main pathways: respiratory (airborne nanoparticles), dermal (cosmetics) and oral (food) routes.

After nanoparticles are absorbed, they may be distributed by the blood and lymphatic system to reach different targets in the body. TiO_2 NPs can reach bone, kidney, pancreas, liver, heart, and the nervous system, where they are retained and can become toxic. Although the toxic mechanisms are not well understood, *in vitro* experimental studies have reported cell membrane damage, production of reactive oxygen species (ROS) that may lead to oxidation of biomolecules (i.e., lipids, DNA, and proteins), genotoxicity, and pro-inflammatory effects. The authors previously showed that TiO_2 NPs exerted different effects on different cell types and induced a strong inflammatory response in human endothelial cells (HUVEC), mediating increases in cellular adhesion (endothelial-monocyte) and adhesion molecule expression. TiO_2 NPs also induce oxidative stress, NF-κB translocation, and cell death, indicating a strong toxic effect.

The aim of this chapter is to present a broad description of the main toxic effects of TiO_2 NPs related to inflammatory responses in various cells types and target organs.

Chapter 11 - Indoor air pollution from toxic gaseous chemicals poses adverse effects on human health, comfort and productivity. This is a serious concern nowadays since people spend more time (over 80%) indoors. In addition, recent studies on indoor air quality show that certain gaseous contaminants have greater concentration levels indoor than in the outdoor environment. Strategies to address this problem include increased ventilation and air cleaning. The former involves the transfer of pollutants from indoor to outdoor while conventional air cleaning methods utilize air filtration and adsorption. Meanwhile, heterogeneous photocatalysis, an advanced oxidation process (AOP), can completely degrade gaseous organics to carbon dioxide and water. In the past two decades, photocatalysis has emerged as a promising technology for the purification of indoor air and exhibited greater effectiveness over conventional air cleaning technologies. A catalyst that has taken center stage in the development of this technology is titanium dioxide (TiO_2), also known as titania, mainly due to these advantages: (1) TiO_2 is relatively inexpensive, non-toxic and chemically stable, (2) it promotes ambient temperature oxidation of the major classes of indoor air pollutants, (3) the complete degradation of a broad class range of pollutants can be achieved under certain operating conditions, and (4) chemical additives are not required for the process to proceed. The photogenerated holes of TiO_2 are highly oxidizing while its photogenerated electrons are reducing enough to produce superoxide from dioxygen. However, TiO_2 has a low solar photoconversion efficiency because of the large band gap of 3.2 and 3.0 eV for the anatase and rutile phases, respectively. Thus, TiO_2 can only be excited by irradiation using ultraviolet (UV) light, which covers only ~5% of the light spectrum. The development of photocatalysts exhibiting high reactivity under visible light ($\lambda > 400$ nm) is of great interest because it would allow the main part of the light spectrum to be utilized and photocatalytic activity would be observed even under poor illumination of interior lighting. This requires engineering the band-gap of TiO_2 to less than 3.2 eV. Studies on TiO_2 modification have been carried out to increase its photocatalytic activity under visible light illumination through doping metals and non-metals onto TiO_2. This chapter offers a review of the status of research related to (1) indoor air pollution by gaseous contaminants such as formaldehyde, benzene, toluene and nitrogen oxide; and (2) the synthesis and application of visible-light active TiO_2 by metal and non-metal doping for the treatment of these gaseous pollutants in indoor air.

Chapter 12 - Synthesis of nanostructured anatase TiO_2 and its composite by ionothermal synthesis (sol–gel method containing ionic liquid (IL) under microwave (MW) irradiation and their water treatment applications were reported in this short review.

In ionothermal synthesis, the ionic liquid plays a hydrothermal role in driving surface crystallization of amorphous TiO_2 to the anatase phase by retaining a suitable amount of water through a dissolution-crystallization mechanism. MW heating can lead to rapid temperature increase of chemical reactions. The effects of MW heating on anatase crystallization are diverse and dependent on the initial water content, MW irradiation mode as well as IL type and its content.

The preparation of two kinds of TiO_2 composite is also reported. Firstly, an innovative one-step immobilization of TiO_2 nano-particles on organic polymer (PMMA) substrate under

IL/MW condition at ambient condition is described. In this technology, a sol–gel reaction and anatase crystallization are simultaneously occurred at the specific sites of the polymer surface. These sites are the tiny cavities of the rough surface resulting from the softening and swelling effect of an alcohol, such as isopropyl alcohol, on the polymer surface under microwave irradiation. The experimental data show that the anatase TiO_2 prepared by this novel method have high immobilization stability and photocatalytic water treatment performance. Secondly, the nanostructured anatase TiO_2/activated carbon (AC) composite material for capacitive deionization (CDI) electrode is prepared by a two-step sol-gel method under IL/MW condition In comparison to the pristine AC electrode, higher specific capacitance was observed for the nanostructured anatase TiO_2/AC composite electrode, especially when the composite was prepared with a molar ratio of titanium tetraisopropoxide/H_2O equal to 1:15. The accompanying capacitive deionization result indicates that the decay of electrosorption capacity of this composite electrode is insignificant after five cycle tests. It means that the ion electrosorption–desorption becomes a reversible process.

In: Titanium Dioxide
Editor: Jerri Brown

ISBN: 978-1-63321-391-3

Chapter 1

ENGINEERED TiO_2 NANOPARTICLES: THEIR FATE AND EFFECTS IN NATURAL AQUATIC ENVIRONMENTS

Xiaojia He and Huey-Min Hwang
Department of Biology, Jackson State University, Jackson, MS, US

ABSTRACT

Engineered nanomaterials (ENMs) have greatly impacted human society as well as natural environments over the past decade, in consequence of their production on a massive scale to support the emerging market for nanotechnology. Among these ENMs, TiO_2 nanoparticles have been extensively employed in the development of novel solutions for photocatalytic applications in environmental remediation. Inevitably, TiO_2 nanoparticles will be released into natural aquatic environments via discharges from sewage treatment, factory releases, and leaks or spills during transportation or use. Thus, they pose potential risks to living organisms in aquatic ecosystems, creating the urgent need for assessing and understanding the consequent environmental hazards. In this chapter, we critically reviewed the physicochemical properties and the fate of TiO_2 nanoparticles in aquatic environments and the subsequent impact on aquatic biota. Their intimate interactions with biotic and abiotic compartments in aquatic ecosystems, also known as nano-bio-eco interfaces, are noted and discussed. On the basis of literature reviewed, the proper assessment of their effects on specific model organisms could provide precise predictions of their adverse effects, as well as concepts for the design of sustainable-use, green nanoparticles.

Keywords: Engineered TiO_2 nanoparticles; fate; effects; aquatic environments

INTRODUCTION

Engineered nanomaterials (ENMs) have shown great potential in novel applications for rapid development of human society. However, they have also been recognized as an emerging class of hazards, with potentially large-scale ecological impacts on aquatic biota. With the massive production in the past decade, engineered titanium dioxide (TiO_2) nanoparticles are the most extensively used ENMs to date, yet their fate and effects in the natural aquatic environments have not been thoroughly investigated or understood. It has been strongly suggested that high concentrations of TiO_2 nanoparticles may occur in sewage and wastewater treatment (Westerhoff et al., 2011). Furthermore, investigation of estimated environmental concentrations suggests that of the most commonly used ENMs in the United States and Europe, TiO_2 nanoparticles have the highest concentrations in surface waters and the annual increase in sludge-treated soil reached 89 μg/kg (Gottschalk et al., 2009). Although the controversy over the safe use of TiO_2 nanoparticles has not subsided, increasing evidence has unveiled their potential risks to aquatic ecosystems, including freshwater, estuarine, and marine ecosystems, the expected destinations of industrial discharges of ENMs.

The knowledgebase regarding fate and behavior of TiO_2 nanoparticles in aquatic ecosystems, their trophic transfer in food webs, and their biomagnification potential is limited. In order to determine the ecotoxicological potential of TiO_2 nanoparticles in the aquatic environment, it is crucial to first determine the actual bioavailability. This, in turn, requires an assessment of their fate and behavior in abiotic and biotic environments, with consideration given to environmentally relevant exposure conditions. Indeed, the quantity and form of TiO_2 nanoparticles released into aquatic environments and their changes over time are recommended to be analyzed in order to assess their ecological hazards and risks. Despite the substantial involvement of microorganisms (Heinlaan et al., 2008, Kasemets et al., 2009, Pathakoti et al., 2013, Simon-Deckers et al., 2009), recent findings suggest that assessments of risk of TiO_2 nanoparticles to pelagic (Dalai et al., 2013, Hall et al., 2009, He et al., 2014a, Ma et al., 2012) and benthic (Barmo et al., 2013, D'Agata et al., 2014, Li et al., 2014) species are urgently needed. More new studies have evidenced the possibility of using model or specific organisms to predict the fate, behavior, and ultimate toxicity of ENMs. In other words, comprehensive research and the collection of ecotoxicological data on the interaction of TiO_2 nanoparticles with such organisms can fill current gaps in the knowledge of the requirements for their design, application, and disposal. Although it is interesting and necessary to know the toxic effects of TiO_2 nanoparticles on certain organisms, the effects of bioaccumulation and biomagnification within an ecosystem cannot be ignored when evaluating hazards and risks from TiO_2 nanoparticles. Thus, acute, as well as long-term exposures, along with more research into bioaccumulation are needed to provide more precise guidance to reveal their actual effects.

While the ongoing debate over potential hazards and risks of TiO_2 nanoparticles is seemingly endless, The International Agency for Research on Cancer (IARC, 2010), U.S. Environmental Protection Agency (EPA, 2010), U.S. National Institute for Occupational Safety and Health (NIOSH, 2011) and the European Commission Scientific Committee on Consumer Safety (SCCS) (2013) have issued some safety concerns about the novel material. Additionally, they have recommended TiO_2 nanoparticles for comprehensive examination to provide data for future assessments of their potential ecological and health implications.

It is worthwhile to note that both IARC and NIOSH recently suggested classifying TiO_2 nanoparticles as potential carcinogens. NIOSH recommends exposure limits of 0.3 mg/m^3 for ultrafine (including engineered nanoscale) TiO_2 as the time-weighted average (TWA) concentration for up to 10 hours per day during a 40-hour work week (U.S. NIOSH 2011).

With uncertainties being as they are, we can neither over-exaggerate nor underestimate the adverse impacts of TiO_2 nanoparticles on human beings or on natural environments. It is important to establish legitimate legislation on this issue to govern their safe use because of their commercial relevance and potentially adverse effects.

In this chapter, we reviewed the progress of advanced research from recent publications. The relationship among the physicochemical properties, the fate and behavior, and toxic consequences of TiO_2 nanoparticles in aquatic ecosystems were elaborated in this review. It is our purpose to reveal the nano-bio-eco interactions and therefore, to provide a paradigm for a credible and reliable predictor of adverse impacts and the design of green nanoparticles, by using existing databases on TiO_2 nanoparticles.

INTERPRETING NANOTOXICITY—THE INTRINSIC PROPERTIES AND THEIR OUTCOMES

Adequate characterization of ENMs is required for understanding nanotoxicity. The relationship between physicochemical properties of ENMs and their toxicity has been reported in numerous studies. These findings apply in general to TiO_2 nanoparticles. The smaller size, larger surface-to-volume ratio, higher surface energy, and surface reactivity enable TiO_2 nanoparticles to have contact with biomolecules and other components in their surrounding microenvironments. Photoactivation by UV radiation, leading to cell damage or dealth, is one of the most recognized properties of TiO_2 nanoparticles. Clearly, the fact that TiO_2 can be photoactivated to function as a photocatalyst by photons having sufficient energy (3.2 eV for anatase) is tantamount to phototoxicity under UV exposure.

Upon excitation, an electron (e^-) can jump to the conduction band from the valence band, leaving a hole (h^+) in the valence band [Equation (1)]. The photoexcited electron-hole pair can then recombine or migrate to the surface of the TiO_2 and could react with water and oxygen [Equation (2-3)], leading to the generation of reactive oxygen species (ROS) [Equation (4-11)]. An overwhelming quantity of ROS could lead to the disruption of cellular homeostasis by interacting with biomolecules such as proteins, lipids, and DNA (Chen et al., 2014). Ultimately, the resultant unbalanced cellular redox condition could result in cell damage and even cell death (Fu et al., 2014). In aquatic environments, planktons are the organisms most vulnerable to photoactivated TiO_2 nanoparticles. For instance, Miller et al. (2012) indicated that TiO_2 nanoparticles could be adsorbed onto the surface of phytoplankton and then participate in a ligand-to-metal charge transfer reaction. Resulting ROS could transfer from the particle surface to cell wall or into cells, leading to cellular oxidation and damage.

$$TiO_2 + hv \rightarrow e^- + h^+ \quad (1)$$

$$e^- + O_2 \rightarrow O_2^{\bullet -} \quad (2)$$

$$H_2O + h^+ \rightarrow H^+ + OH^- \quad (3)$$

$$OH^- + h^+ \rightarrow {}^{\bullet}OH \quad (4)$$

$$O_2^{\bullet-} + H^+ \rightarrow HO_2^- \quad (5)$$

$$HO_2^- + O_2^{\bullet-} + H^+ \rightarrow H_2O_2 + O_2 \quad (6)$$

$$2HO_2^- \rightarrow H_2O_2 + O_2 \quad (7)$$

$$HO_2^- + O_2^{\bullet-} + e^- \rightarrow H_2O_2 \quad (8)$$

$$H_2O_2 + O_2^{\bullet-} \rightarrow OH^- + {}^{\bullet}OH + O_2 \quad (9)$$

$$H_2O_2 + e^- \rightarrow OH^- + {}^{\bullet}OH \quad (10)$$

$$H_2O_2 + O_2^{\bullet-} + h^+ \rightarrow {}^1O_2 + {}^{\bullet}OH + H_2O \quad (11)$$

However, the modification of physicochemical properties of TiO_2 nanoparticles could also alter their intrinsic toxicity along with their novel applications. A change in the band gap can extend the light absorbance of TiO_2 into the visible light spectrum, enabling photoactivation to occur under visible light. Band gap, also called energy gap, refers to the energy difference between the valence band and the conduction band. Band gap tuning is a common approach to enhance the photocatalytic activity of TiO_2 nanoparticles, a class of semiconductors, to facilitate electron–hole separation and promote interfacial electron transfer that in turn increases the surface reactivity (Paramasivam et al., 2008). It is possible to tailor the band gap through size engineering (Baskoutas and Terzis 2006, Robel et al., 2007) and surface constructing (Pathakoti et al., 2013, Smith and Nie 2010, Yun et al., 2010). It has been highly recommended that the working region of TiO_2 nanoparticles should be extended to visible light in the future in order to improve the utilization of solar energy for field applications, thus inevitably leading to greater hazards and risk to exposed biota (Lan et al., 2013).

Band gap narrowing could be induced by the electronic perturbations caused by change of the lattice parameters and/or the presence of the trap states within conduction and valence bands. Consequently, the photons of lower energy can induce electron–hole pairs in nanoparticles with smaller band gaps. This is especially useful and feasible in solar cells and other applications for photocatalysts operating beyond UV range (Hamal and Klabunde 2007, Paramasivam et al., 2008, Pathakoti et al., 2013). Furthermore, these photo-generated electrons and holes may be further transferred to the surface of nanoparticles, acting as reducing and oxidizing agents in the subsequent redox reactions. Additionally, these photo-induced charge-carriers last longer and predominate over the fast charge-recombination process, resulting in a powerful visible light photocatalytic activity (Hamal and Klabunde 2007). The band gap change of TiO_2 nanoparticles could also alter their intrinsic toxicity, along with their novel applications. For example, Pathakoti and coworkers reported that sulfur-doped TiO_2 nanoparticles with a relatively smaller band gap indicated higher toxic effects on zebrafish embryos than did conventional TiO_2 (He et al., 2014a, Pathakoti et al., 2013).

Among their many characteristics, the crystal structure of nanoparticles is critical in determining their toxicity. The specific spatial arrangement of atoms is closely relevant to their potential oxidative reactivity. Nanoparticles demonstrate different chemical reactivity with respect to various crystal faces (Mclaren et al., 2009, Vankayala et al., 2013). For instance, the anatase TiO_2 nanocrystals with well-faceted (101) surfaces exhibit enhanced photocatalytic activity, while the rutile (100) and (110) facets exhibit lower photoreduction activity (Mor et al., 2006b). This could be attributed to the higher aptitude of anatase to adsorb oxygen in the form of O_2^- and O^- and the dissociative adsorption of water molecules in the form of H^+ and OH^- (Sclafani and Herrmann 1996, Selloni et al., 1998, Vittadini et al., 1998). Analysis from available data implies that physicochemically stable nanoparticles tend to show insignificant cellular toxicity, whereas chemically-reactive nanoparticles are usually toxic to living organisms or cells. Indeed, the anatase TiO_2 nanoparticles were reported to be more biologically toxic than rutile in terms of cytotoxicity, oxidative stress, or neurological lesion (Hirakawa et al., 2004, Jin et al., 2011, Sato and Taya 2006, Wang et al., 2008).

Besides the aforementioned characteristics, crystal morphology is also related to TiO_2 nanoparticle toxicity. Tong et al., (2013) reported that nanotubes and nanosheets of TiO_2 are less phototoxic than their rod- and sphere-shaped counterparts under simulated solar irradiation. Interestingly, they found that TiO_2 nanotubes and nanosheets exhibited equal or even higher photoactivity than other shapes. They suggested that the photoactivity, aggregation, and the nano-bio interactions should be considered in assessing the toxicity of TiO_2 nanoparticles.

Noticeably, crystallites are known to be highly sensitive to phase transformation as the size changes, especially in the case of TiO_2 nanoparticles. This phenomenon was suggested to be involved in both chemical activity and toxicity towards cellular organisms (Auffan et al., 2009, Jang et al., 2007). Based on valid investigations, Jiang et al. (2009) found that for a fixed total surface area, the generation of ROS per unit surface area as a function of particle size, is described an S-shaped curve (4–195 nm with the same crystal phase). Other than primary size, the aggregation of particles, especially the metal-based nanoparticles, could play a critical role as well (Braydich-Stolle et al., 2009, Keller et al., 2010).

RELEASE OF TIO_2 NANOPARTICLES INTO AQUATIC ENVIRONMENTS

TiO_2 nanoparticles can be released into aquatic environments via three major ways. First, discharges directly from consumer products. It is well known that TiO_2 nanoparticles serve as color additive in toothpaste and capsules, products that are regulated by the U. S. Food and Drug Administration (U.S. FDA 2013). The primary purpose of using TiO_2 herein is as a white pigment in paints, paper, fibers, and sunscreens, owing to its highly stable property and brightness. Manufacture, processing, direct and indirect use, and disposal of these consumer products can lead to their release into domestic sewage systems. For instance, in 2010, the production of TiO_2 was 5000 tons, and it is expected to continue to increase with greater reliance upon nanosized TiO_2 (Weir et al., 2012). Consequently, a large amount of discharges can be expected. Indeed, Kaegi et al. (2008) reported and confirmed the release of synthetic TiO_2 from domestic paints as whitening pigments into aquatic environments.

Secondly, the intentional distribution of TiO_2 into aquatic systems for remediation purposes is an additional entry route into the environment (Kamat and Meisel 2003, Karn et al., 2009, Obare and Meyer 2004). It is well known that TiO_2 is one of the most promising photocatalysts and has been extensively used in environmental purification, e.g., water (Bai et al., 2010, Ryu and Choi 2008) and air purification (El-Roz et al., 2012, Li et al., 2005). Their low expense and photostable properties enable the large-scale application in industrial processes. In the presence of sunlight, TiO_2 has been shown to be more efficient in the decomposition of organic compounds and dyes, such as bisphenol A (Kim et al., 2008), rhodamine B (Zhou et al., 2010), orange G (Vinu et al., 2010), and reactive red 195 (Tzikalos et al., 2013), in aqueous solutions than conventional photocatalysts (Vinu and Madras 2011). Besides, it has been demonstrated that TiO_2 nanoparticles are highly efficient for remediation of a variety of aquatic contaminants, including pesticides (Senthilnathan and Philip 2010, Yu et al., 2007), heavy metal ions (Chen et al., 1997a, Chen et al., 1997b), and pharmaceuticals (Liang et al., 2013, Shahmoradi et al., 2010). Not surprisingly, at one waste water treatment plant studied in detail, raw sewage contained 100 to nearly 3000 μg Ti/L (Kiser et al., 2009). Consequently, as a potent candidate for large-scale *in situ* remediation, it must be critically studied since its environmental impacts remain largely unknown.

Thirdly, application in renewable energy is another potential source that accounts for TiO_2 discharges. Solar cell nanotechnology has been recognized as the most critical technology to resolve the shortage and crisis of energy in the future. Nano-sized TiO_2 demonstrates impressive overall light-to-electricity conversion efficiencies through the improvement of electron transport (Kuang et al., 2008, Mor et al., 2006a, Zhu et al., 2007). Large scale manufacturing can be expected in the future. TiO_2 photocatalytic water-splitting allows us to produce hydrogen in a promising way, owing to its clean, low-cost and environmentally friendly advantages (Ni et al., 2007). Thus, considerable research and efforts have been invested in its development (Cho et al., 2011, Khan et al., 2002), in order to further improve the solar-to-hydrogen conversion efficiency.

Other nano-sized TiO_2 uses including antimicrobial applications (Kong et al., 2010), medical applications (Narayan et al., 2009), and sensor applications (Wang et al., 2006) might discharge massive amounts into aquatic environments as well (Macwan et al., 2011, Weir et al., 2012).

As discussed above, there is need for a more extensive knowledge base on the sources (point and nonpoint sources), occurrence, and ultimate fate of TiO_2 nanoparticles, once released into the aquatic environment. Upon entry into the aquatic environment, TiO_2 nanoparticles may be taken up by living organisms. They could also remain intact as suspended particles or become immobilized by dissolved organic matter (DOM). Further sedimentation of TiO_2 nanoparticles from surface water may occur at a later stage, with the potential to affect organisms in benthic habitats. In addition, it has been suggested that toxic responses to TiO_2 nanoparticles can be species-specific, resulting in the loss of TiO_2 disturbance-sensitive species. Together, these processes and interactions could interrupt an entire aquatic ecosystem.

Recent studies have indicated that TiO_2 nanoparticles express a higher level of complexity towards aquatic biota than conventionally hazardous pollutants (Clemente et al., 2012, Moore 2006). He and co-workers critically suggested that using a holistic approach concerning different trophic levels and species is required to recognize and understand the potential hazards and toxicity mechanisms of ENMs, including TiO_2 nanoparticles (He et al., 2014b). Microorganisms serve essential ecological functions as decomposers in aquatic ecosystems. A microbial community and its function could be altered by TiO_2 nanoparticles via direct contact with the suspended nanoparticles, or by indirect contact through decomposing detritus (Binh et al., 2013). Zheng and colleagues reported that 50 ppm of TiO_2 nanoparticles significantly decreased the diversity of a microbial community in activated sludge (Zheng et al., 2011). Their investigation revealed that the exposure of TiO_2 nanoparticles caused serious deterioration of ammonia oxidation, which is an important step in the nitrogen cycle. Consequently, a deteriorated ecological food chain could lead to severe population loss, community degradation, and ultimately, disruption of ecological systems. In addition, the adverse effects of TiO_2 nanoparticles on phytoplankton have been extensively investigated in recent years (Cardinale et al., 2012, Clément et al., 2013, Hund-Rinke and Simon 2006, Metzler et al., 2011, Miller et al., 2012). The dependence of phytoplankton on sunlight for photosynthesis makes them more vulnerable to the photo-induced toxicity of TiO_2 nanoparticles. Phytoplanktons are responsible for oxygen production in the Earth's atmosphere, and the global carbon cycle. As the basis of most aquatic food chains, phytoplanktons are critical to the survival of other species in aquatic food webs. Clearly, the effects of any decline in their population due to contamination by TiO_2 nanoparticles could be profound. Now, abundant research has shown that zooplanktons are also affected by TiO_2 nanoparticles (Dabrunz et al., 2011, Hund-Rinke and Simon 2006, Kim et al., 2010, Ma et al., 2012, Zhu et al., 2010). *Daphnia magna*, a key group in the aquatic food web structure, has been used extensively as a representative of zooplanktons. It is also noteworthy that TiO_2 nanoparticles could adversely affect zooplanktons through enhancing other toxic contaminants, such as Cu^{2+} (Fan et al., 2011). Fan et al. (2011) reported that Cu^{2+}could be adsorbed onto the TiO_2 nanoparticles, ingested by and accumulated in the organisms, thereby resulting in toxic injury. They also suggested that TiO_2 nanoparticles may compete for free Cu^{2+} with sulfhydryl groups, preventing the detoxification by metallothioneins.

Besides microscopic organisms, macro-invertebrate and vertebrate organisms are impacted by TiO_2 nanoparticles as well. Freshwater and marine bivalve (Canesi et al., 2013, D'Agata et al., 2014) and gastropod (Zhu et al., 2011a, Zhu et al., 2011b) mollusks are largely employed in the biomonitoring of environmental perturbations (Canesi et al., 2012); therefore, the investigation of the adverse impact of TiO_2 nanoparticles on those species is crucial for aquatic ecosystems. Vertebrates, mainly including fish and amphibians, are another important group in aquatic ecosystems. Assessment of the ecotoxicological effects of TiO_2 nanoparticles on fish is largely available for goldfish (*Carassius auratus*) (Ates et al., 2013, Diniz et al., 2013), fathead minnow (*Pimephales promelas*) (Hall et al., 2009, Jovanović et al., 2011), zebrafish (*Danio rerio*) (Bar-Ilan et al., 2012, He et al., 2014a), rainbow trout (*Oncorhynchus mykiss*) (Al-Jubory and Handy 2013, Boyle et al., 2013), and carp (*Cyprinus carpio*) (Hao et al., 2009, Lee et al., 2012). Furthermore, amphibians are extremely vulnerable to xenobiotic chemicals, enabling them to serve as appropriate bioindicators for potential hazards owing to their lives in both terrestrial and aquatic ecosystems. The African clawed frog (*Xenopus laevis*), as a representative species of amphibians, is widely utilized in assessing the ecotoxicity of TiO_2 nanoparticles (Bacchetta et al., 2012, Bİrhanli et al., 2014).

CASE STUDY

In this section, we present selected case studies to illustrate the interactions among TiO_2 nanoparticles, abiotic environments, and biotic environments that TiO_2 nanoparticles might encounter once they are released into aquatic environments. It is also worth mentioning that most experts have noted the need for the consideration of micro- and macro-environments, and for the elaboration of nano-bio-eco interaction regarding this subject. Readers are encouraged to retrieve further information on this topic from the review report by He et al. (2014b).

Table 1. Averaged zeta potential and pH for TiO_2 nanoparticles with and without humic acid (HA) at various concentrations. Data are collected from the publication by Dasari and Hwang (2013)

Groups	Zeta potential (mV)	pH
25 ppm TiO_2	–(20.70 ± 0.40)	7.63
25 ppm TiO_2 + 10 ppm HA	–(24.33 ± 0.51)	7.71
25 ppm TiO_2 + 40 ppm HA	–(24.33 ± 0.51)	8.05
75 ppm TiO_2	–(21.07 ± 0.32)	7.68
75 ppm TiO_2 + 10 ppm HA	–(21.63 ± 0.38)	7.71
75 ppm TiO_2 + 40 ppm HA	–(23.23 ± 0.58)	7.93

Nano-Eco Interaction

Nano–eco interaction refers to the interactions between nanoparticles and their surrounding environments. Amongst the various surrounding environments, DOM has been considered the most important factor in determining the behavior and fate of nanoparticles in aquatic environments. Another critical factor however, is sunlight, owing to their highly photocatalytic activities. To investigate the interaction between TiO_2 nanoparticles and their abiotic environments, Dasari and Hwang (2013) comprehensively studied the effect of humic acid (HA) and sunlight on their cytotoxicity to a river bacterial assemblage. The results indicated that the viability of natural bacteria was significantly affected by HA. Generally, the supplementation with HA reduced the toxicity of TiO_2 nanoparticles. The results (Table 1) indicate that HA increased the zeta potential of nanoparticle suspensions in a concentration-dependent manner. In the presence of HA, the physicochemical properties of TiO_2 nanoparticle surfaces were changed and the suspension stability was increased (Yang et al., 2009). HA could introduce the barrier of steric hindrance between TiO_2 nanoparticles and cells, leading to diminishing bioavailability and toxicity (Jiang et al., 2011). Co-precipitation experimentation and transmission electron microscopy (TEM) observation revealed that HA prevented the adhesion of TiO_2 nanoparticles to cells due to the increased electrosteric repulsion (Dasari and Hwang 2013, Lin et al., 2012). The data analysis also indicated that HA could attenuate the sunlight and the oxidative stress generated by TiO_2 nanoparticles. It is in agreement with other observations that HA could retard the photodegradation rates of TiO_2 nanoparticles (Lin and Lin 2007), resulting in decreased generation of intracellular ROS (Lin et al., 2012).

Nano-Bio Interaction

Nano–bio interactions occur at the time when nanomaterials come into contact with living organisms directly. Complex biological microenvironments enable direct interactions between nanomaterials and biomolecules (e.g., lipids, proteins, and DNA). These interactions could lead to denaturation or particle wrapping, followed by biocatalytic reactions that may result in adverse effects. Nanomaterials per se could also go through phase transformations, free energy releases, particle aggregation, restructuring, and dissolution at its surface (Nel et al., 2009). It is critically important to fully examine the biochemical reactions at the contacting interfaces and to clarify their physiochemical interactions. With sufficient knowledge, the adverse impacts of nanomaterials could be diminished through proper design or surface modification.

An investigation was conducted to reveal the interaction between TiO_2 nanoparticles and cytoplasmic proteins (Gheshlaghi et al., 2008). In their study, 0−50 ppm of TiO_2 nanoparticle solution was used for investigation. A 50% decrease of 2 mg/ml tubulin protein activity was observed in the presence of 35 ppm TiO_2 nanoparticles. The turbidimetric assay of microtubules (MT) and tubulin indicated that TiO_2 nanoparticles could inhibit tubulin polymerization and the MT nucleation rate in a concentration- dependent manner. Data showed TiO_2 nanoparticles could affect both soluble tubulin and tubulin in MT structures, suggesting that the decreased tubulin polymerization ability resulted from tubulin conformational changes. In addition, the results suggested that changes in protein conformation were involved in protein function alteration. In this case, TiO_2 nanoparticles altered the polarity in the vicinity of tryptophan residues. Evidence also demonstrated that TiO_2 nanoparticles could cause changes in tryptophan positioning, moving them towards guanosine-5'-triphosphate (GTP) binding sites in the protein structures. Additional data showed that these changes result in the suppression of tubulin polymerization.

In vitro/in vivo study

For a comprehensive and reliable toxicological assessment, both *in vitro* and *in vivo* studies on the biological effects of TiO_2 nanoparticles need to be performed (Maurer-Jones and Haynes 2012, Sahu and Casciano 2009). The use of *in vitro* models is the most common approach that allows scientists to evaluate nanotoxicity in a less expensive and affordable way (Arora et al., 2012). They can also provide a rapid and effective pathway to obtain extensive toxicological data regarding various biological endpoints. Thus, the successful development of valid and reproducible *in vitro* assays for assessing the toxicity of nanomaterials is critical for early stage nanomaterial design. The ultimate goal of these assays is to establish a quantitative structure-activity relationship (QSAR) involving nanomaterials, meanwhile enabling high-throughput screening of large numbers of new engineered nanomaterials.

Research teams at the Interdisciplinary Nanotoxicity Center, at Jackson State University, have endeavored to establish quantitative structure–activity relationship (QSAR) models involving TiO_2 nanoparticles via *in vitro* assessment (Puzyn et al., 2011, Pathakoti et al., 2014), as illustrated in Figure 1. Puzyn and coworkers first established a QSAR model regarding the cytotoxicity of 17 different types of metal oxide nanoparticles to the bacteria *Escherichia coli* [Equation (12)].

$$\log(1/EC_{50})=2.59-0.50\cdot\Delta H_{Me+} \quad (12)$$

where the ΔH_{Me+} represents the enthalpy of formation of a gaseous cation having the same oxidation state as that in the metal oxide structure.

Pathakoti et al. (2014), established another QSAR model in regard to their photo-induced toxicity, also using *E. coli* [Equation (13)].

$$-\log(LC_{50})=-4.950\mathrm{Log}(Cp)+45.023\mathrm{ALZLUMO}+18.185 \quad (13)$$

where the LC_{50} is in units of mol/L, and *Cp* is the literature molar heat capacity of the metal oxide at 298.15 K; ALZLUMO is the average of the alpha and beta LUMO energies of the metal oxide.

The validity of these QSAR models and their reliability in predicting the nanotoxicity of nanomaterials are thus established by extensive testing. Both strongly suggest that molecular descriptors are useful in elucidating the mechanisms of nanotoxicity and in predicting the environmental risk associated with the release of the ENMs. The quantitative methodology is hence expected to provide the basis for the design of safe nanomaterials in the future.

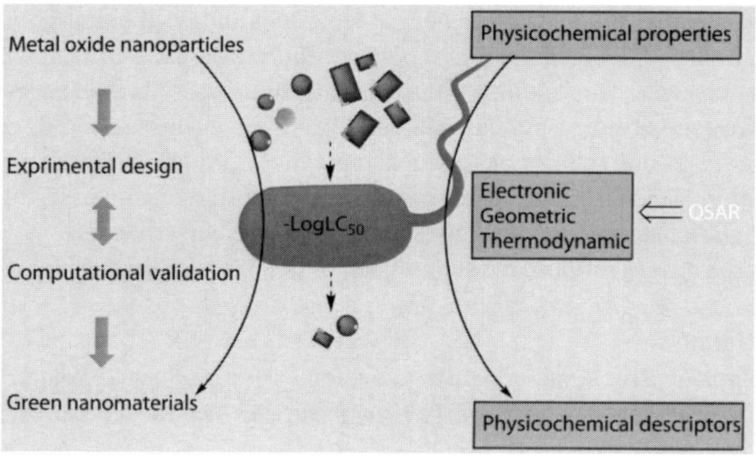

Figure 1. Schematic representation of the QSAR procedure for the risk assessment and prediction of nanotoxicity.

However, *in vitro* study needs validation by *in vivo* experimentation. *In vivo* nanotoxicity screening could provide high reliability and predictability compared to its *in vitro* analogue. It is employed to observe the overall effects of the selected chemicals on a living organism, instead of a partial tissue or a single organ/cell. Though it is relatively expensive and time-consuming, it remains irreplaceable until a highly reliable and reproducible *in vitro* candidate is established. Hence, the optimal design for assessing TiO_2 nanotoxicity is to have both *in vivo* and *in vitro* experiments, as elaborated in Figure 2.

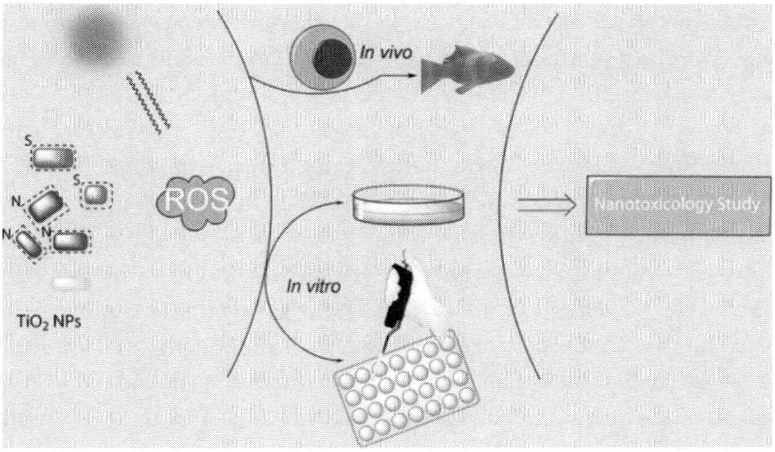

Figure 2. A particular *in vitro/in vivo* design of nanotoxicology study with TiO_2 nanoparticles. Reprinted with permission from Pathakoti et al. (2013), Copyright (2013) American Chemical Society.

The zebrafish has been used extensively as an *in vivo* model for nanotoxicological studies and for the development of high-throughput screening (HTS) platform (George et al., 2011, Pardo-Martin et al., 2010). Transparent zebrafish larvae enable scientists to develop imaging platform and analysis systems for phenotypic and developmental abnormalities in embryos and larvae. This is particularly essential for evaluating the photo-induced toxicity of TiO_2 nanoparticles. Moreover, current accomplishment of the zebrafish genome study offers scientists another important way to thoroughly investigate the adverse affects from nanomaterials at the genetic level. Researchers from one of the aforementioned research groups at Jackson State University, reported photo-enhanced toxicities of TiO_2 nanoparticles to zebrafish embryos (*Danio rerio*) in terms of malformation, mortality, rheotaxis dysfunction, and DNA damage (He et al., 2014a). They highlighted the advantages of using the transparent zebrafish embryo as an *in vivo* model for investigating photo-induced nanotoxicity, *via* both bright-field and fluorescence imaging analysis. It is also highly useful to use fluorescent tracing for monitoring real-time localization, distribution, and possible translocation of TiO_2 nanoparticles in aquatic ecosystems. Noticeably, the use of *in vivo* study in conjunction with *in vitro* screening and *in silico* (performed on computer or *via* computer simulation) data handling can constitute a paradigm for rapid and integrated ENM toxicological screening (George et al., 2011).

Direction for Environmental Risk Assessment

Both *in vitro* and *in vivo* study are at a fairly rudimentary stage in the quest to develop a low-cost, less labor-intensive, highly reliable and reproducible, and data accessible paradigm for the toxicological assessment of TiO_2 nanoparticles. QSAR or HTS will be needed to integrate toxicological data with nanoparticle design. However, there are many other issues to address as well. The lack of comprehensive protocols limits the cross-comparison of available toxicological data from different research institutes or groups. Controversies in the research findings from those research groups may mislead the research direction. There are ongoing debates concerning the effect of particle size on the chemical activity and toxicity of ENMs. Some claim that size plays a critical role in engineering green nanoparticles (Karlsson et al., 2009, Sohaebuddin et al., 2010, Wang et al., 2009); whereas others don't (Vecchio et al., 2012, Warheit et al., 2006). For instance, Warheit and coworkers suggested that considerations of primary particle size and surface area may sometimes be misleading, and they concluded that the toxicity of nanoparticles is not dependent upon particle size and surface area (Warheit et al., 2006). However, the effect of particle size should be compared within a given crystal structure and morphology (He et al., 2014b), while taking into account the secondary size effect (Menard et al., 2011). The investigated size ranges should also be sufficient for any further conclusion and comparison. Another important issue is that false positive results may lead to incorrect conclusions. Rajapakse et al. (2013), reported that Comet results from the direct interactions between nanoparticles and DNA intracellularly may mislead scientists to conclude it is caused by ROS generated from nanoparticles. Thus, one possible solution is to develop experimental protocols, and those critically evaluated protocols should be available from a recognized institute or agency for all scientists.

Conclusion

The promises of nanotechnology to human society are enormous benefits and advances as it is developed further. However, the thriving nanotechnology is still in transition to a well underpinned discipline with environmental impact as well as fields of application established. During its development and uses, scientific research and governmental regulations should be applied to minimize its adverse impacts and to ensure its safe use in the long-term. The critical literature review on the fate and effects of TiO_2 nanoparticles in this chapter leads to three conclusions. First, the physicochemical properties of TiO_2 nanoparticles are closely related to their chemical activities and subsequent toxicity. The QSAR platform can be efficiently used to formulate a paradigm for the prediction of toxicity of TiO_2 nanoparticles. A thorough characterization of TiO_2 nanoparticles can ensure scientists a more reliable comparison and comprehensive conclusion. Second, nano-bio-eco interactions should be thoroughly considered for assessing their behavior, fate, and toxicity in aquatic ecosystems. The immediate micro- and macro-environments must be considered with respect to the biological and biochemical interferences. In addition, multiple trophic levels, and species variety must be considered for the screening of species sensitive to TiO_2 nanoparticles. Third, the sources and routes of TiO_2 nanoparticles that are released into aquatic environments must be controlled and regularly monitored. Hazard prevention is always better than remediation. Ultimately, the adverse effects of TiO_2 nanoparticles can be controlled and diminished in the future, thus a sustainable development of green nanotechnology can be achieved as well.

Acknowledgments

This study was supported by a grant received from NSF-CREST program (National Science Foundation-Centers of Research Excellence in Science and Technology) with grant #HRD-0833178.

REFERENCES

Al-Jubory, AR; Handy, RD. Uptake of titanium from TiO_2 nanoparticle exposure in the isolated perfused intestine of rainbow trout: nystatin, vanadate and novel CO_2-sensitive components. *Nanotoxicology*, 2013, 7 (8), 1282-1301.

Arora, S; Rajwade, JM; Paknikar, KM. Nanotoxicology and *in vitro* studies: The need of the hour. *Toxicol Appl Pharm*, 2012, 258(2), 151-165.

Ates, M; Demir, V; Adiguzel, R; Arslan, Z. Bioaccumulation, subacute toxicity, and tissue distribution of engineered titanium dioxide nanoparticles in goldfish (*Carassius auratus*). *J Nanomater*, 2013, 2013, 460518.

Auffan, M; Rose, J; Wiesner, MR; Bottero, J-Y. Chemical stability of metallic nanoparticles:A parameter controlling their potential cellular toxicity *in vitro*. *Environ Pollut*, 2009, 157(4), 1127-1133.

Bacchetta, R; Santo, N; Fascio, U; Moschini, E; Freddi, S; Chirico, G; et al. Nano-sized CuO, TiO_2 and ZnO affect *Xenopus laevis* development. *Nanotoxicology*, 2012, 6(4), 381-398

Bai, H; Liu, Z; Sun, DD. Hierarchically multifunctional TiO_2 nano-thorn membrane for water purification. *Chem Commun*, 2010, 46, 6542-6544.

Bar-Ilan, O; Louis, KM; Yang, SP; Pedersen, JA; Hamers, RJ; Peterson, RE; et al. Titanium dioxide nanoparticles produce phototoxicity in the developing zebrafish. *Nanotoxicology*, 2012, 6(6), 670-679.

Barmo, C; Ciacci, C; Canonico, B; Fabbri, R; Cortese, K; Balbi, T; et al. *In vivo* effects of n-TiO_2 on digestive gland and immune function of the marine bivalve *Mytilus galloprovincialis*. *Aquat Toxicol*, 2013, 132-133, 9-18.

Baskoutas, S; Terzis, AF. Size-dependent band gap of colloidal quantum dots. *J Appl Phys*, 2006, 99, 013708.

Binh, CTT; Tong, T; Gaillard, J-F; Gray, KA; Kelly, JJ. Common freshwater bacteria vary in their responses to short-term exposure to nano-TiO_2. *Environ Toxicol Chem*, 2013, Epub ahead of print. doi: 10.1002/etc.

Bİrhanli, A; Emre, FB; Sayilkan, F; Güngördü, A. Effect of nanosized TiO_2 particles on the development of *Xenopus laevis* embryos. *Turk J Biol*, 2014, 38. 10.3906/biy-1307-37.

Boyle, D; Al-Bairuty, GA; Ramsden, CS; Sloman, KA; Henry, TB; Handy, RD. Subtle alterations in swimming speed distributions of rainbow trout exposed to titanium dioxide nanoparticles are associated with gill rather than brain injury. *Aquat Toxicol*, 2013, 126, 116-127.

Braydich-Stolle, LK; Schaeublin, NM; Murdock, RC; Jiang, J; Biswas, P; Schlager, JJ; et al. Crystal structure mediates mode of cell death in TiO_2 nanotoxicity. *J Nanopart Res*, 2009, 11(6), 1361-1374.

Canesi, L; Ciacci, C; Fabbri, R; Marcomini, A; Pojana, G; Gallo, G. Bivalve molluscs as a unique target group for nanoparticle toxicity. *Marine Environmental Research*, 2012, 76, 16-21.

Canesi, L; Frenzilli, G; Balbi, T; Bernardeschi, M; Ciacci, C; Corsolini, S; et al. Interactive effects of n-TiO_2 and 2,3,7,8-TCDD on the marine bivalve *Mytilus galloprovincialis*. *Aquat Toxicol*, 2013, In press. 10.1016/ j.aquatox.2013.11.002.

Cardinale, BJ; Bier, R; Kwan, C. Effects of TiO_2 nanoparticles on the growth and metabolism of three species of freshwater algae. *J Nanopart Res*, 2012, 14, 913.

Chen, LX; Rajh, T; Mićic, O; Wang, Z; Tiede, DM; Thurnauer, M. Photocatalytic reduction of heavy metal ions on derivatized titanium dioxide nano-particle surface studied by XAFS. *Nucl Instr Meth Phys Res*, 1997a, 133(1-4), 8-14.

Chen, LX; Rajh, T; Wang, Z; Thurnauer, MC. Xafs studies of surface structures of TiO_2 nanoparticles and photocatalytic reduction of metal ions. *J Phys Chem B*, 1997b, 101(50), 10688-10697.

Chen, T; Yan, J; Li, Y. Genotoxicity of titanium dioxide nanoparticles. *J Food Drug Anal*, 2014, 22(1), 95–104.

Cho, IS; Chen, Z; Forman, AJ; Kim, DR; Rao, PM; Jaramillo, TF; et al. Branched TiO_2 nanorods for photoelectrochemical hydrogen production. *Nano Lett*, 2011, 11(11), 4978-4984.

Clément, L; Hurel, C; Marmier, N. Toxicity of TiO_2 nanoparticles to cladocerans, algae, rotifers and plants--Effects of size and crystalline structure. *Chemosphere*, 2013, 90(3), 1083-1090.

Clemente, Z; Castro, VL; Jonsson, CM; Fraceto, LF. Ecotoxicology of nano-TiO_2: An evaluation of its toxicity to organisms of aquatic ecosystems. *Int J Environ Res*, 2012, 6(1), 33-50.

D'Agata, A; Fasulo, S; Dallas, LJ; Fisher, AS; Maisano, M; Readman, JW; et al. Enhanced toxicity of 'bulk' titanium dioxide compared to 'fresh' and 'aged' nano-TiO_2 in marine mussels (*Mytilus galloprovincialis*). *Nanotoxicology*, 2014, 8 (5), 549-558.

Dabrunz, A; Duester, L; Prasse, C; Seitz, F; Rosenfeldt, R; Schilde, C; et al. Biological surface coating and molting inhibition as mechanisms of TiO_2 nanoparticle toxicity in *Daphnia magna*. *PLoS ONE*, 2011, 6(5), e20112.

Dalai, S; Pakrashi, S; Nirmala, MJ; Chaudhri, A; Chandrasekaran, N; Mandal, AB; et al. Cytotoxicity of TiO_2 nanoparticles and their detoxification in a freshwater system. *Aquat Toxicol*, 2013, 138-139, 1-11.

Dasari, TP; Hwang, H-M. Effect of humic acids and sunlight on the cytotoxicity of engineered zinc oxide and titanium dioxide nanoparticles to a river bacterial assemblage. *J Environ Sci*, 2013, 25(9), 1925-1935.

Diniz, MS; de, Matos, APA; Lourenço, J; Castro, L; Peres, I; Mendonça, E; et al. Liver alterations in two freshwater fish species (*Carassius auratus* and *Danio rerio*) following exposure to different TiO_2 nanoparticle concentrations. *Microsc Microanal*, 2013, 19(5), 1131-1140.

El-Roz, M; Kus, M; Cool, P; Thibault-Starzyk, F. New operando IR technique to study the photocatalytic activity and selectivity of TiO_2 nanotubes in air purification: Influence of temperature, uv intensity, and VOC concentration. *J Phys Chem C*, 2012, 116(24), 13252-13263.

Fan, W; Cui, M; Liu, H; Wang, C; Shi, Z; Tan, C; et al. Nano-TiO_2 enhances the toxicity of copper in natural water to *Daphnia magna*. *Environ Pollut*, 2011, 159(3), 729-734.

Fu, P; Xia, Q; Hwang, H-M; Ray, PC; Yu, H. Mechanisms of nanotoxicity: Generation of reactive oxygen species. *J Food Drug Anal*, 2014, 22(1), 64–75.

George, S; Xia, T; Rallo, R; Zhao, Y; Ji, Z; Lin, S; et al. Use of a high-throughput screening approach coupled with *In Vivo* zebrafish embryo screening to develop hazard ranking for engineered nanomaterials. *ACS Nano*, 2011, 5(3), 1805-1817.

Gheshlaghi, ZN; Riazi, GH; Ahmadian, S; Ghafari, M; Mahinpour, R. Toxicity and interaction of titanium dioxide nanoparticles with microtubule protein. *Acta Biochim Biophys Sin*, 2008, 40(9), 777-782.

Gottschalk, F; Sonderer, T; Scholz, RW; Nowack, B. Modeled environmental concentrations of engineered nanomaterials (TiO_2, ZnO, Ag, CNT, fullerenes) for different regions. *Environ Sci Tech*, 2009, 43(24), 9216-9222

Hall, S; Bradley, T; Moore, JT; Kuykindall, T; Minella, L. Acute and chronic toxicity of nano-scale TiO_2 particles to freshwater fish, cladocerans, and green algae, and effects of organic and inorganic substrate on TiO_2 toxicity. *Nanotoxicology*, 2009, 3(2), 91-97.

Hamal, DB; Klabunde, KJ. Synthesis, characterization, and visible light activity of new nanoparticle photocatalysts based on silver, carbon, and sulfur-doped TiO_2. *J Colloid Interface Sci*, 2007, 311(2), 514-522.

Hao, L; Wang, Z; Xing, B. Effect of sub-acute exposure to TiO_2 nanoparticles on oxidative stress and histopathological changes in Juvenile Carp (*Cyprinus carpio*). *J Environ Sci*, 2009, 21(10), 1459-1466.

He, X; Aker, WG; Hwang, H-M. An *in vivo* study on the photo-enhanced toxicities of S-doped TiO_2 nanoparticles to zebrafish embryos (*Danio rerio*) in terms of malformation, mortality, rheotaxis dysfunction, and DNA damage. *Nanotoxicology*, 2014a, Accepted

He, X; Aker, WG; Leszczynski, J; Hwang, H-M. Using a holistic approach to assess the impact of engineered nanomaterials inducing toxicity in aquatic systems. *J Food Drug Anal*, 2014b, 22(1), 128–146.

Heinlaan, M; Ivask, A; Blinova, I; Dubourguier, H-C; Kahru, A. Toxicity of nanosized and bulk ZnO, CuO and TiO_2 to bacteria *Vibrio fischeri* and crustaceans *Daphnia magna* and *Thamnocephalus platyurus*. *Chemosphere*, 2008, 71(7), 1308-1316.

Hirakawa, K; Mori, M; Yoshida, M; Oikawa, S; Kawanishi, S. Photo-irradiated titanium dioxide catalyzes site specific DNA damage via generation of hydrogen peroxide. *Free Radical Res*, 2004, 38(5), 439-447.

Hund-Rinke, K; Simon, M. Ecotoxic effect of photocatalytic active nanoparticles (TiO_2) on algae and daphnids (8 pp). *Environ Sci Pollut Res*, 2006, 13(4), 225-232.

IARC. Monographs on the evaluation of carcinogenic risks to humans. Volume 93. carbon black, titanium dioxide and talc. In: *International Agency for Research on Cancer* Who, editor. Lyon, France, 2010.

Jang, H; Kim, S; Kim, S. Effect of particle size and phase composition of titanium dioxide nanoparticles on the photocatalytic properties. *J Nanopart Res*, 2007, 3, 141-147.

Jiang, G-X; Shen, Z-Y; Niu, J-F; Bao, Y-P; Chen, J; He, T-D. Toxicological assessment of TiO_2 nanoparticles by recombinant *Escherichia coli* bacteria. *J Environ Monitor*, 2011, 13(1), 42-48.

Jin, C; Tang, Y; Yang, FG; Li, XL; Xu, S; Fan, XY; et al. Cellular toxicity of TiO_2 nanoparticles in anatase and rutile crystal phase. *Biol Trace Elem Res*, 2011, 141(1-3), 3-15.

Jovanović, B; Anastasova, L; Rowe, EW; Zhang, Y; Clapp, AR; Palić, D. Effects of nanosized titanium dioxide on innate immune system of fathead minnow (*Pimephales promelas* Rafinesque, 1820). *Ecotox Environ Safe*, 2011, 74(4), 675-683.

Kaegi, R; Ulrich, A; Sinnet, B; Vonbank, R; Wichser, A; Zuleeg, S; et al. Synthetic TiO_2 nanoparticle emission from exterior facades into the aquatic environment. *Environ Pollut*, 2008, 156(2), 233-239.

Kamat, PV; Meisel D. Nanoscience opportunities in environmental remediation. *Comptes Rendus Chimie*, 2003, 6(8-10), 999-1007.

Karlsson, HL; Gustafsson, J; Cronholm, P; Möller, L. Size-dependent toxicity of metal oxide particles—A comparison between nano- and micrometer size. *Toxicol Lett*, 2009, 188(2), 112-118.

Karn, B; Kuiken, T; Otto, M. Nanotechnology and *in situ* remediation: A review of the benefits and potential risks. *Environ Health Perspect*, 2009, 117(12), 1813-1831.

Kasemets, K; Ivask, A; Dubourguier, H-C; Kahru, A. Toxicity of nanoparticles of ZnO, CuO and TiO_2 to yeast *Saccharomyces cerevisiae*. *Toxicol in Vitro*, 2009, 23(6), 1116-1122.

Keller, AA; Wang, H; Zhou, D; Lenihan, HS; Cherr, G; Cardinale, BJ; et al. Stability and aggregation of metal oxide nanoparticles in natural aqueous matrices. *Environ Sci Technol*, 2010, 44(6), 1962-1967.

Khan, SUM; Al-Shahry, M; Ingler, Jr. WB. Efficient photochemical water splitting by a chemically modified n-TiO_2. *Science*, 2002, 297(5590), 2243-2245.

Kim, J; Lee, J; Choi, W. Synergic effect of simultaneous fluorination and platinization of TiO_2 surface on anoxic photocatalytic degradation of organic compounds. *Chem Commun*, 2008, 2008, 756-758.

Kim, KT; Klaine, SJ; Cho, J; Kim, S-H; Kim, SD. Oxidative stress responses of *Daphnia magna* exposed to TiO_2 nanoparticles according to size fraction. *Sci Total Environ*, 2010, 408(10), 2268-2272.

Kiser, MA; Westerhoff, P; Benn, T; Wang, Y; Pérez-Rivera, J; Hristovski, K. Titanium nanomaterial removal and release from wastewater treatment plants. *Environ Sci Technol*, 2009, 43(17), 6757-6763.

Kong, H; Song, J; Jang, J. Photocatalytic antibacterial capabilities of TiO_2 biocidal polymer nanocomposites synthesized by a surface-initiated photopolymerization. *Environ Sci Technol*, 2010, 44(14), 5672-5676

Kuang, D; Brillet, J; Chen, P; Takata, M; Uchida, S; Miura, H; et al. Application of highly ordered TiO_2 nanotube arrays in flexible dye-sensitized solar cells. *ACS Nano*, 2008, 2(6), 1113-1116.

Lan, Y; Lu, Y; Ren, Z. Mini review on photocatalysis of titanium dioxide nanoparticles and their solar applications. *Nano Energy*, 2013, 2(5), 1031-1045.

Lee, BC; Kim, KT; Cho, JG; Lee, JW; Ryu, TK; Yoon, JH; et al. Oxidative stress in juvenile common carp (*Cyprinus carpio*) exposed to TiO_2 nanoparticles. *Mol Cell Toxicol*, 2012, 8(4), 357-366.

Li, D; Haneda, H; Hishita, S; Ohashi, N. Visible-light-driven n-f-codoped TiO_2 photocatalysts. 2. Optical characterization, photocatalysis, and potential application to air purification. *Chem Mater*, 2005, 17(10), 2596-2602.

Li, S; Wallis, LK; Ma, H; Diamond, SA. Phototoxicity of TiO_2 nanoparticles to a freshwater benthic amphipod: Are benthic systems at risk? *Sci Total Environ*, 2014, 466-467, 800-808.

Liang, R; Hu, A; Li, W; Zhou, YN. Enhanced degradation of persistent pharmaceuticals found in wastewater treatment effluents using TiO_2 nanobelt photocatalysts. *J Nanopart Res*, 2013, 15, 1990.

Lin, C; Lin, K-S. Photocatalytic oxidation of toxic organohalides with TiO_2/UV: The effects of humic substances and organic mixtures. *Chemosphere*, 2007, 66(10), 1872-1877.

Lin, D; Ji, J; Long, Z; Yang, K; Wu, F. The influence of dissolved and surface-bound humic acid on the toxicity of TiO_2 nanoparticles to *Chlorella sp.Water Research*, 2012, 46(14), 4477-4487.

Ma, H; Brennan, A; Diamond, SA. Phototoxicity of TiO_2 nanoparticles under solar radiation to two aquatic species: *Daphnia magna* and *Japanese medaka*. *Environ Toxicol Chem*, 2012, 31(7), 1621-1629.

Macwan, DP; Dave, PN; Chaturvedi, S. A review on nano-TiO_2 sol-gel type syntheses and its applications. *J Mater Sci*, 2011, 46(11), 3669-3686.

Maurer-Jones, MA; Haynes, CL. Toward correlation in *in vivo* and *in vitro* nanotoxicology studies. *J Law Med Ethics*, 2012, 40(4), 795-801.

Mclaren, A; Valdes-Solis, T; Li, G; Tsang, SC. Shape and size effects of ZnO nanocrystals on photocatalytic activity. *J Am Chem Soc*, 2009, 131(35), 12540-12541.

Menard, A; Drobne, D; Jemec, A. Ecotoxicity of nanosized TiO_2. Review of *in vivo* data. *Environ Pollut*, 2011, 159(3), 677-684.

Metzler, DM; Li, M; Erdem, A; Huang, CP. Responses of algae to photocatalytic nano-TiO_2 particles with an emphasis on the effect of particle size. *Chem Eng J*, 2011, 170(2-3), 538-546.

Miller, RJ; Bennett, S; Keller, AA; Pease, S; Lenihan, HS. TiO_2 nanoparticles are phototoxic to marine phytoplankton. *PLoS ONE*, 2012, 7(1), e30321.

Moore, MN. Do nanoparticles present ecotoxicological risks for the health of the aquatic environment? *Environ Int*, 2006, 32(8), 967-976.

Mor, GK; Shankar, K; Paulose, M; Varghese, OK; Grimes, CA. Use of highly-ordered TiO_2 nanotube arrays in dye-sensitized solar cells. *Nano Lett*, 2006a, 6(2), 215-218.

Mor, GK; Varghese, OK; Paulose, M; Shankar, K; Grimes, CA. A review on highly ordered, vertically oriented TiO_2 nanotube arrays: Fabrication, material properties, and solar energy applications. *Sol Energ Mat Sol C*, 2006b, 90(14), 2011-2075.

Narayan, RJ; Monteiro-Riviere, NA; Brigmon, RL; Pellin, MJ; Elam, JW. Atomic layer deposition of TiO_2 thin films on nanoporous alumina templates: Medical applications. *JOM*, 2009, 61(6), 12-16.

Nel, AE; Mädler, L; Velegol, D; Xia, T; Hoek, EMV; Somasundaran, P; et al. Understanding biophysicochemical interactions at the nano-bio interface. *Nat Mater*, 2009, 8, 543-557.

Ni, M; Leung, MKH; Leung, DYC; Sumathy K. A review and recent developments in photocatalytic water-splitting using TiO_2 for hydrogen production. *Renew Sust Energ Rev*, 2007, 11(3), 401-425.

Obare, SO; Meyer, GJ. Nanostructured materials for environmental remediation of organic contaminants in water. *Journal of Environmental Science and Health, Part A: Toxic/Hazardous Substances and Environmental Engineering*, 2004, 39(10), 2549-2582

Paramasivam, I; Macak, JM; Schmuki, P. Photocatalytic activity of TiO_2 nanotube layers loaded with Ag and Au nanoparticles. *Electrochem Commun*, 2008, 10(1), 71-75.

Pardo-Martin, C; Chang, T-Y; Koo, BK; Gilleland, CL; Wasserman, SC; Yanik, MF. High-throughput *in vivo* vertebrate screening. *Nat Methods*, 2010, 7, 634-636.

Pathakoti, K; Huang, M-J; Watts, JD; He, X; Hwang, H-M. Using experimental data of *Escherichia coli* to develop a QSAR model for predicting the photo-induced cytotoxicity of metal oxide nanoparticles. *J Photochem Photobiol B*, 2014, 130, 234-240.

Pathakoti, K; Morrow, S; Han, C; Pelaez, M; He, X; Dionysiou, DD; et al. Photo-inactivation of *Escherichia coli* by sulfur and nitrogen-fluorine-codoped TiO_2 nanoparticles under solar simulated light and visible light irradiation. *Environ Sci Technol*, 2013, 47(17), 9988-9996.

Puzyn, T; Rasulev, B; Gajewicz, A; Hu, X; Dasari, TP; Michalkova, A; et al. Using nano-QSAR to predict the cytotoxicity of metal oxide nanoparticles. *Nat Nanotech*, 2011, 6, 175-178.

Rajapakse, K; Drobne, D; Kastelec, D; Marinsek-Logar, R. Experimental evidence of false-positive Comet test results due to TiO_2 particle--assay interactions. *Nanotoxicology*, 2013, 7, 1043-1051.

Robel, I; Kuno, M; Kamat, PV. Size-dependent electron injection from excited CdSe quantum dots into TiO_2 nanoparticles. *J Am Chem Soc*, 2007, 129(14), 4136-4137.

Ryu, J; Choi, W. Substrate-specific photocatalytic activities of TiO_2 and multiactivity test for water treatment application. *Environ Sci Technol*, 2008, 42(1), 294-300.

Sahu, SC; Casciano, DA. *Nanotoxicity, From in vivo and in vitro models to health risks*. Chichester, West Sussex, UK: John Wiley & Sons, 2009.

Sato, T; Taya, M. Enhancement of phage inactivation using photocatalytic titanium dioxide particles with different crystalline structures. *Biochem Eng J*, 2006, 29(3), 303-308.

SCCS. Opinion on titanium dioxide (nano form). In: *Scientific Committee on Consumer Safety* EC, editor., 2013.

Sclafani, A; Herrmann, JM. Comparison of the photoelectronic and photocatalytic activities of various anatase and rutile forms of titania in pure liquid organic phases and in aqueous solutions. *J Phys Chem*, 1996, 100, 13655-13661

Selloni, A; Vittadin, iA; Gratzel, M. The adsorption of small molecules on the TiO_2 anatase(101) surface by first-principles molecular dynamics. *Surf Sci*, 1998, 404, 219-222.

Senthilnathan, J; Philip, L. Removal of mixed pesticides from drinking water system using surfactant-assisted nano-TiO_2. *Water Air Soil Pollut*, 2010, 210(1-4), 143-154.

Shahmoradi, B; Ibrahim, IA; Sakamoto, N; Ananda, S; Row, TNG; Soga, K; et al. *In situ* surface modification of molybdenum-doped organic-inorganic hybrid TiO_2 nanoparticles under hydrothermal conditions and treatment of pharmaceutical effluent. *Environ Technol*, 2010, 31(11).

Simon-Deckers, A; Loo, S; Mayne-L'hermite, M; Herlin-Boime, N; Menguy, N; Reynaud, C; et al. Size-, composition- and shape-dependent toxicological impact of metal oxide nanoparticles and carbon nanotubes toward bacteria. *Environ Sci Technol*, 2009, 43(21), 8423-8429.

Smith, AM; Nie, S. Semiconductor nanocrystals: Structure, properties, and band gap engineering. *Acc Chem Res*, 2010, 43(2), 190-200.

Sohaebuddin, SK; Thevenot, PT; Baker, D; Eaton, JW; Tang, L. Nanomaterial cytotoxicity is composition, size, and cell type dependent. *Part Fibre Toxicol*, 2010, 7, 22. 10.1186/1743-8977-7-22.

Tong, T; Shereef, A; Wu, J; Binh, CTT; Kelly, JJ; Gaillard, J-F; et al. Effects of material morphology on the phototoxicity of nano-TiO_2 to bacteria. *Environ Sci Technol*, 2013, 47(21), 12486-12495.

Tzikalos, N; Belessi, V; Lambropoulou, D. Photocatalytic degradation of Reactive Red 195 using anatase/brookite TiO_2 mesoporous nanoparticles: Optimization using response surface methodology (RSM) and kinetics studies. *Environ Sci Pollut Res*, 2013, 20(4), 2305-2320.

U.S. EPA. Nanomaterial case studies, Nanoscale titanium dioxide in water treatment and in topical sunscreen. In: Agency USEP, editor. Washington, DC, 2010.

U.S. FDA. Title 21--food and drugs. Chapter I--food and drug administration. Part 73 -- Listing of color additives exempt from certification. Subpart A--Foods. Sec. 73.575 Titanium dioxide. In: Department Of Health And Human Services USFaDA, editor., 2013.

U.S. NIOSH. Current intelligence bulletin 63: Occupational exposure to titanium dioxide. In: National Institute for Occupational Safety and Health CfDCaP, Department of Health and Human Services, editor., CreateSpace Independent Publishing Platform, 2011.

Vankayala, R; Kuo, C-L; Sagadevan, A; Chen, P-H; Chiang, C-S; Hwang, KC. Morphology dependent photosensitization and formation of singlet oxygen ($^1\Delta_g$) by gold and silver nanoparticles and its application in cancer treatment. *J Mater Chem B*, 2013, 1, 4379-4387.

Vecchio, G; Galeone, A; Brunetti, V; Maiorano, G; Sabella, S; Cingolani, R; et al. Concentration-dependent, size-independent toxicity of citrate capped AuNPs in *Drosophila melanogaster. PLoS ONE*, 2012, 7(1), e29980

Vinu, R; Akki, SU; Madras, G. Investigation of dye functional group on the photocatalytic degradation of dyes by nano-TiO_2. *J Hazard Mater*, 2010, 176(1-3), 765-773.

Vinu, R; Madras, G. Photocatalytic degradation of water pollutants using nano-TiO_2. In: Zang L, editor. *Energy Efficiency and Renewable Energy Through Nanotechnology Green Energy and Technology*. e-book, Springer, 2011, 625-677.

Vittadini, A; Selloni, A; Rotzinger, FP; Gratzel, M. Structure and energetics of water adsorbed at TiO_2 anatase (101) and (001) surfaces. *Phys Rev Lett*, 1998, 81, 2954-2957.

Wang, C; van, der, Vliet, D; Chang, K-C; You, H; Strmcnik, D; Schlueter, JA; et al. Monodisperse Pt_3Co nanoparticles as a catalyst for the oxygen reduction reaction: Size-dependent activity. *J Phys Chem C*, 2009, 113(45), 19365-19368.

Wang, G; Wang, Q; Lu, W; Li, J. Photoelectrochemical study on charge transfer properties of TiO_2-B nanowires with an application as humidity sensors. *J Phys Chem B*, 2006, 110(43), 22029-22034.

Wang, J; Chen, C; Liu, Y; Jiao, F; Li, W; Lao, F; et al. Potential neurological lesion after nasal instillation of TiO_2 nanoparticles in the anatase and rutile crystal phases. *Toxicol Lett*, 2008, 183(1-3), 72-80.

Warheit, DB; Webb, TR; Sayes, CM; Colvin, VL; Reed, KL. Pulmonary instillation studies with nanoscale TiO_2 rods and dots in rats: Toxicity is not dependent upon particle size and surface area. *Toxicol Sci*, 2006, 91(1), 227-236.

Weir, A; Westerhoff, P; Fabricius, L; Hristovski, K; Goetz, Nv. Titanium dioxide nanoparticles in food and personal care products. *Environ Sci Technol*, 2012, 46(4), 2242-2250.

Westerhoff, P; Song, G; Hristovski, K; Kiser, MA. Occurrence and removal of titanium at full scale wastewater treatment plants: implications for TiO_2 nanomaterials. *J Environ Monit*, 2011, 13(5), 1195-1203.

Yang, K; Lin, D; Xing, B. Interactions of humic acid with nanosized inorganic oxides. *Langmuir*, 2009, 25(6), 3571-3576.

Yu, B; Zeng, J; Gong, L; Zhang, M; Zhang, L; Chen, X. Investigation of the photocatalytic degradation of organochlorine pesticides on a nano-TiO_2 coated film. *Talanta*, 2007, 72(5), 1667-1674.

Yun, HJ; Lee, H; Joo, JB; Kim, ND; Yi, J. Tuning the band-gap energy of $TiO_{2-x}C_x$ nanoparticle for high performance photo-catalyst. *Electrochem Commun*, 2010, 12(6), 769-772.

Zheng, X; Chen, Y; Wu, R. Long-term effects of titanium dioxide nanoparticles on nitrogen and phosphorus removal from wastewater and bacterial community shift in activated sludge. *Environ Sci Technol*, 2011, 45(17), 7284-7290.

Zhou, W; Pan, K; Qu, Y; Sun, F; Tian, C; Ren, Z; et al. Photodegradation of organic contamination in wastewaters by bonding TiO_2/single-walled carbon nanotube composites with enhanced photocatalytic activity. *Chemosphere*, 2010, 81(5), 555-561.

Zhu, K; Neale, NR; Miedaner, A; Frank, AJ. Enhanced charge-collection efficiencies and light scattering in dye-sensitized solar cells using oriented TiO_2 nanotubes arrays. *Nano Lett*, 2007, 7(1), 69-74.

Zhu, X; Chang, Y; Chen, Y. Toxicity and bioaccumulation of TiO_2 nanoparticle aggregates in *Daphnia magna*. *Chemosphere*, 2010, 78(3), 209-215.

Zhu, X; Zhou, J; Cai, Z. TiO_2 nanoparticles in the marine environment: Impact on the toxicity of tributyltin to abalone (*haliotis diversicolor supertexta*) embryos. *Environ Sci Technol*, 2011a, 45(8), 3753-3758.

Zhu, X; Zhou, J; Cai, Z. The toxicity and oxidative stress of TiO_2 nanoparticles in marine abalone (*Haliotis diversicolor supertexta*). *Mar Pollut Bull*, 2011b, 63(5-12), 334-338.

In: Titanium Dioxide
Editor: Jerri Brown

ISBN: 978-1-63321-391-3

Chapter 2

INDOOR ENVIRONMENT CONTROL WITH FUNCTIONAL MORTARS

S. S. Lucas*

University of Aveiro CICECO/Department of Civil Engineering, Campus Santiago, Portugal

ABSTRACT

In developed countries, people spend most of the time inside buildings, either their offices or houses. Indoor air quality has a strong impact on wellbeing since unhealthy or uncomfortable environments can cause illness and low productivity. The selection of materials that contribute to improve indoor air quality improves significantly the building quality and sustainability. Mortars, because they are one of the most commonly used materials, represent a strong environmental impact in construction. Preparing mortars with new and innovative functionalities, that promote energy savings and increase air quality, help to develop buildings with better environmental performance. However, adding a new functionality comprises the introduction of new additives into the mortar matrix, causing microstructural modifications that influence fresh and hardened state properties. Adding new additives always imply a compromise between functionality and durability. In this chapter, was investigated the introduction of a photocatalytic additive and a phase change material, in single and binary formulations. Latent heat storage, self-cleaning and depolluting properties were studied and the impact of the additives in the mortar final properties was assessed. This chapter proved that functional mortars maintain good fresh and hardened state properties improving sustainability in new and historical buildings.

Keywords: Construction, materials, mortars, photocatalysis, phase change materials

* Email: sandra@sandralucas.pt.

1. Introduction

The process of selecting materials it is one of the most important tasks related with the building project. To achieve sustainability the choice should be based in specific criteria, assuring that the building materials and products will contribute for:

- Improve indoor air quality
- Increase energy efficiency
- Lowering the incorporated energy
- Improve building durability and reduce maintenance costs
- Increasing the use of recycled or recyclable materials
- Improve the use of materials from renewable sources
- Lowers pollutants emission during the building life cycle

Lee et al. conducted a study in 2008 that demonstrated how indoor air quality could affect building occupants. In commercial buildings, were people spend most of their working hours lower quality levels can decrease productivity and increase absence by health problems [1]. Jones et al., refer thermal comfort and indoor air quality as playing a very important role in quality perception by users. This chapter encompasses the study and development of mortars able to reduce pollutant concentration inside buildings and at the same time, to control the temperature by thermal storage [2].

Indoor air can be influenced by both internal and external factors, like:

- Pollutants released by finishing and furniture
- Pollutants released by construction materials
- Human activities (cooking, smoking, fireplaces)
- Contamination from external air (from windows and ventilation)

Several different contaminants can be present in the indoor air:

- Substances released in combustion and human activities – carbon dioxide, carbon monoxide, nitrogen oxides, sulfur oxides.
- Substances released by furniture, paints, cleaning products – VOC (volatile organic compounds) like formaldehyde or BTEX (benzene-toluene-ethylbenzene and xylene) [3]

The application of wall coatings (mortars or paints) with the capability to degrade these pollutants by photocatalysis, will contribute for better indoor air quality [4].

2. Photocatalytic Degradation of Indoor Pollutants

The photocatalytic effect is well known since 1950, some authors reported studies from even earlier (1920) [4]. Two different catalysers were investigated, zinc oxide (ZnO) and titanium dioxide (TiO_2), the last one in two different crystalline phases: rutile and anatase [5-

8]. TiO_2 has been by far, the most studied due to its high chemical stability, good efficiency and low cost. The photocatalytic efficiency was improved when researchers started to produce TiO_2 in nanosized particles [9, 10] When TiO_2 is applied in coatings, for instance, as a thin film, these are able to decompose organic compounds in the presence of UV or solar light. This behaviour, combined with the hydrophobicity of the surface, is responsible for the self-cleaning capabilities [11, 12]. Another property of titanium dioxide is the ability to react with some common pollutants like VOC, NOx and SOx, among others. Surfaces prepared with this photocatalytic additive are able to destroy the pollutants in the surrounding area. The application in construction materials has been deeply explored in the last decades [13-15].

Titanium dioxide can be found in 3 different crystalline forms: rutile, anatase and brookite. The first two are easier to produce in nanoparticles and only anatase shows good photocatalytic properties [16]. If TiO_2 is irradiated with light from a wavelength higher than the energy of the valence band, electrons from this level are excited and switch to the conduction band, leaving a hole.

The hole exhibits a strong oxidation capacity, and electrons have strong reduction potential. Consequently, the electron-hole pair can induce catalytic reactions at the surface. The holes react with the water molecules around TiO_2 particles producing a hydroxyl radical. The electrons react with oxygen creating a superoxide radical [17]. Equations 1 to 3 resume the reactions involved in the photocatalytic process, with 2 and 3 occurring simultaneously. Where hf stands for energy demanded to cause the electron to switch from the valence band to the conduction band, e^- matches the electron from the conduction band and h^+ the electron hole.

$$\mathrm{TiO_2 + hf \rightarrow TiO_2 + e^- + h^+} \quad (1)$$

$$\mathrm{e^- + O_2 \rightarrow O_2^-} \quad (2)$$

$$\mathrm{h^+ + H_2O \rightarrow HO + H^+} \quad (3)$$

These free radicals react with air pollutants and organic matter, converting it into harmless substances [18]. The reaction products lay on the surface until gravity, wind or water removes them, which reactivates the catalyser. The efficiency of this process can be evaluated by:

- Colorimetric tests
- Degradation of one or more aromatic and organic compounds (VOC, formaldehyde)
- Degradation tests with inorganic pollutants

2.1. Photocatalysis of Indoor Contaminants

Although VOC is one of the preferred contaminants for testing photocatalytic degradation, its application with mortars can constitute a problem [19]. Carbon dioxide (CO_2) is one of the reaction products and because mortars, especially the ones prepared with aerial binders endure carbonation, this product can influence the properties of the mortar and affect the results [20]. Nitrogen oxides (NOx) are a suitable alternative to solve the carbonation problem. Equations 4 to 6 describe the degradation process where OH and O_2^- radicals created during the photocatalytic activation, reacts with NO_2 to produce HNO_3 [21].

$$NO + OH^- \rightarrow NO_2 + H^+ \tag{4}$$

$$NO_2 + OH^- \rightarrow HNO_3 \tag{5}$$

$$NO + O_2^- + \rightarrow NO_3^- \tag{6}$$

Using the initial and final concentration of the gas, $C_{initial}$ and C_{final} respectively, photocatalytic efficiency (η) it can then be calculated according to Equation 7.

$$\eta = 1 - \frac{C_{final}}{C_{initial}} \tag{7}$$

Titanium dioxide has demonstrated good degradation efficacy for concentrations between 0.01 and 10 ppmv [22, 23].

2.2. Self-Cleaning

Titanium dioxide ability to degrade inorganic matter provides surfaces with self-cleaning capabilities. Building facades, especially in historic buildings are often attacked by mould or damaged with graffiti. TiO_2 nanoparticles confer to mortars and concrete self-cleaning properties guaranteeing higher durability with lower maintenance costs. The efficiency of the cleaning process is directly related with the surface decontamination rate. The decontamination, provided by an effective removal of the polluting agent (ink, mould or other) needs to take place to assure the catalyser active sites became again available. The faster the decontamination occurs the sooner the surface will be active and ready to degrade the contaminant. External facades, exposed to wind and rain can be rapidly cleaned, improving self-cleaning efficiency[11].

Self-cleaning is strongly related with the hydrophobicity (water repulsion) of TiO_2 surfaces resultant from the extremely low contact angle of water molecules when titania is activated (Figure 1). This phenomenon occurs simultaneously with photocatalysis and contributes to the removal of the decomposed organic matter [24]. Testing conditions can influence photocatalytic activity and self-cleaning efficiency. Factors affecting photocatalysis will be discussed with more detail in the next sub-section.

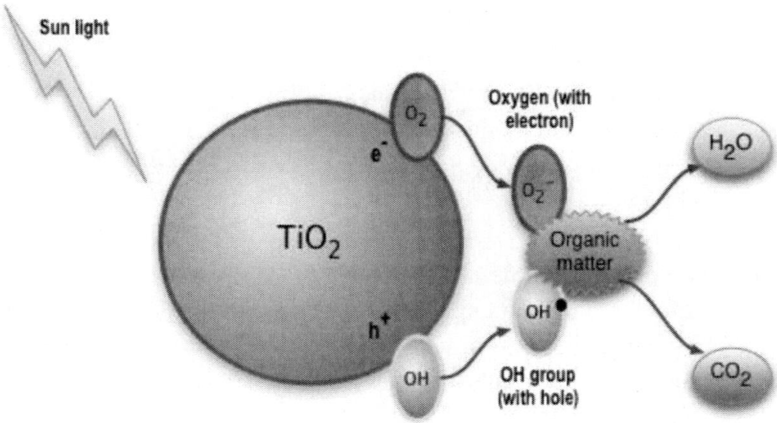

Figure 1. Photocatalytic self-cleaning in surfaces with TiO2.

2.3. Factors Affecting the Photocatalytic Process

2.3.1. Light Intensity

Light conditions are fundamental to ensure that heterogeneous photocatalysis occur since the activity is a direct result from exposure to a wavelength of 385 nm or less. However, light intensity does not seem to influence the process as photocatalysis still takes place in lower intensities [25]. This means that photocatalysis can be performed in confined spaces such as offices or houses.

2.3.2. Catalyzer Particle Size

Because photocatalytic reactions only occur at the surface, the efficiency will increase with the area exposed, which explains the interest in using nanoparticles as an additive. Though reducing the particle size will improve photocatalysis this is only effective to a certain extent. Below a certain minimum size (10 nm), the efficiency starts to decay due to the high recombination rate of electron-hole pairs leading to the inactivation of the catalyser [9].

The titanium dioxide selected for this chapter exhibits a particle size of 21 nm, small enough to ensure effective surface area and sufficient to avoid the catalyst deactivation.

Xie et al., presented an alternative route to enhance photocatalytic activity – metal doping [26]. The process is still too complex, expensive and inefficient to be applied in a large-scale production [27].

2.3.3. Temperature

Ideally, photocatalysis should be conducted at room temperature, and the process will not be affected by small deviations. For extreme conditions, above 80 °C, the degradation rate can diminish considerably because of lower catalyzer activation [28, 29].

2.3.4. Humidity

High values of humidity lead to an excess of water molecules that start to compete with the pollutants for the occupation of the active sites. Several studies indicate values around 40% to 50% as the most favourable [30, 31].

2.3.5. Catalyser Content

Photocatalytic activity improves with the increasing amount of catalyser. Above what was established as the optimum content, the electron-hole pairs reach a recombination rate that leads to the reduction of active sites and surface deactivation [17]. The ideal additive content differs with the type of substrate, in mortars can vary depending on the type of binder [32].

2.3.6. Pollutant Concentration and Intermediate Products

The products resulting from simultaneous oxidation and reduction reactions are adsorbed at the surface. For high concentrations of a pollutant, the active sites are rapidly captured, and the degradation rate decreases [21]. Intermediate products deposited at the surface can create a barrier between the pollutant and the photocatalytic additive causing a temporary deactivation until their removal [17].

2.3.7. Pollutant Flow Rate (Gaseous Substances)

Besides pollutant concentration, the flow rate can also affect the process. If it is too high, the time that the gas remains in contact with the catalytic additive will be so short that degradation does not take place. When the goal is to assess the photocatalytic efficiency indoors, the residence time for the gas inside the reactor should simulate accurately the real service conditions. This means that flow rate needs to be adjusted to the reactor configuration to simulate the renovation rate for indoor conditions [33].

2.4. Combining Photocatalysis with Other Indoor Control Systems

2.4.1. Using Phase Change Materials for Heat Storage in Mortars

Several factors can influence indoor comfort, not only air quality levels, also temperature can be an important factor. To maintain an adequate indoor temperature for human in buildings, HVAC systems for heating and/or cooling or other active strategies can be applied. This always implies energy demand, and researchers look for new and passive methods to achieve thermal comfort [34]. Phase change materials (PCM) may be applied as a passive storage media for heat.

The latent heat transfer, caused by the PCM fusion, occurs at a temperature designated as phase change temperature. Latent heat storage (ΔQ) is therefore, dependent of the enthalpy change and can be determined from Equation 8.

$$\Delta Q = m \times \Delta H \tag{8}$$

where ΔH, enthalpy variation (J); and m, material mass (g). When the temperature rises, the PCM stores a certain amount of heat that is able to release as soon as the temperature drops below the phase change point [35]. This effect can be advantageous when the capsules are mixed with a mortar, creating walls capable of thermal indoor control.

These two additives have been studied separately, mostly because combining more than one additive in the same mortar needs a careful investigation of microstructural properties to select the best combination for the dual system. Functionalities such as heat storage and self-cleaning have been always considered separately, so the products available so far can only include one of these options. With the increasing demand for new and more innovative materials, future research will necessarily pass by including several functionalities in one product [36]. This chapter includes the development of a multifunctional mortar combining energy storage, self-cleaning and air depolluting capabilities. This goal was accomplished by the manipulation of mortars internal structure, studying the microstructural modifications with different amounts of PCM (phase change material) and nanoparticles of titanium dioxide. This chapter demonstrates that it is possible to develop new advanced mortars for building applications, more complex and with multiple functions, opening a new field of opportunities for the construction sector.

3. EXPERIMENTAL

3.1. Materials

The raw materials used for the mortars preparation consists of commercial aerial lime (Calcidrata), Portland cement CEM II 32.5 N, gypsum (SIVAL) as binders and siliceous sand as the aggregate. The titanium dioxide, used as a photocatalytic additive, is a commercial product (Degussa P25), mainly consisting of anatase with a specific surface area of 50 ± 15 m2 g-1. A superplasticizer (Glenium 51) was added as an admixture to all compositions (0.5%[1]) in order to control the workability and water content.

Mortars were prepared with those different binders and compositions, which are summarized in Table 1. For each formulation, mortars with 0, 0.5, 1, 2.5 and 5% of TiO_2 added to the total solid content were tested. The mortar workability was adjusted using the flow table. The amount of water added in relation to the total weight of solid material varies in order to keep constant the slump value for each composition.

Table 1. Binder, aggregate and additive content for the photocatalytic compositions

Composition		0% TiO_2	0.5% TiO_2	1% TiO_2	2.5% TiO_2	5% TiO_2
L	Lime	25	24.9	24.8	24.4	23.8
	Sand	75	74.6	74.3	73.2	71.4
	TiO_2	0	0.5	1	2.4	4.8
LC	Lime	9	9	8.9	8.8	8.6
	Cement	13	12.9	12.9	12.7	12.4
	Sand	78	77.6	77.2	76.1	74.3
	TiO_2	0	0.5	1	2.4	4.8
LG	Lime	12	11.9	11.9	11.7	11.4
	Gypsum	8	8	7.9	7.8	7.6
	Sand	80	79.6	79.2	78	76.2
	TiO_2	0	0.5	1	2.4	4.8
C	Cement	20	19.9	19.8	19.5	19
	Sand	80	79.6	79.2	78	76.2
	TiO_2	0	0.5	1	2.4	4.8
G	Gypsum	20	19.9	19.8	19.5	19
	Sand	80	79.6	79.2	78	76.2
	TiO_2	0	0.5	1	2.4	4.8

[1] Within this chapter, whenever % is used for TiO_2 or PCM content, it always refer to weight percentage also designated as wt%.

Table 2. Formulations for the multifunctional mortars

	2.5% TiO_2 20% PCM	2.5% TiO_2 25% PCM	2.5% TiO_2 30% PCM	5% TiO_2 20% PCM	5% TiO_2 25% PCM	5% TiO_2 30% PCM
Lime	20	20	19	20	19	19
Sand	61	59	57	60	58	56
TiO2	2.0	2.0	1.9	4.0	3.8	3.7
PCM	16	20	23	16	19	22

The phase change material (Micronal DS 5008) comprises a paraffin mixture encapsulated in a polymethylmethacrylate (PMMA) shell, with an average particle size of 6 μm, transition temperature of 23°C and enthalpy 135 kJ/kg. For the binary compositions, the author selected the lime mortar with a 2:1 sand/binder ratio and combined amounts of titanium dioxide and PCM as described in Table 2.

To guarantee the tests reproducibility a methodology for the preparation of mortars was defined, and was followed throughout the study.

- Solid materials weighing (raw materials and additives)
- Manual homogenization in a sealed plastic bag
- Weighing of superplasticizer and mixing water
- Addition of superplasticizer to the mixing water
- Mixing of solids with the liquid
- Homogenization in a blender for 60 seconds
- A final manual mixing for 60s followed by homogenization during 90 seconds, in the automatic blender.

3.2. Fresh State Properties

Flow table measurements when complemented with rheology tests are useful to determine the mixture formulations helping to establish the additive content. It is a quick and convenient way to estimate the workability of mortars, and it is sensitive to identify the effect of several additions [37]. However, its accuracy is not enough to define the rheological properties of mortars, because two compositions with distinct rheological profiles can exhibit similar spread values on the flow table. The rheological profile of mortars has a significant influence on some hardened state properties, such as porosity, bulk density, mechanical strength and durability [38, 39]. The rheological behaviour of mortars is described according to the Bingham model, as:

$$\tau = \tau 0 + \eta \gamma \tag{9}$$

where, τ represents the shear stress, τ_0 is the yield stress and η the plastic viscosity.

The rheology tests were made with a specific equipment, Viskomat NT. Torque variation, T, as a function of rotation speed, N, follows the Bingham model:

$$T = g + h \times N \tag{10}$$

where, g is related to the yield stress and h to the plastic viscosity. The maximum rotation speed was kept constant (100 rpm) with programmed pauses each 5 minutes for lime mortars and 15 minutes for cement, which shaped the flow curves for the calculation of g and h.

3.3. Hardened State Properties

The samples for the mechanical strength tests were prepared following the procedure described in the European Standard EN 1015-11. The tests were carried out after a curing time of 90 days in storage conditions with a temperature of 20 °C ± 2 °C and 65% ± 5% of relative humidity. The pore size and pore distribution were determined using a mercury intrusion porosimeter (AutoPore IV Micromeritics) working in the pressure range from 4 kPa to 228 MPa and intrusion in pores between 5.5 nm and 360 μm.

3.4. Photocatalytic Tests

To conduct the photocatalytic test, a specific setup using NOx as a pollutant was developed. NOx is a common pollutant gas present inside buildings and urban areas, in concentrations around 1 ppmv [40]. This value was selected as the initial concentration for the laboratory tests. Several authors also used this amount and it would facilitate comparisons with related studies [13, 41-44]. NOx (10 ppmv) was diluted in standardised air, to obtain the desired concentration. Gas inlet measured with mass flow controllers was injected to the chamber at a constant flow rate of 1 l/min. The Portuguese legislation was the reference for the requirements of indoor air exchange rate, 0.6 renovations per hour [45]. The reaction took place inside the stainless steel chamber. The cylinder with a 35 L capacity is completely sealed and the glass window on the top allowed the light from the 300-Watt solar lamp (OSRAM UltraVitalux), placed at 1m from the sample, to irradiate the mortar. During the tests, a thermocouple and a humidity sensor monitored temperature and humidity, and the values remained steady at 20 0C and 40% respectively. Excessive temperature can affect the photocatalytic measurements making difficult to distinguish the degradation effect from an overheating phenomenon [46]. A chemiluminescence analyser (AC-30 M, Environment SA) measured the concentration of the outlet gas.

A blank reference and TiO_2 containing samples were tested according to the following process:

- Placement of the sample in the holder inside the chamber with the glass window covered.
- The inlet gas mixture starts to flow until NOx reaches 1 ppmv measured by the analyser. This stabilisation step assures that sample is completely saturated, preventing an absorption effect from the sample and reactor walls.
- The solar lamp is turned on, the reactor glass is uncovered, and the reaction starts. During this phase, the NOx concentration is monitored continuously. When the concentration reaches a constant minimum level, it is considered that the reaction stopped and the sample is unable to degrade more pollutant.

- The reaction is completed.

The photocatalytic efficiency is determined by equation 11

$$NOx_{removed} = \frac{[NOx]_{TiO_2} - [NOx]_{blank}}{[NOx]_{TiO_2}} \times 100 \quad (11)$$

where $[NOx]_{TiO2}$ is the final pollutant concentration after the irradiation test and $[NOx]_{blank}$ is the NOx concentration for the blank test (0% TiO_2).

3.5. Latent Heat Storage

The test developed to assess the heat storage capability was designed to simulate as closely as possible the real application conditions. Test cells placed inside a climatic chamber were submitted to a specific temperature cycle. The minimum temperature of 10 °C was the starting point and the maximum value of 40°C was reached at a constant rate. The highest and lowest temperatures were kept constant for 10 minutes, and this cycle was repeated for 24 hours. The test cells made with an insulating material (extruded polystyrene) were coated on the inside with a mortar layer of 3 mm. During the test, thermocouples placed inside the boxes controlled the temperature inside the cells with PCM and the reference sample. The temperature cycle forces the PCM to change phase (around 23 to 25 ºC) and it is possible to determine the impact of the heat storage capability of the mortar [47]. The multifunctional mortars were submitted to cyclic thermal tests to determine if the use of multiple additives can influence the thermal behaviour.

4. Discussion

4.1. Mortars with Titanium Dioxide

4.1.1. Effect of TiO_2 on Mortar Fresh State Properties

The rheology tests were performed using the two most important binders for this chapter, lime and cement. In Figure 2, it is visible the effect of incorporating nanoparticles in lime mortars, causing a strong increase in the torque and a reduction of the testing time. Separation of aerial lime aggregates and the high surface area of titanium dioxide nanoparticles cause the reduction of the water, promoting interparticle collision that leads to the mixture thickening [48]. Mortars with 5% of titanium dioxide exhibit higher torque showing low plasticity after only 15 minutes. Because of intermolecular forces (Van der Walls), this additive tends to agglomerate causing unstable flow during rheology tests [49]. According to Figure 3, increasing the photocatalytic additive increases plastic viscosity, a result from the separation of nanoparticle agglomerates however, after constant stirring, goes down again. Lower workability for lime mortars with higher content of TiO_2 with reduction of free water inside the system, increases yield stress (Figure 4) [50].

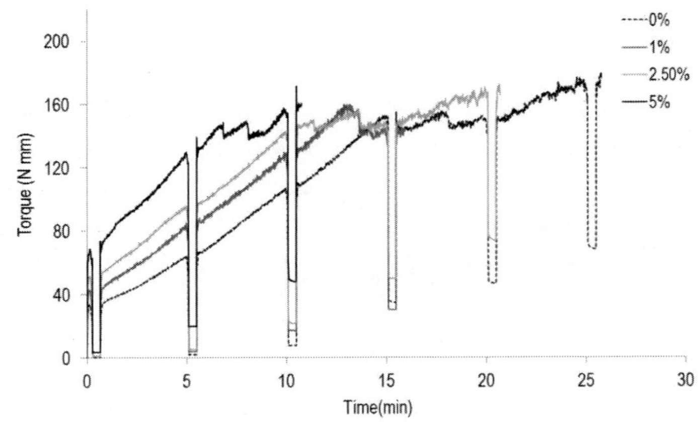

Figure 2. Torque variation with time for the lime mortars.

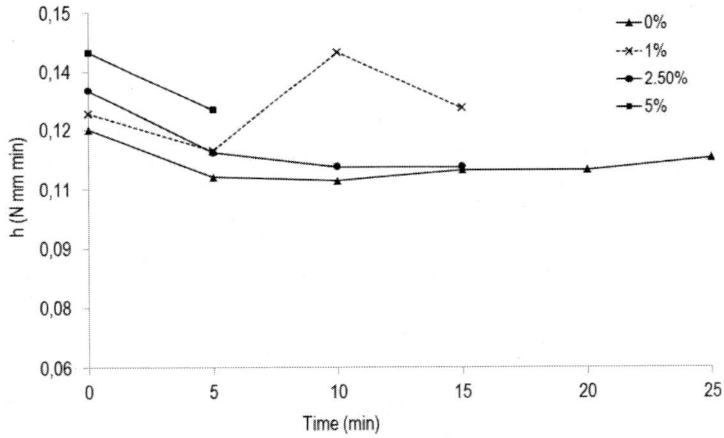

Figure 3. Plastic viscosity of the lime mortars.

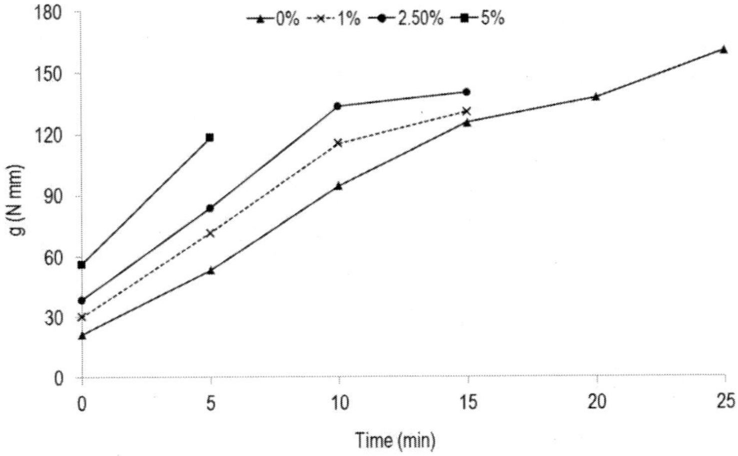

Figure 4. Yield stress for the lime compositions.

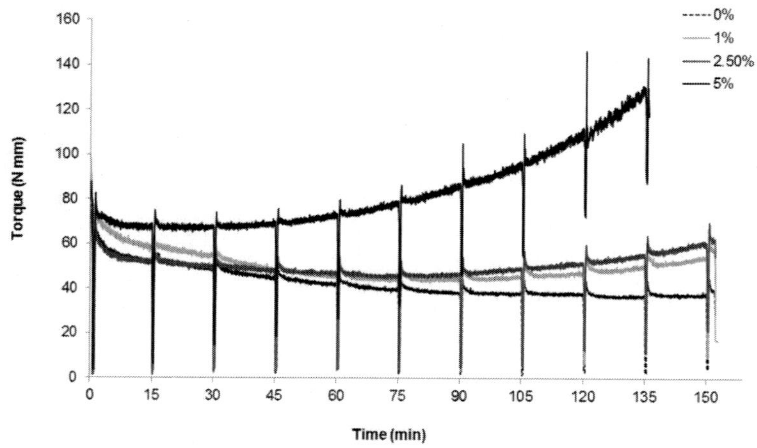

Figure 5. Torque variation with time for the cement formulations.

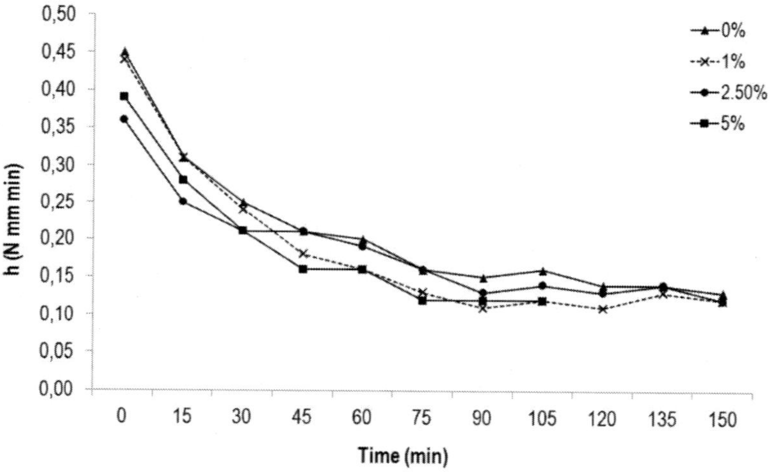

Figure 6. Plastic viscosity of the cement mortars.

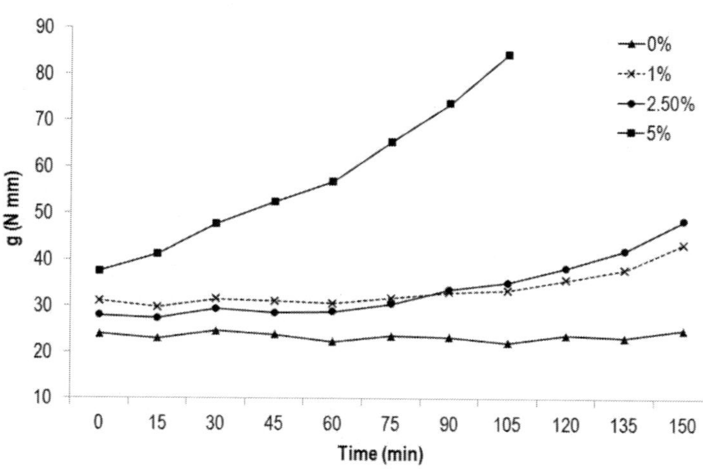

Figure 7. Yield stress for the cement compositions.

Cement mortars show different rheological profile due to the absence of lime agglomerates and the effect of free water reduction appears later during the test (Figure 5).

The tests can be conducted for more than 120 minutes even for 5% TiO_2. The effect of nanoparticles in the system is no longer detectable by measuring plastic viscosity (Figure 6). Only by monitoring yield stress (Figure 7) one can see the effect of nanoparticle agglomeration for the composition with 5%.

Adding up to 2.5% of titanium dioxide to lime and cement compositions does not compromise fresh state properties, making viable its preparation as wall coatings.

4.1.2. Effect on Hardened State Properties

Figure 8 and 9 show the mechanical strength of the compositions tested in this chapter whereas the lime mortars are designated by L, blended lime and gypsum mortars are LG, lime and cement are designated by LC. The cement mixture is C and G represents gypsum compositions. The incorporation of titanium dioxide nanoparticles in lime mortars (L) reduces both flexural and compression strength though decrease is non-linear with the introduction of more additive.

Adding just 0.5% of TiO_2 reduces the flexural and compression strength, however increasing amounts up to 2.5% cause a slight improvement in the flexural strength while the compressive strength remains unaffected. It is also noticeable that more than 2.5% affect the mechanical properties even though 5% strength is close to the 0.5% values. Up to 2.5% the mechanical strength remains higher than 1 MPa, only when the additive exceeds this amount the strength falls below this limit.

For the purpose of clarification in the pore distribution analysis, pores above 10 μm are designated as macropores, pores between 10 and 1 μm are considered as micropores and below 1 μm nanopores. The pore size analysis (Figure 10) shows a bimodal distribution for all compositions with a clear separation in two distinct pore size classes, an event highlighted by the introduction of TiO_2 in the mortar.

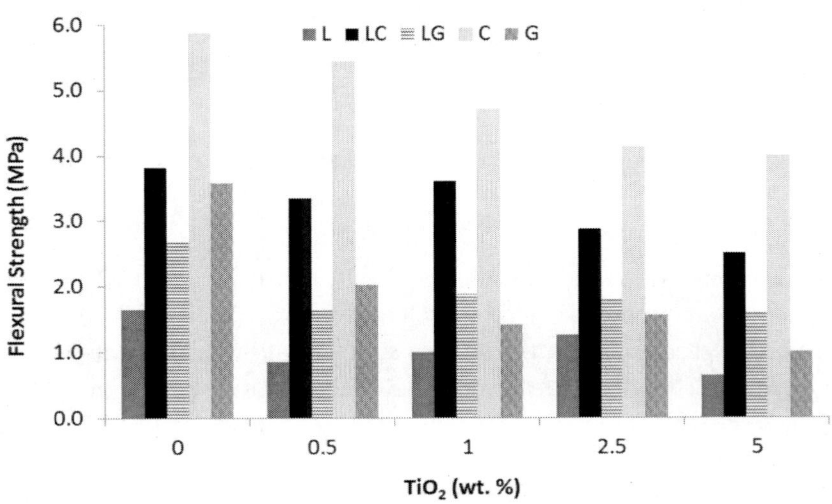

Figure 8. Flexural strength of the compositions tested.

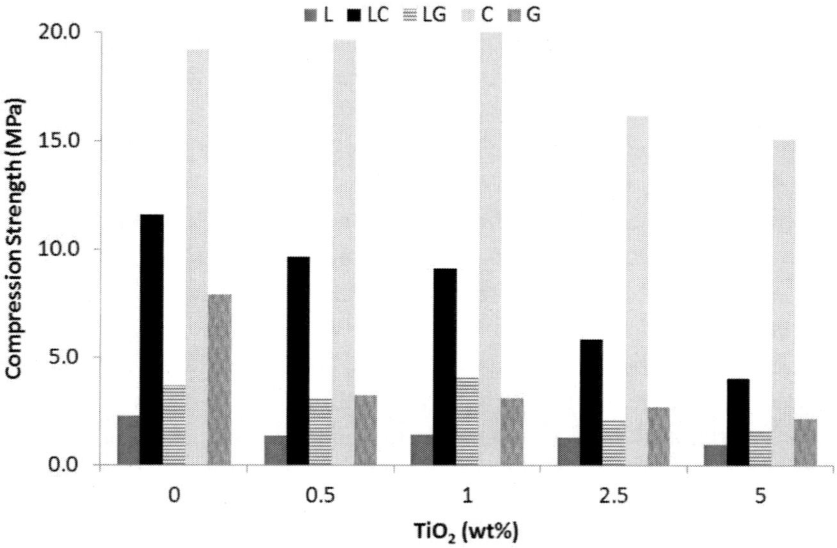

Figure 9. Compression strength of the compositions tested.

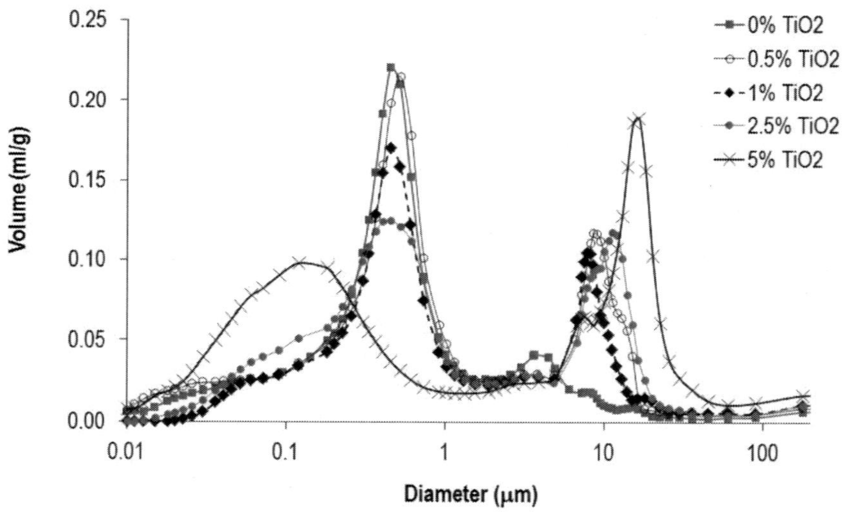

Figure 10. Pore distribution for the lime mortars (L).

The reference composition (0%) exhibits a porosity distributed mostly between 0.1 and 1 μm, with a small amount of pores in the range of 5 to 100 μm. With the introduction of 0.5% TiO_2 the presence of macropores in the internal structure of the mortar causes a decrease in the flexural strength. The effect of an increased macroporosity (> 10 μm) reflects in the evolution of the mechanical properties. Titanium dioxide additions of 1 to 2.5% cause a slight reduction in both macropores and micropores which explains the flexural strength recovery. The increase of larger pores accounts for the higher fragility of the mortar structure, for 5% TiO_2, regardless of a strong presence of nanopores.

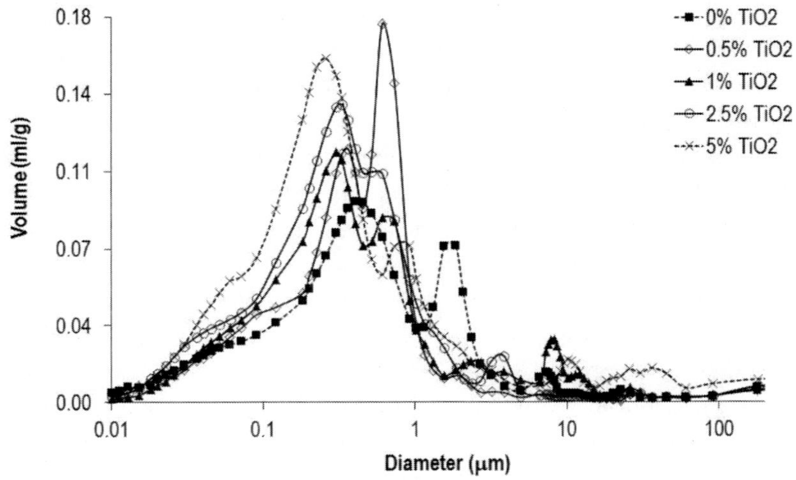

Figure 11. Pore distribution for the lime-cement mortars (LC).

The partial substitution for cement binder contributes to a higher mechanical strength in aerial lime and cement composition (LC) compared to a lime-based mortar (L). The cement-lime mortar shows a progressive reduction of the mechanical strength with the incorporation of TiO_2 nanoparticles, more pronounced above 2.5%. The reference composition (0%) exhibits an internal porosity distributed in two major intervals corresponding to the highest peaks in Figure 11, 2-3 μm and 0.05-1 μm (micropores and nanopores respectively), there is also a residual amount of macropores in the range of 9 to 15 μm. The addition of 0.5% TiO_2 causes an approximation of the higher peaks, to a range between 0.06 and 1 μm. Although the total amount of pores increases, the residual macroporosity disappears. Establishing a comparison between the lime-based and the cement-lime mortar it can be concluded that the partial substitution with cement reduces the average pore size and the macroporosity above 1 μm. This effect explains the better hardened state performance of the LC mortar. The microstructure of cement-lime mortars is more complex because of the evolution of the carbonation process. The pore size distribution tends to evolve to lower size ranges (micro and nanopores). Karatasios et al., reported that the incorporation of TiO_2 nanoparticles might accelerate the carbonation process which can contribute to the increase of micropores, in formulations with TiO_2 additions above 1% [51]. On the other hand, the water content rise with the increasing amount of nanoparticles contributes to a higher porosity so there is a joint effect of pore size reduction and increased porosity for the compositions with higher amounts of TiO_2 [32].

The third mortar composition tested (LG) was a lime-gypsum plaster, commonly used in old buildings before the development of Portland cement and the consequent widespread of cement-based mortars [52]. This composition shows a reduction in the mechanical strength for 0.5% TiO_2 and recovery when 1% of additive is added. This performance is common to the blended cement pastes (LC and LG) and the lime mortars, indicating that the presence of the lime binder can favour the accommodation of up to 1% nanoparticles and stabilizes the mechanical properties. The titanium dioxide incorporation above 2.5% leads to a strength decrease, resulting from the higher nanoparticles content. The additive excessive content overcomes the benefit from the aerial lime binder incorporation, nevertheless, the mechanical strength remains above 1 MPa. For the LG composition, the pore size distribution exhibits similar curves for the different amounts of titanium dioxide (Figure 12). This single-mode pore size distribution differs only in the porosity range for the additive content. For the reference mortar (without TiO_2) the pores are distributed in the interval between 1 and 3 μm, the increase up to 1% additive reduces the pore size to an interval between 0.2 and 1.5 μm. The porosity distribution from 0% to 1% TiO_2 follows a very similar pattern. The variation observed in the mechanical properties results mainly from the difference in pores size and quantity. When the amount of TiO_2 in the mortar varies between 2.5% and 5% the tendency is inverted, the pore size increases causing a decay in the mechanical strength.

The cement based mortar (C) shows a mechanical strength reduction with the increasing in additive content, however, even for the maximum content (5% TiO_2), the reduction is lower than the observed in previous compositions, and remain above 15 MPa. Some authors report a mechanical strength increase with the partial substitution of cement binder for up to 1% TiO_2 [18], though this percentage corresponded to less than 0.3% on the total weight, which might be insufficient to ensure photocatalytic activity.

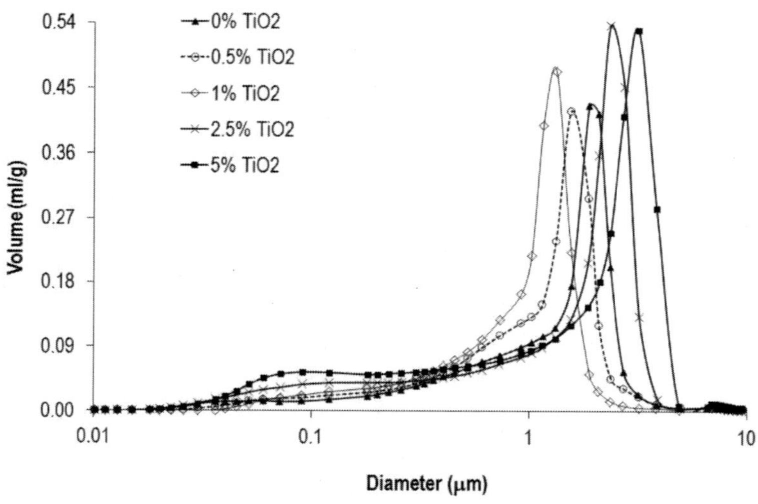

Figure 12. Pore distribution for the lime-gypsum mortars (LG).

In the present study, the titanium dioxide content was determined considering the total mass instead of a partial substitution of the binder, as a result it was used a higher additive content, and consequently, a better photocatalytic performance was achieved. Essawy et al., indicate a strong mechanical strength improvement for a partial replacement of cement binder with up to 11% titanium dioxide, however they do not evaluate the photocatalytic activity of such mortars [53]. The porosity distribution in Figure 13 shows the difference between the mortar without and with TiO_2. In the reference or base composition, the porosity distribution is divided into two intervals, a set of pores of larger size, ranged between 10 and 60 μm and another group between 0.02 and 1 μm. With 0.5 - 1% titania added to the cement matrix, the larger pores completely disappear remaining solely pores between 0.02 and 1 μm, the total porosity is reduced and the compressive strength increases. With 2.5% TiO_2, a set of nanopores begins to emerge that, combined with the disappearance of the macropores, clarify why the mechanical strength does not decrease so significantly for the cement paste as for the lime-cement composition. For the maximum TiO_2 content, the nanoporosity increases notably and even with the presence of residual micropores (1.5 to 2.7 μm), the mechanical strength remains stable.

The gypsum mortar (G) is the one that registers the lowest mechanical strength, where only 0.5% TiO_2 addition results in a 60% decrease. Figure 14 shows the pore distribution for the 0, 2.5 and 5% TiO_2 gypsum based compositions. In the absence of nanoparticles, the pores are distributed in two ranges: 10 – 45 μm and 1-10 μm. The incorporation of 2.5 - 5% TiO_2 put in evidence the presence of pores below 10 μm, a significant size reduction, however not sufficient to prevent the mechanical strength decrease. Cement and blended lime-cement mortars (C and LC compositions), that combine the effect of diminishing content of macropores (> 10 μm) with increasing amount of nanopores (< 1 μm), are able to maintain a higher mechanical strength. On the contrary, gypsum mortars that exhibit only macro and micropores, above 1 μm, show a significant decrease.

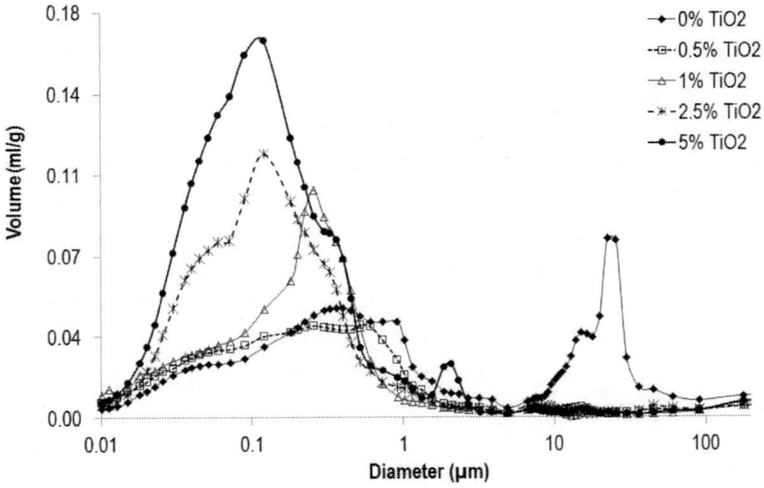

Figure 13. Pore distribution for the cement mortars (C).

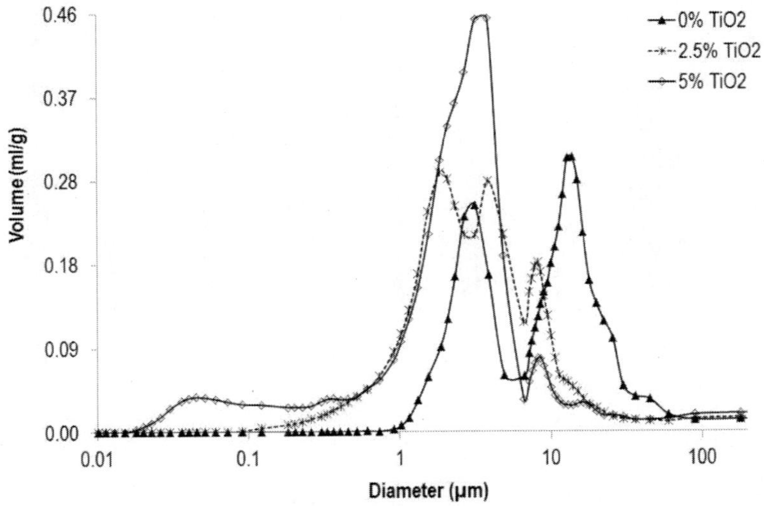

Figure 14. Pore distribution for the gypsum mortars (G).

4.1.3. Photocatalytic Efficiency

The NOx reduction was evaluated for all the mortar's formulations according to the procedure previously described and the degradation rate is presented in Figure 15. The lime mortar (L) exhibits high photocatalytic efficiency even for the minimum additive content (0.5%). The microstructural modifications observed for the different titanium dioxide contents, discussed in the prior section does not seem to determine the photocatalytic efficiency. Because the pollutant is in the gaseous state, it can reach even the micropores close to the surface. Hence, the amount of catalyser exposed will be a more important factor to determine the degradation efficiency. The decrease of the degradation rate for the samples with 2.5 and 5% can be related to the excess of catalyser exposed meaning that a higher porosity does not necessarily imply a higher photoactivity. Carp et al., mention that an excessive amount of titanium dioxide can lead to a reduction in the active sites leading to a partial deactivation. This is produced by the effect of electron-hole recombination [17]. The photocatalytic activity improves with the increasing catalyst load, although there is an optimum content.

For the lime-cement mortar (LC), the sample with only 0.5% of TiO_2, the highest pore size range (1- 0.5 μm) has the maximum degradation rate. For 1-5% titanium dioxide, the NOx degradation rate progressively decreases, even with an increasing additive content and higher porosity. This behaviour can result from one of the following factors: strong prevalence of pores below 1 μm (nanopores) which restrains the gas diffusion into the mortar matrix, reducing the TiO_2 available for the reaction [54], or elevated porosity that causes an excessive exposure of titanium dioxide, exceeding the optimum content and leading to a partial deactivation of active sites [32]. The results demonstrate that for this particular formulation (LC) there is no benefit resulting from adding more than 0.5% TiO_2.

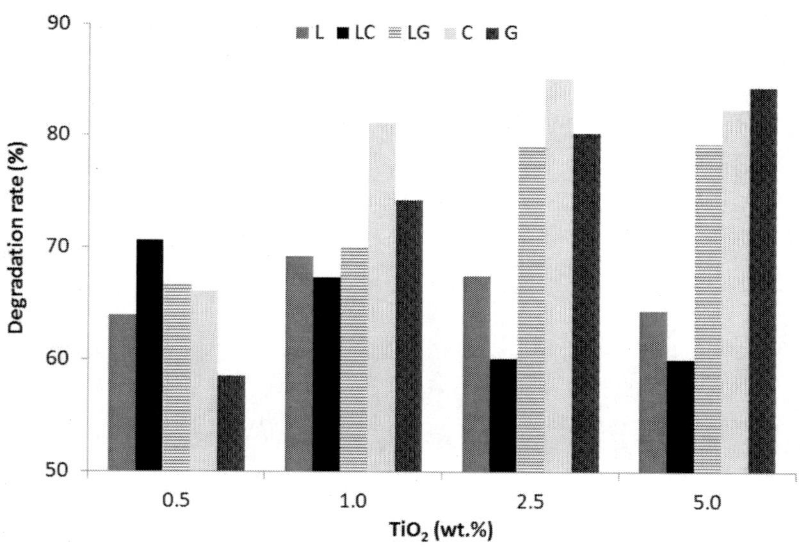

Figure 15. Degradation rate of the photocatalytic mortars.

The lime and gypsum mortar (LG) exhibit higher degradation rates than the L and LC mortars (lime and blended lime-cement compositions). The effect of higher pores sizes and higher porosity, promotes the contact between the catalyser and the pollutant gas, not only at the surface but also into the bulk. The titanium dioxide exposure increases and consequently, the photocatalytic efficiency improves. Adding more than 1% titanium dioxide represents a 9% increase in the NOx degradation rate (from 70% to 79%) which is not advantageous enough to counterbalance the 50% loss for the compression strength.

The overall degradation rate for the cement mortar (C) is the highest amongst all the compositions tested, reaching approximately 80% with only 1% TiO_2. From 0.5 to 2.5% TiO_2 a higher total porosity increases the NOx degradation rate. Folli et al., in a study conducted with photocatalytic cements reported that the use of titanium dioxide nanoparticles favours the degradation of gaseous pollutants even for the small size internal pores, since there is a higher pore surface area and hence, more TiO_2 exposed [55]. However, in the present study, the 5% TiO_2 composition has the maximum total porosity and exhibits a slight reduction of the photocatalytic activity. In this mixture, the porosity increase results from a higher amount of nanopores (below 1 μm). It can be concluded that the photocatalytic activity is favoured by a higher porosity if not accompanied by a strong prevalence of nanopores, which appear to pose an obstacle for the pollutant diffusion into the cementitious matrix [56].

The mortar with gypsum (G) shows the most important increase in degradation rate from 0.5 to 1% TiO_2, for higher amounts, the difference is less significant. Despite the reduction in the pore sizes (from approximately 20 μm to 5 μm), porosity is high (above 34%) helping internal gas diffusion and increasing the catalyst surface area.

4.2. Mortars with Multiple Additives – Combining TiO_2 and PCM

4.2.1. Effect on Final Properties

The mechanical strength is higher for the maximum content of PCM (30%) and 2.5% of titanium dioxide, as shown in Figure 16. If we compare the results when additives are used separately with the combined formulations, it can be concluded that generally the two additives combinations exhibit better results.

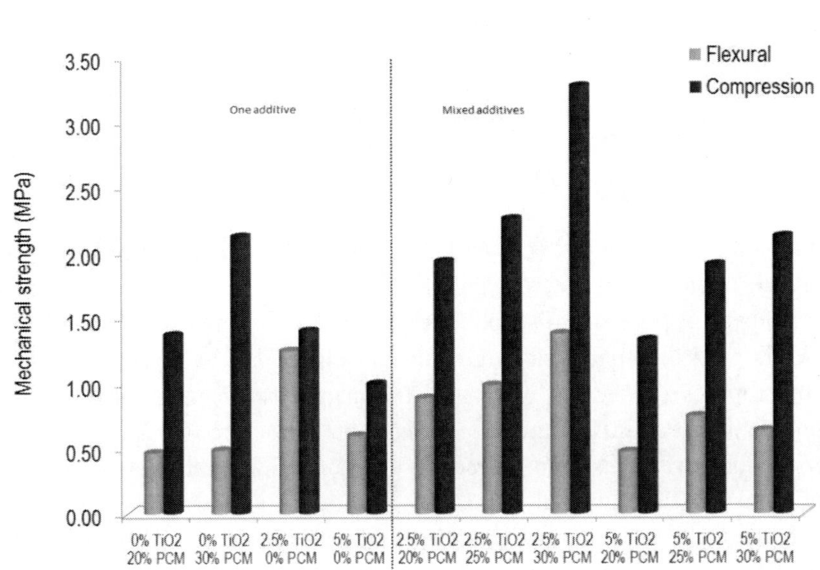

Figure 16. Mechanical strength of the compositions tested.

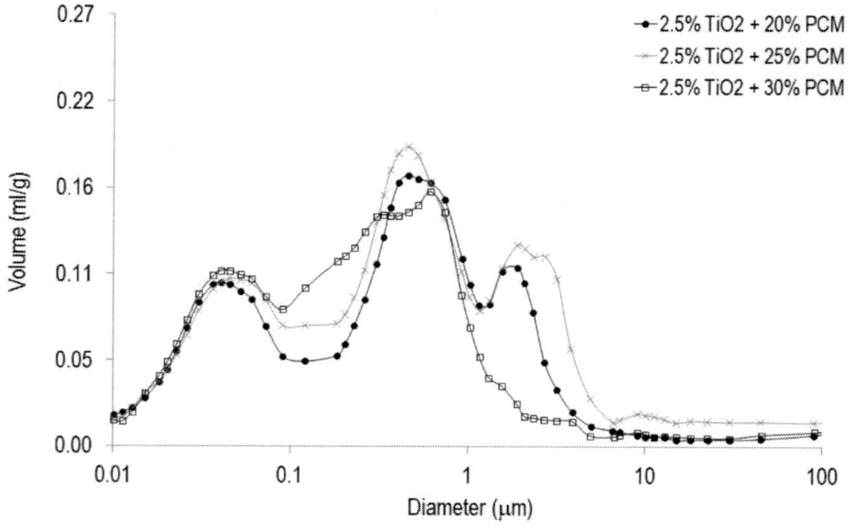

Figure 17. Porosity distribution of the 2.5% TiO2 mortars combined with 20 to 30% of PCM.

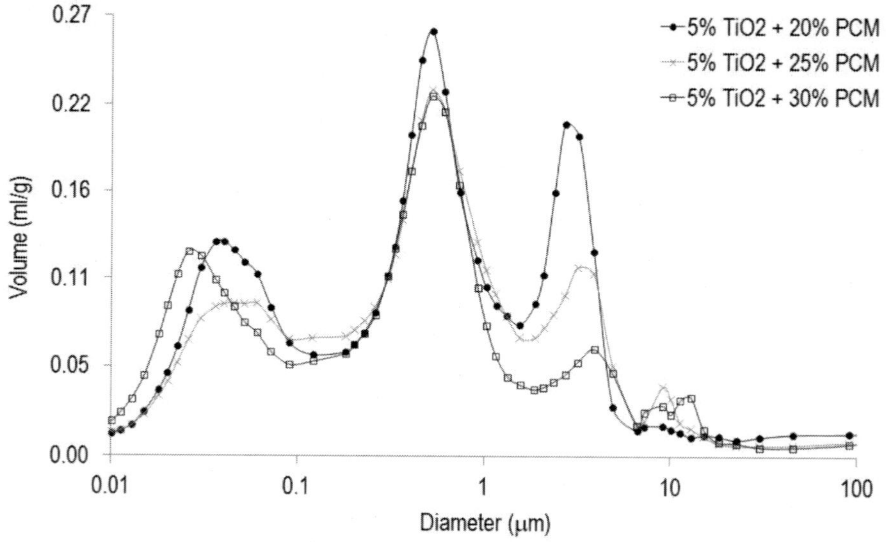

Figure 18. Porosity distribution of the 5% TiO2 mortars combined with 20 to 30% of PCM.

This can be a consequence of better particle packaging inside the mortar matrix. Because combining PCM microcapsules with TiO_2 nanoparticles results in a complex internal structure, the analysis of pore size distribution contributes for better understanding the mortar microstructure. For 2.5% of photocatalytic additive (Figure 17), increasing the PCM content reduces the macropores (above 2.5 μm), lead to an increase in micro and nanopores, with better accommodation of additives into the internal microstructure.

Lucas et al., observed this same effect in PCM containing mortars, and PCM act as counterbalance for the detrimental effect of titanium dioxide nanoparticles [47]. The compositions containing only titanium dioxide exhibits a high content of pores above 10 μm however when PCM is added, these pores are eliminated. With 5% of TiO_2, the compensation from the PCM microcapsules is still evident, though less effective, due to the excess of nanoparticles (Figure 18). The high volume of macropores present in the 20% PCM formulation disappeared when PCM content increases, and the volume of nanopores become higher. However, a parallel effect occurs with 5% titanium dioxide that was not present with 2.5%, the presence of a high porosity, above 8 μm. These bigger pores can explain why the composition with 2.5% TiO_2 and 30% PCM performs better compared to the 5% TiO_2 and 30% PCM combination, since total porosity and average pore size are similar in both. The ideal combination that can achieve better mechanical strength will be the mortar with 30% of PCM and 2.5% of titanium dioxide.

4.2.2. Photocatalytic Activity

The photocatalytic tests demonstrate that an increase in the titanium dioxide content does not imply an increase in performance (Figure 19). The excess of catalyst contributes to a decrease in the degradation rate from 2.5% to the 5% compositions [57]. For the same amount of photocatalytic additive, the increase in PCM reduces the photocatalytic efficiency. The strong decrease in pores above 1 μm for compositions with 30% of PCM reduces the amount of catalytic agent exposed to the NOx and explains the difference for the 20 and 30% PCM, when the TiO_2 content remains the same.

4.2.3. Latent Heat Storage

Figure 20 and Figure 21 shows the behaviour of the PCM/ TiO_2 mortars when submitted to the test in the climatic chamber. For mortars containing 2.5% of TiO_2, when the temperature increases, the efficiency is similar for 20 and 30% of PCM. During the cooling cycle, corresponding to the night period in this simulation, the composition with 30% is slightly more effective. A change in the internal porosity does not play an important role, because the amount of nanopores (below 0.1 μm) is the same for both mortars.

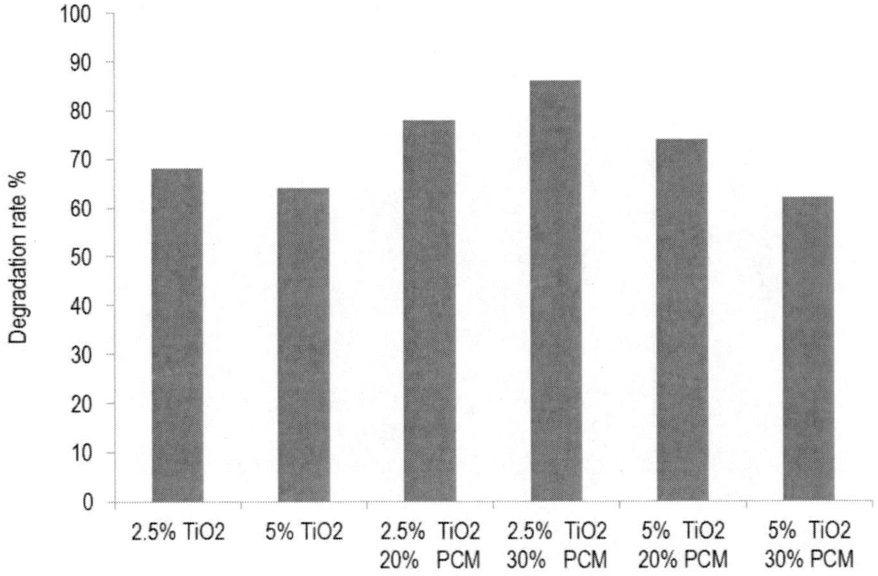

Figure 19. Photocatalytic tests of the TiO2/PCM mortars.

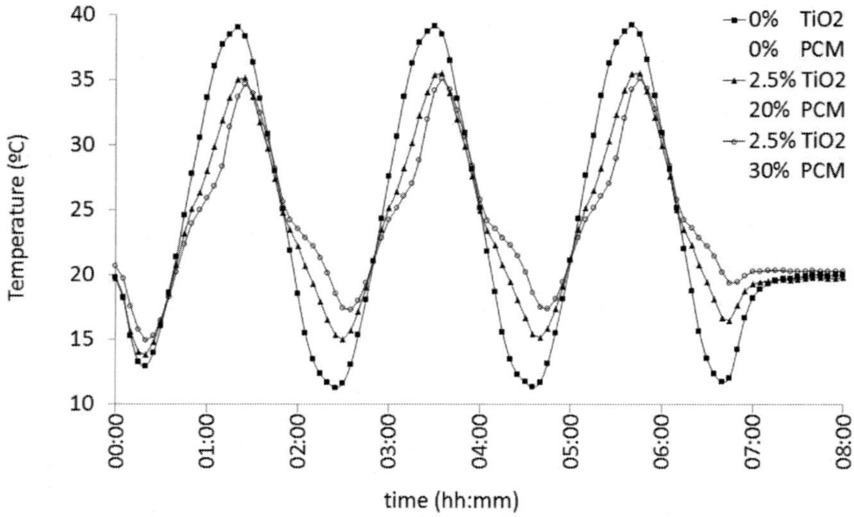

Figure 20. Thermal tests for the 2.5% TiO2 mortars with 20 to 30% of PCM.

The porosity between 0.5 and 1 μm present in the 30% PCM mortar is high enough to ensure an effective heat exchange, since the critical limit to delay the heat transfer in PCM mortars is below 0.5 μm.

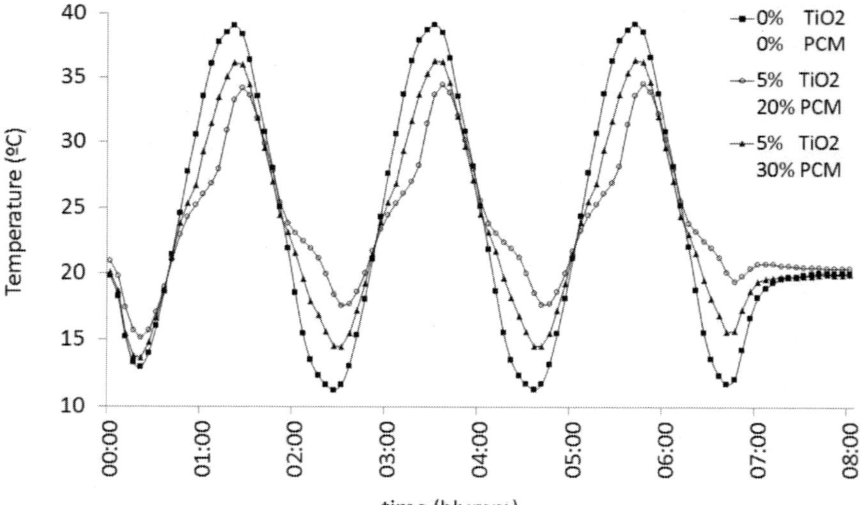

Figure 21. Thermal tests for the 5% TiO2 mortars with 20 to 30% of PCM.

When the amount of titanium dioxide nanoparticles increases to 5%, the mortar with 20% PCM is more efficient than the 30% formulation. This event shows the importance that the internal pore distribution play in the heat transfer. The strong reduction in macropores for the 30%PCM-5%TiO_2 composition, when the content of nanopores (below 0.1 μm) and micropores (0.5 to 1 μm) remains the same, determines the heat efficiency differences between this composition and the 30%PCM-2.5%TiO_2 formulation. However, an improvement in latent heat storage performance for the mortars with 5% of TiO_2 is not enough to compensate the loss in mechanical strength. Clearly, a good compromise between hardened state properties and heat transfer would be advantageous for the durability of the mortar.

CONCLUSION

The photocatalytic additive used in this chapter is a nanomaterial consisting of titanium dioxide, with two crystalline phases, anatase (86%) and rutile (14%). This additive has a high surface area, which increases the mixing water content in order to maintain the mortar's workability.

The cement and lime-cement mortars (C and LC) exhibit mechanical strength loss for additions of more than 1% TiO_2, due to the porosity increase. For the lime-gypsum mortar (LG), the mechanical strength reduction (to an additive content above 1%) is accompanied by total porosity and pore size increase. The gypsum mortar (G) is the one with the greatest mechanical strength reduction, which indicates a higher difficulty for the incorporation of the nanoparticles. In general, the presence of low size pores, particularly in the range between 1 and 0.1 μm helps to minimize the detrimental effect of the introduction of nanoadditives.

Comparing the NOx degradation rates, all the compositions exhibit high photocatalytic activity even in lower additions (0.5 and 1%). Considering the loss in mechanical strength, the slight increase in the degradation rate for higher amounts of additive (2.5 and 5%), is not enough to consider this incorporation content advantageous. The lime and lime-cement compositions (L, LC) have the highest degradation of NOx to 0.5% and 1% TiO_2 respectively and increasing the additive content reduces the degradation rate. The remaining compositions (LG, C, G) show better photocatalytic activity with higher nanoparticles content.

The photocatalytic activity study of the mortars tested in this chapter, confirm that the decrease in photocatalytic activity can result from an excessive concentration of catalyst on the mortar surface. The excess of catalyst causes a partial electron-hole pair deactivation, which can reduce the efficiency for some compositions. It was found that the pollutant gas could infiltrate the mortar microstructure even when the compositions present a low average pore size, which indicates that low porosity together with a strong presence of micropores (> 1μm) is not an obstacle for the degradation process. However, it was also verified that mortars with a high prevalence of nanopores (< 0.1 μm) exhibit an efficiency decrease. This restricted the gas intrusion in these samples, reducing the amount of catalytic agent exposed.

There is an optimal catalyst content for each mortar composition, which does not exceed, generally 2.5% TiO_2. The most favourable compositions combine a proper amount of catalyst (that does not cause deactivation), with a pore size distribution and total porosity that favour the pollutant access into the internal structure of the mortar. It was established that it is possible to develop photocatalytic mortars, capable of reducing the pollutant concentration in indoor air, contributing to a more sustainable construction.

The results presented in this chapter also confirmed that combining micro-sized (PCM) and nano-size (TiO_2) components opens new possibilities to develop mortars with multiple functionalities such as self-cleaning and heat storage. The microstructural modification does not necessarily mean that hardened state properties will be compromised. Understanding microstructural changes involved in the development of new and innovative mortars can lead to an optimization of compositions without compromising their functionality.

ACKNOWLEDGMENTS

This chapter is part of the results from the author PhD thesis (project PTDC/ECM/72104/2006), funded by the Foundation for Science and Technology (FCT) and supervised by Professor Victor Ferreira from University of Aveiro and Professor Barroso de Aguiar from the University of Minho.

REFERENCES

[1] Lee, Y.S. and S.K. Kim, Indoor Environmental Quality in LEED-Certified Buildings in the US. *Journal of Asian Architecture and Building Engineering,* 2008. 7(2): p. 293-300.

[2] Jones, A.P., Indoor air quality and health. *Atmospheric Environment,* 1999. 33(28): p. 4535-4564.

[3] Yu, C.W.F. and Jeong Tai Kim, Building Environmental Assessment Schemes for Rating of IAQ in Sustainable Buildings. *Indoor and Built Environment,* 2011. 20(1): p. 5-15.

[4] Fujishima, A., X.T. Zhang, and D.A. Tryk, TiO2 photocatalysis and related surface phenomena. *Surface Science Reports*, 2008. 63(12): p. 515-582.

[5] Inglis, H.S. and D. Taylor, Decomposition of formic acid on titanium, vanadium, chromium, manganese, iron, cobalt, nickel, and copper. *Journal of the Chemical Society A: Inorganic, Physical, Theoretical*, 1969: p. 2985-2987.

[6] Kennedy, D.R., M. Ritchie, and J. Mackenzie, The photosorption of oxygen and nitric oxide on titanium dioxide. *Transactions of the Faraday Society,* 1958. 54: p. 119-129.

[7] McLintock, I.S. and M. Ritchie, Reactions on Titanium Dioxide; Photo-adsorption and Oxidation of Ethylene and Propylene. *Transactions of the Faraday Society*, 1965. 61.

[8] Terenin, A. and Y. Solonitzin, Action of light on the gas adsorption by solids. *Discussions of the Faraday Society,* 1959. 28.

[9] Beydoun, D., et al., Role of Nanoparticles in Photocatalysis. *Journal of Nanoparticle Research*, 1999. 1(4): p. 439-458.

[10] Henglein, A., Small-particle research: physicochemical properties of extremely small colloidal metal and semiconductor particles. *Chemical Reviews,* 1989. 89(8): p. 1861-1873.

[11] Wang, R., et al., Photogeneration of Highly Amphiphilic TiO2 Surfaces. *Advanced Materials*, 1998. 10(2): p. 135-138.

[12] Guan, K., Relationship between photocatalytic activity, hydrophilicity and self-cleaning effect of TiO2/SiO2 films. *Surface and Coatings Technology,* 2005. 191(2-3): p. 155-160.

[13] Ballari, M.M., et al., Modelling and experimental study of the NOx photocatalytic degradation employing concrete pavement with titanium dioxide. *Catalysis Today*, 2010. 151(1-2): p. 71-76.

[14] Bolte, G., Innovative building materials - reduction of pollutants with TioCem (R). *Zkg International*, 2009. 62(1): p. 63-70.

[15] Toma, F.L., et al., Photocatalytic removal of nitrogen oxides via titanium dioxide. *Environmental Chemistry Letters*, 2004. 2(3): p. 117-121.

[16] Fujishima, A., X. Zhang, and D.A. Tryk, Heterogeneous photocatalysis: From water photolysis to applications in environmental cleanup. *International Journal of Hydrogen Energy*, 2007. 32(14): p. 2664-2672.

[17] Carp, O., C.L. Huisman, and A. Reller, Photoinduced reactivity of titanium dioxide. *Progress in Solid State Chemistry*, 2004. 32(1-2): p. 33-177.

[18] Dalton, J.S., et al., Photocatalytic oxidation of NOx gases using TiO2: a surface spectroscopic approach. *Environmental Pollution*, 2002. 120(2): p. 415-422.

[19] Hunger, M., G. Husken, and J. Brouwers, Photocatalysis applied to concrete products - Part 1: Principles and test procedure. *Zkg International,* 2008. 61(8): p. 77-85.

[20] Lawrence, R.M., et al., Effects of carbonation on the pore structure of non-hydraulic lime mortars. *Cement and Concrete Research*, 2007. 37(7): p. 1059-1069.

[21] Hunger, M., G. Husken, and J. Brouwers, Photocatalysis applied to concrete products - Part 2 : Influencing factors and product performance. *Zkg International*, 2008. 61(10): p. 76-84.

[22] Gunschera, J., et al., Surface-catalysed reactions on pollutant-removing building products for indoor use. *Chemosphere*, 2009. 75(4): p. 476-482.

[23] Salthammer, T. and F. Fuhrmann, Photocatalytic surface reactions on indoor wall paint. *Environmental Science & Technology*, 2007. 41(18): p. 6573-6578.

[24] Agrios, A. and P. Pichat, State of the art and perspectives on materials and applications of photocatalysis over TiO2. *Journal of Applied Electrochemistry,* 2005. 35(7): p. 655-663.

[25] Fujishima, A., T.N. Rao, and D.A. Tryk, Titanium dioxide photocatalysis. *Journal of Photochemistry and Photobiology, C: Photochemistry Reviews*, 2000. 1: p. 1-21.

[26] Xie, Y., et al., Low temperature preparation and characterization of n-doped and n-s-codoped TiO2 by sol-gel route. *Catalysis Letters,* 2007. 118(3-4): p. 231-237.

[27] Shi, H.W., et al., Influence of Fe3+ and Ho3+ co-doping on the photocatalytic activity of TiO2. *Materials Chemistry and Physics*, 2007. 106(2-3): p. 247-249.

[28] Herrmann, J.M., Heterogeneous photocatalysis: state of the art and present applications. *Topics in Catalysis*, 2005. 34(1): p. 49-65.

[29] Yamazaki-Nishida, S., et al., An Experimental and Theoretical Study of the Reaction Mechanism of the Photoassisted Catalytic Degradation of Trichloroethylene in the Gas Phase. *The Journal of Physical Chemistry,* 1995. 99(43): p. 15814-15821.

[30] Pichat, P., Some views about indoor air photocatalytic treatment using TiO2: Conceptualization of humidity effects, active oxygen species, problem of C1-C3 carbonyl pollutants. *Applied Catalysis B: Environmental,* 2010. 99(3-4): p. 428-434.

[31] Maggos, T., et al., Application of photocatalytic technology for NOx removal. *Applied Physics a-Materials Science & Processing*, 2007. 89(1): p. 81-84.

[32] Chen, J. and C.-s. Poon, Photocatalytic Cementitious Materials: Influence of the Microstructure of Cement Paste on Photocatalytic Pollution Degradation. *Environmental Science & Technology*, 2009. 43(23): p. 8948-8952.

[33] Sleiman, M., et al., Photocatalytic oxidation of toluene at indoor air levels (ppbv): Towards a better assessment of conversion, reaction intermediates and mineralization. *Applied Catalysis B-Environmental,* 2009. 86(3-4): p. 159-165.

[34] Cabeza, L.F., et al., Materials used as PCM in thermal energy storage in buildings: A review. *Renewable and Sustainable Energy Reviews*, 2011. 15(3): p. 1675-1695.
[35] Tyagi, V.V., et al., Development of phase change materials based microencapsulated technology for buildings: A review. *Renewable and Sustainable Energy Reviews*, 2011. 15(2): p. 1373-1391.
[36] Machida, M., et al., The effect of SiO2 addition in super-hydrophilic property of TiO2 photocatalyst. *Journal of Materials Science*, 1999. 34(11): p. 2569-2574.
[37] Wallevik, J.E., Relationship between the Bingham parameters and slump. *Cement and Concrete Research*, 2006. 36(7): p. 1214-1221.
[38] Banfill, P.F.G., The rheology of fresh cement and concrete - a review, in 11th International Cement Chemistry Congress. 2003: Durban.
[39] Kaci, A., et al., Adhesive and Rheological Properties of Mortar. *Applied Rheology*, 2009. 19(5).
[40] Jantunen, M., Effect of outdoor generated pollutants on indoor air quality and health, in REHVA World Congress Clima 2007 Wellbeing Indoors, FINVAC, Editor. 2007: Helsinki, Finland.
[41] Ballari, M.M., et al., NOx photocatalytic degradation employing concrete pavement containing titanium dioxide. *Applied Catalysis B: Environmental,* 2010. 95(3-4): p. 245-254.
[42] Cassar, L., et al., White Cement for Architectural Concrete, Possessing Photocatalytic Properties, in 11th Int. Congr. on the Chemistry of Cement. 2003: Durban.
[43] Maggos, T., et al., Photocatalytic degradation of NOx in a pilot street canyon configuration using TiO2-mortar panels. *Environmental Monitoring and Assessment*, 2008. 136(1-3): p. 35-44.
[44] Diamanti, M.V., M. Ormellese, and M. Pedeferri, Characterization of photocatalytic and superhydrophilic properties of mortars containing titanium dioxide. *Cement and Concrete Research*, 2008. 38(11): p. 1349-1353.
[45] RCCTE, Regulamento das Características de Comportamento Térmico dos Edifícios (RCCTE), in Decreto-Lei n.o 80/2006, de 4 de Abril, ERSE (Entidade reguladora dos serviços energéticos), Editor. 2006: Portugal.
[46] Akpan, U.G. and B.H. Hameed, Parameters affecting the photocatalytic degradation of dyes using TiO2-based photocatalysts: A review. *Journal of Hazardous Materials*, 2009. 170(2-3): p. 520-529.
[47] Lucas, S.S., V.M. Ferreira, and J.L.B.d. Aguiar, Latent heat storage in PCM containing mortars—Study of microstructural modifications. *Energy and Buildings*, 2013. 66(0): p. 724–731.
[48] Westerholm, M., et al., Influence of fine aggregate characteristics on the rheological properties of mortars. *Cement & Concrete Composites,* 2008. 30(4): p. 274-282.
[49] Artelt, C. and E. Garcia, Impact of superplasticizer concentration and of ultra-fine particles on the rheological behaviour of dense mortar suspensions. *Cement and Concrete Research*, 2008. 38(5): p. 633-642.
[50] Flatt, R.J., D. Larosa, and N. Roussel, Linking yield stress measurements: Spread test versus Viskomat. *Cement and Concrete Research,* 2006. 36(1): p. 99-109.
[51] Karatasios, I., et al., Photo-induced carbonation of lime-TiO2 mortars. *Applied Catalysis B: Environmental*, 2010. 95(1-2): p. 78-86.

[52] Iordanidis, A., et al., Thermal, mineralogical and spectroscopic study of plasters from three post-Byzantine churches from Kastoria (northern Greece). *Journal of Thermal Analysis and Calorimetry*, 2011. 103(2): p. 577-586.

[53] Essawy, A.A. and S. Abd El. Aleem, Physico-mechanical properties, potent adsorptive and photocatalytic efficacies of sulfate resisting cement blends containing micro silica and nano-TiO2. *Construction and Building Materials*, 2014. 52: p. 1-8.

[54] Care, S. and F.ß. Derkx, Determination of relevant parameters influencing gas permeability of mortars. *Construction and Building Materials,* 2011. 25(3): p. 1248-1256.

[55] Folli, A., et al., TiO2 photocatalysis in cementitious systems: Insights into self-cleaning and depollution chemistry. *Cement and Concrete Research*, 2012. 42(3): p. 539-548.

[56] Nazari, A. and S. Riahi, The effects of TiO2 nanoparticles on properties of binary blended concrete. *Journal of Composite Materials*, 2010.

[57] Lucas, S.S., V.M. Ferreira, and J.L.B.d. Aguiar, Incorporation of titanium dioxide nanoparticles in mortars — Influence of microstructure in the hardened state properties and photocatalytic activity. *Cement and Concrete Research*, 2013. 43: p. 112–120.

In: Titanium Dioxide
Editor: Jerri Brown
ISBN: 978-1-63321-391-3

Chapter 3

TITANIUM DIOXIDE-BASED PHOTOCATALYSTS FOR THE CONVERSION OF WATER POLLUTANTS

C. G. Silva[1,*], S. A. C. Carabineiro[1], M. J. Lima[1], G. Dražić[2,3], J. L. Figueiredo[1] and J. L. Faria[1]

[1]LCM – Laboratory of Catalysis and Materials – Associate Laboratory LSRE-LCM, Faculdade de Engenharia, Universidade do Porto, Porto, Portugal

[2]Department of Nanostructured Materials, Jožef Stefan Institute, Ljubljana, Slovenia

[3]Laboratory for Materials Chemistry, National Institute of Chemistry, Ljubljana, Slovenia

ABSTRACT

Heterogeneous photocatalytic processes have been gaining increasing commercial interest worldwide mostly in the fields of environmental cleanup and energy. In particular, photocatalytic water treatment is considered a cost-effective and environmentally benign technology enabling the destruction of a great variety of hazardous emerging pollutants. One of the most challenging topics in heterogeneous photocatalysis is the development of efficient catalysts working under solar light irradiation. Titanium dioxide (TiO_2), the benchmark catalyst for this type of applications, requires UV-light excitation, which only accounts for 5% of the incoming solar energy. Loading TiO_2 with carbon materials and/or noble metal nanoparticles are effective techniques used to increase the photonic efficiency and to extend light absorption of TiO_2 towards the visible spectral range. In this chapter, we will describe some of our recent results on the development of TiO_2-based photocatalysts, namely TiO_2-activated carbon fiber composites and gold-loaded TiO_2, for the degradation of the antibiotic ciprofloxacin in water.

* Corresponding author: C. G. Silva. LCM – Laboratory of Catalysis and Materials – Associate Laboratory LSRE-LCM, Faculdade de Engenharia, Universidade do Porto, Rua Dr. Roberto Frias, 4200-465 Porto, Portugal. Tel: + 351 22 508 1779, fax: + 351 22 508 1447, e-mail: cgsilva@fe.up.pt.

1. INTRODUCTION

Water is present in multiple human activities and is used for diverse purposes like domestic use, agriculture, industry and energy production. The exponential growth of human population and the intensification of agricultural and industrial activities led to a continuous increase in the demand for earth's limited supply of freshwater. Protection of natural water resources and development of new technologies for water and wastewater treatment became key environmental issues of the 21st century.

It has been frequently observed that pollutants not amenable to biological treatments may also be characterized by high chemical stability and/or by strong difficulty to be completely mineralized. In these cases, it is necessary to use reactive systems much more effective than those adopted in conventional purification processes. A lot of research work has been done in this direction in the last decades, highlighting the prominent role of a special class of oxidation techniques defined as Advanced Oxidation Processes (AOPs) [1].

These technologies are based on the generation of the extremely reactive hydroxyl radical ($HO^{\bullet}$), which can oxidize a great variety of pollutant molecules found in wastewaters. Known as efficient, relatively low cost and easy to operate technologies, AOPs could be combined with biological or traditional physical-chemical processes for enhanced wastewater treatment [2, 3]. AOPs include water treatment techniques such as ozonation [4], Fenton and photo-Fenton [5] and photocatalysis [6, 7].

Due to the fast development of light-assisted technologies, photocatalytic processes have emerged as one of the most attractive wastewater treatment routes, since these processes require mild operation conditions of temperature and pressure, and also due to the possibility of using solar light for the photo-activation of semiconductor catalysts [8, 9]. Titanium dioxide is the most used catalyst for photocatalytic applications due to important properties such as high chemical stability, low cost and high photocatalytic activity [10].

According to the usually proposed mechanism, irradiation of TiO_2 particles with suitable photons (of energy $h\nu \geq Eg$ = 3.2 eV) generates a bound electron/hole pair, which can dissociate in competition with recombination to produce a conduction band electron and a valence band hole (Figure 1).

Both charge carriers migrate to the TiO_2 surface to reduce adsorbed electron acceptors and to oxidize adsorbed electron donors. For photocatalytic oxidation processes in aqueous systems, oxygen is often present as the electron acceptor to form the superoxide anion radical ($O_2^{\bullet -}$), while H_2O and OH^- are available as electron donors to yield the hydroxyl radical ($HO^{\bullet}$).

Both radicals are very reactive and strongly oxidizing, capable of totally mineralizing most of the organic compounds.

The major limitation of using TiO_2 is due to its relatively large bandgap energy, meaning that it mostly absorbs UV light, which accounts for only 5% of the solar spectrum. In order to improve its photocatalytic efficiency, a great deal of effort has been devoted in recent years to extend the photocatalytic response to the visible range [11].

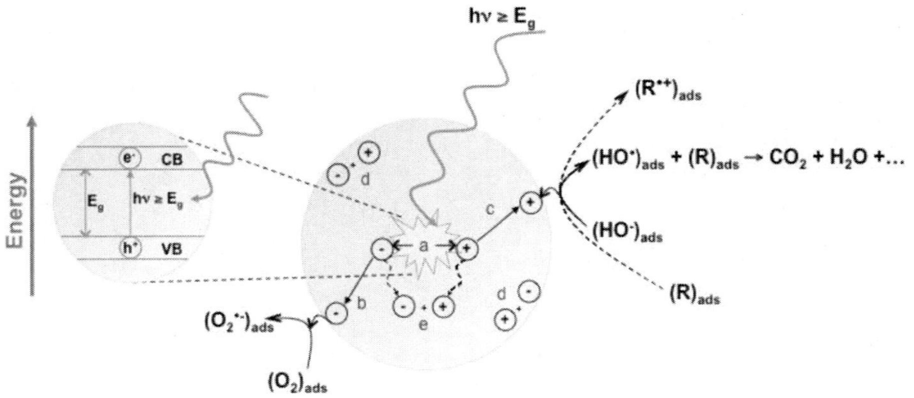

Figure 1. Main processes occurring on a semiconductor particle upon photo-excitation: (a) photogeneration of electron/hole pair; (b) diffusion of the electron acceptor and reduction at the surface of the semiconductor; (c) oxidation of the electron donor on the surface of the semiconductor particle; (d) and (e) electron/hole recombination at the surface and in the bulk, respectively.

The major practices involve catalyst modification by metal and non-metal doping [12, 13], metal loading [14, 15], dye photosensitization [16], mixing with other semiconductors [17], or addition of carbon materials [18-20].

The use of TiO_2 hybrids containing carbon materials such as activated carbon, carbon nanotubes, activated carbon fibers and graphene, has been reported as an effective approach for the enhancement of the semiconductor activity both under UV and visible/solar light conditions [21-23]. The main effects of these materials have been ascribed to their role as co-adsorbents, as TiO_2 dispersing media and/or as photo-sensitizers.

Another important technique used to increase TiO_2 performance consists in loading it with noble metals, such as Pt, Au, Ir, Pd, Ru and Rh [15, 24, 25]. A variety of explanations have been advanced to rationalize the observed improvement in the photoefficiency of TiO_2 by addition of noble metal nanoparticles, including: i) increased absorption due to surface plasmons; ii) improved charge separation as a result of localized electromagnetic field; iii) promotion of electron transfer to adsorbed species; or iv) electron storage effects that can drive the Fermi level to more negative potentials [26].

TiO_2-based photocatalytic processes have been applied for the treatment of contaminated waters containing a wide variety of organic pollutants such as pesticides, dyes and pharmaceuticals [27-29]. In particular, the search for effective technologies addressing the treatment of water streams containing pharmaceutical and personal care products (PPCP) is increasing, due to their low biodegradability and generalized use [30, 31].

Therapeutic drugs (e.g., antibiotics, anti-inflammatory drugs, steroids), fragrances, cosmetics, sun-screen products, diagnostic agents, nutraceuticals (e.g., vitamins) are classified as PPCPs [32]. The specific case of antibiotics is of major concern due to the increasing resistance of bacteria to this class of drugs. As a consequence of being discharged along the years, antibiotics may increase bacterial resistance and in the future, public health might recede to stages equivalent to pre-penicillin years [33-35].

Indeed, Cek et al. studied the resistance status of several bacteria, including *E. coli*, to antibiotics used for urinary tract infections, one of the most common diseases, and found bacterial resistance rates higher than 30% for most of the studied antibiotics [36].

Included in the top of resistance rates, we can find ciprofloxacin (Figure 2), a broad-spectrum synthetic antibiotic (a 2nd generation fluoroquinolone) widely employed for the treatment of bacterial infections [37], which can be found in pharmaceutical wastewaters as well as in hospital effluents [38, 39] and surface waters [39-43].

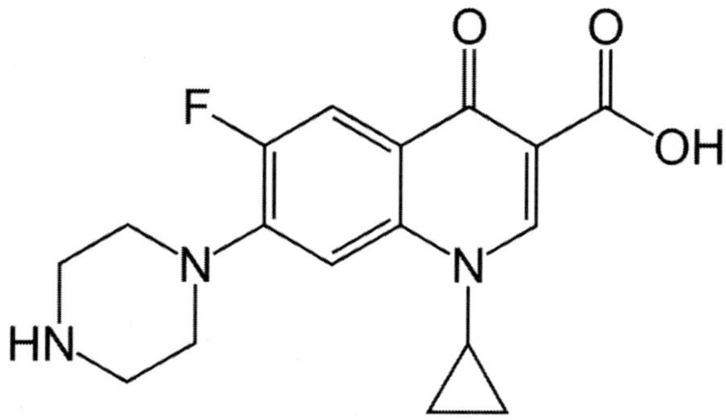

Figure 2. Molecular structure of ciprofloxacin.

This chapter describes some recent results on the degradation of ciprofloxacin in aqueous medium using TiO_2-based photocatalysts, namely TiO_2-activated carbon fiber composites and gold-loaded TiO_2.

2. EXPERIMENTAL

2.1. Synthesis of Activated Carbon Fiber-TiO_2 Hybrid Catalysts

TiO_2 was prepared by means of an acid-catalyzed sol-gel method starting from an alkoxide precursor. The preparation was performed at room temperature as follows: 0.1 mol of $Ti(OC_3H_7)_4$ (Aldrich 97%) was dissolved in 200 mL of ethanol. The solution was magnetically stirred for 30 min and then 1.56 mL of nitric acid (65%) was added.

The solution was loosely covered and kept under stirring until a homogeneous gel formed. The gel was aged in air for several days. Then, the obtained xerogel was ground into fine powder and dried at room temperature. The powder was calcined at different temperatures in nitrogen flow for two hours to obtain TiO_2 samples.

TiO_2-based hybrid catalysts were prepared by a similar sol-gel procedure modified by the introduction of activated carbon fibers (ACF). The fibers were previously washed with distilled water to remove impurities and dried at 373 K for 24 h. A given amount of commercial activated carbon fiber Kuractive FR-15 (Kuraray Chemical) was added to the $Ti(OC_3H_7)_4$ ethanol solution. The mixture was kept stirring until an ACF-containing gel was formed. The gel was then aged, ground, dried and calcined to obtain the hybrid materials.

Catalysts were designated as X-ACF/TiO_2-Y, where the number X (17, 33 and 44) corresponds to the weight percentage of ACF in the composite material and Y is the calcination temperature (673, 723 and 873 K).

2.2. Synthesis of Au-Loaded TiO_2 Catalysts

Gold nanoparticles (1% wt.) were loaded on commercial TiO_2 Evonik P25 by two different techniques: double impregnation (DIM) and liquid phase reductive deposition (LPRD). The DIM method is similar to traditional impregnation (the support is impregnated with a solution of $HAuCl_4$ using sonication), but using a second impregnation step with addition of an aqueous solution of Na_2CO_3 (1 M), under constant ultrasonic stirring [44]. LPRD consists of mixing a solution of the gold precursor with a solution of NaOH (with a ratio of 1:4 in weight) with stirring at room temperature [45]. The resulting solution is aged for 24 h in the dark at room temperature, to complete the hydroxylation of Au^{3+} ions. Then the appropriate amount of support is added to the solution and, after ultrasonic dispersion for 30 min, the suspension is aged in the oven at 373 K overnight. In both cases, the resulting solid is washed repeatedly with distilled water for chloride removal, which is well known to cause sintering of Au nanoparticles, thus rendering them inactive [46], and dried in the oven at 373 K overnight.

2.3. Catalysts Characterization

The thermal behavior of the TiO_2 xerogel and the ACF/TiO_2 hybrid materials was analyzed using a Mettler TA 4000 system from 298 to 1073 K at a heating rate of 10 K min^{-1} under N_2 or air. The carbon content in the composite catalysts was determined from weight loss in air by using the same conditions. The specific surface area (S_{BET}) of the powder materials was determined by multipoint analysis of N_2 adsorption isotherms at 77 K using the BET method, with a Quantachrome NOVA 4200e multi-station apparatus.

Powder X-Ray diffraction (XRD) patterns were recorded in the range $2\theta = 20 - 60°$ on a Philips X'Pert MPD rotatory target diffractometer using Cu Kα radiation ($\lambda = 0.15406$ nm) as X-Ray source, operated at 40 kV and 50 mA. Scanning electron microscopy (SEM) was carried in a Jeol JSM-6301F instrument, which was equipped with an energy dispersive X-ray (EDX) detector.

Transmission electron microscopy (TEM) experiments for the TiO_2 samples were carried out in a JEOL 1011 transmission electron microscope operating at 100 keV. The samples were dispersed in ethanol using an ultrasonic bath for 1 min. A drop of the dispersion was placed on a holey carbon film deposited on a copper grid and left to dry at ambient temperature. A JEOL 2010F analytical electron microscope, equipped with a field-emission gun, was used for high-resolution transmission electron microscopy (HRTEM) investigations of the Au/TiO_2 materials. The microscope was operated at 200 keV, and an energy dispersive X-ray spectrometer (EDXS) LINK ISIS–300 from Oxford Instruments with an UTW Si–Li detector employed for the chemical analysis.

The samples for TEM were prepared from a diluted suspension of nanoparticles in ethanol. A drop of suspension was placed on lacey carbon–coated Ni grid and allowed to dry in air. Z–contrast images were collected using a high-angle annular dark field detector (HAADF), in scanning transmission mode (STEM). Particle sizes were measured from TEM images, using the ImageJ program. Histograms were drawn from measurements made on 100–500 particles, depending on the sample. Average particle sizes were calculated for all samples. The UV-Vis spectra of the solid powder materials were measured on a JASCO V-560 UV-Vis spectrophotometer equipped with an integrating sphere attachment (JASCO ISV-469). The powder was not diluted in any matrix to avoid a decrease in the absorbance. The spectra were recorded in diffuse reflectance mode and transformed by the instrument software to equivalent absorption Kubelka-Munk units.

2.4. Photocatalytic Degradation of Ciprofloxacin Using ACF/TiO_2 Hybrid Catalysts under UV Irradiation

Photodegradation tests using ACF-containing catalysts were performed in a glass batch reactor of 1000 mL of capacity, charged with 800 mL of solution/suspension. UV light was supplied by a Heraeus TNN 15/32 low-pressure mercury vapor lamp with an emission line at 253.7 nm (3 W of radiant flux), located axially and held in a quartz immersion tube. The initial ciprofloxacin (Fluka) concentration in aqueous solutions was set to 20 mg L^{-1}.

The pH of the aqueous solutions was set to 7 using HCl 0.1M and NaOH 0.1M solutions and the catalyst load was maintained at 1 g L^{-1}. Before the illumination was turned on, the suspensions were magnetically stirred in the absence of light for 30 min to establish adsorption-desorption equilibrium.

Then the suspension was irradiated with ultraviolet light at constant stirring speed. A sample was taken at the end of the dark adsorption period, just before the light was turned on, in order to determine the antibiotic concentration in solution, which was hereafter considered as the initial concentration (C_0).

Samples were withdrawn regularly at controlled times and centrifuged immediately for separation of the suspended solids. The clean solution was analyzed by UV-Vis spectroscopy. The full UV-Vis spectrum (200-800 nm) of each sample was recorded and the absorbance at a selected wavelength (271 nm) registered to determine the concentration of ciprofloxacin. Repeated tests were made to assess the reproducibility. For the purpose of kinetic analysis, each sample taken from the reactor was divided into three vials and the final absorption was given by the arithmetic average over three measurements. In order to determine the mineralization of the organic molecule, total organic carbon (TOC) determination was performed using a Shimadzu 5000A TOC analyzer.

2.5. Photocatalytic Degradation of Ciprofloxacin Using Au-TiO_2 Catalysts under Visible Light Irradiation

The experiments were carried out in a glass cylindrical reactor filled with 7.5 mL of a 20 mg L^{-1} ciprofloxacin aqueous solution. The solution was magnetically stirred and continuously purged with an oxygen flow.

The irradiation source, located 2.5 cm from the photocatalytic reactor, consisted of a Heraeus TQ 150 medium pressure mercury vapor lamp (main emission lines at excitation wavelength λ_{exc} = 254, 313, 365, 436 and 546 nm). A DURAN® glass cooling jacket was used for irradiation in the near-UV to visible light range (λ_{exc}= 365, 436 and 546 nm).

When needed, a cut-off filter was used to eliminate irradiation below 400 nm (visible light conditions). In this case, the samples withdrawn from the reactor were analyzed by HPLC using a Hitachi Elite LaChrom liquid chromatograph equipped with an L-2450 diode array detector.

The stationary phase consisted of a YMC-Pack ODS-AQ endcapped column (250 mm x 4.6 mm, 5 μm particles) working at room temperature. The mobile phase was a mixture of a 20 mM NaH_2PO_4/H_3PO_4 (pH = 2.8) buffer solution and acetonitrile with a gradient concentration at a flow rate of 1 mL min^{-1}. Total organic carbon (TOC) measurements were also performed.

3. Results and Discussion

3.1. Materials Characterization

Nitrogen adsorption isotherms at 77 K were used for the determination of the surface area of the catalyst samples. The results, listed in Table 1, show that there is a dependence of the surface area on the TiO_2 calcination temperature for the materials prepared by sol-gel. The higher the temperature, the lower the surface area, which is due to the increase in the TiO_2 crystallite dimensions, as can be observed in Figures 3a and 3b.

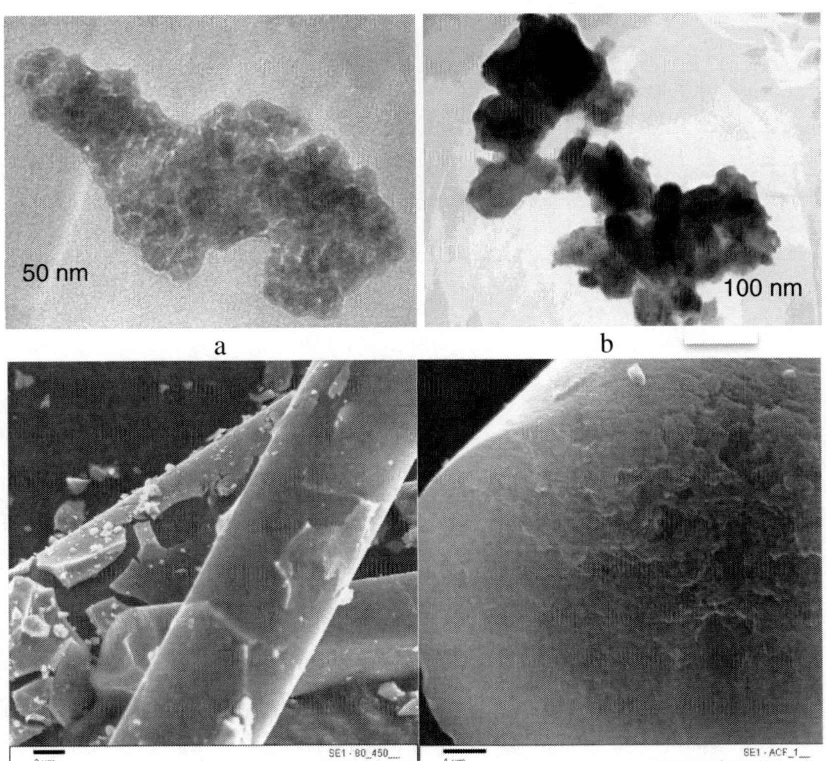

c d

Figure 3. TEM images of TiO_2-573 (a) and TiO_2-873 (b), and SEM micrographs of 44-ACF/TiO_2-723 (c) and ACF (d).

The adsorption isotherm of ACF (not shown) can be ascribed to type I [47], indicating the presence of a microporous structure. ACF shows an extremely high specific surface area of 1280 m^2 g^{-1}. The surface area of the ACF/TiO_2 hybrid materials increases with the ACF content. Nevertheless, the values of S_{BET} were lower than those estimated from the AC and TiO_2 contents, indicating the existence of a contact surface between both phases that is not available for adsorption, as can be observed in Figure 3c. Figure 3d shows the aspect of the top of one fiber. For the materials synthesized at 573 K, only the anatase phase can be found. It has been previously reported that for neat sol-gel TiO_2, transition from anatase to rutile occurs between 723 and 873 K [48]. At 873 K, TiO_2 is constituted by 94% of rutile. However, upon calcination at 723 K, the material with the highest ACF load (44-ACF/TiO_2-723), presented only 63% of anatase, i.e., there is a fraction of TiO_2 that has been converted to rutile. This may result from the creation of higher inner temperature sites in the hybrid catalysts during calcination, due to the presence of a high amount of ACF, thus leading to phase conversion from anatase to rutile. The crystallite sizes of anatase TiO_2 ($d_{anatase}$), estimated from the line broadening of the (101) XRD reflection plane (not shown), are also given in Table 1. It was observed that the inclusion of a carbon phase did not show any significant effect on the dimensions of the anatase crystallites.

Table 1. BET surface area (S_{BET}), carbon content determined by themogravimetry (C_{TG}), anatase content and dimensions of the anatase crystallites ($d_{anatase}$), ciprofloxacin concentration after the dark adsorption period (C_0) and TOC removal at the end of the photocatalytic reactions

Catalyst	S_{BET} (m^2g^{-1})	C_{TG} (%)	Anatase (%)	$d_{anatase}$ (nm)	C_0 (mg L^{-1})	*TOC removal* (%)
TiO_2-573	134	-	100	6	17.6	71
TiO_2-723	84	-	100	10	14.8	36
Catalyst	S_{BET} (m^2g^{-1})	C_{TG} (%)	Anatase (%)	$d_{anatase}$ (nm)	C_0 (mg L^{-1})	*TOC removal* (%)
TiO_2-873	3	-	6	27	18.9	21
17-ACF/TiO_2-573	306	16	100	5	11.5	58
17-ACF/TiO_2-723	143	14	100	10	14.2	50
17-ACF/TiO_2-873	103	11	74	24	13.6	59
33-ACF/TiO_2-573	463	32	100	5	7.6	72
33-ACF/TiO_2-723	374	29	100	11	4.4	75
33-ACF/TiO_2-873	319	29	6	21	4.4	76
44-ACF/TiO_2-573	590	44	100	5	4.8	80
44-ACF/TiO_2-723	551	44	63	8	3.1	74
44-ACF/TiO_2-873	511	43	7	26	3.0	70
TiO_2 P25	45	-	77	24	17.6	71

In the case of the Au-loaded materials, the TiO_2 used was a commercial material, normally used as the benchmark for photocatalytic applications: TiO_2 Evonik P25 (former Degussa P25). This material consists of 77% anatase and 23% rutile (Table 1). In this case, the introduction of gold nanoparticles did not affect the textural or crystalline properties of the TiO_2 material. The dimensions of the anatase and rutile crystallites of TiO_2 P25, estimated from the XRD data, were 24 and 33 nm, respectively. The dimensions of the crystallites were confirmed by TEM analysis of the Au/TiO_2 materials (Figure 4). Dark field TEM analysis allowed the detection of gold particles of very small dimensions deposited on the surface of TiO_2 (Figures 4b and 4d).

The dimensions of the Au particles prepared by LPRD and DIM, determined by TEM analysis, were 3.9 and 9.3 nm [49], respectively, which indicate that the preparation method has a determining influence on the properties of the catalyst. Diffuse reflectance UV-Vis spectra of TiO_2 P25 and Au/TiO_2 materials prepared by LPRD and DIM are shown in Figure 5. As expected, two main bands are observed due to the presence of gold nanoparticles: one in the visible region between 500 and 650 nm, typical of the gold surface plasmon band, and another in the UV region, with an edge rising at 400 nm, due to the bandgap transition of the TiO_2 semiconductor.

It can be noticed that the plasmon resonance Au absorption band of the material prepared by LPRD is much more intense than that of Au/TiO_2 obtained by DIM, which can be ascribed to the lower dimensions of the gold nanoparticles.

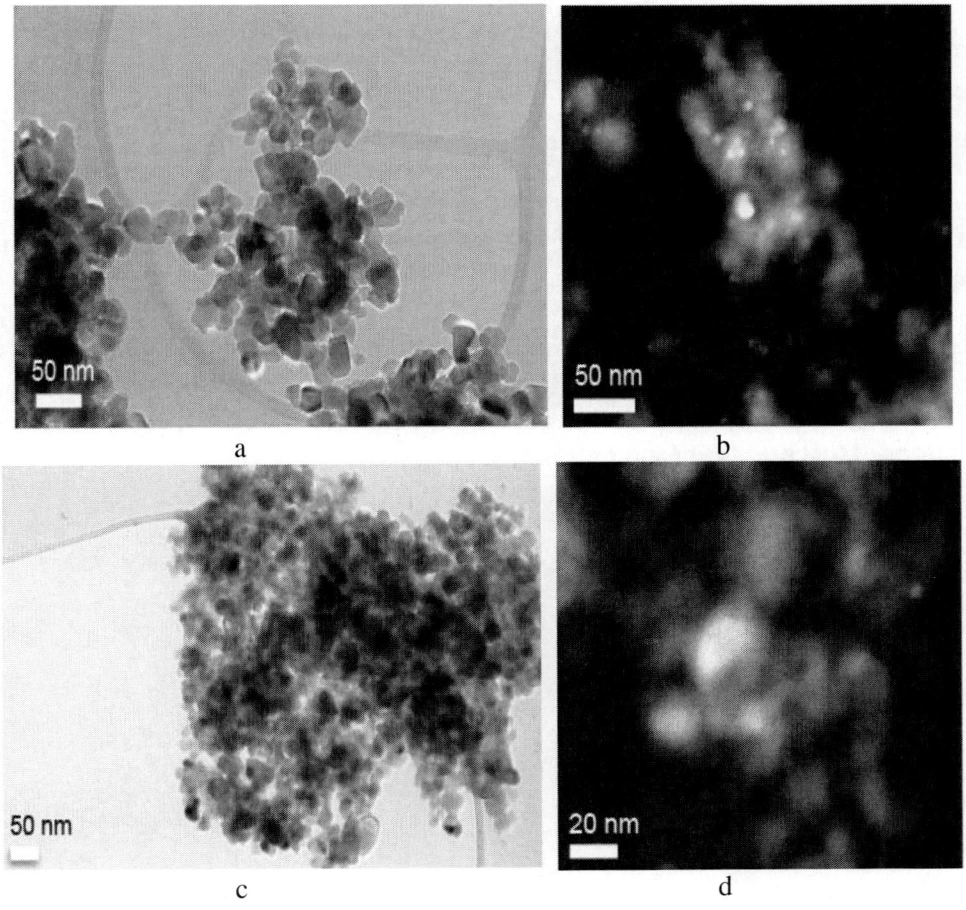

Figure 4. Images of Au/TiO_2 materials prepared by LPRD (a-TEM and b-HAADF) and DIM (c-TEM and d-HAADF) methods.

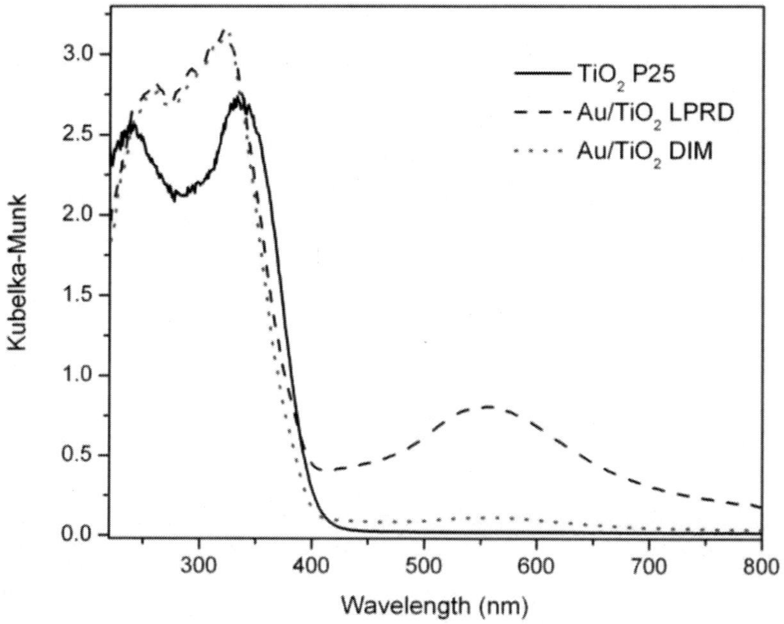

Figure 5. Diffuse reflectance UV-Vis spectra of TiO_2 and Au/TiO_2 catalysts.

3.2. Photocatalytic Degradation of Ciprofloxacin Using ACF-TiO_2 Materials under UV Irradiation

The degradation of ciprofloxacin in aqueous medium was studied using TiO_2 and ACF/TiO_2 catalysts under UV (254 nm) irradiation. Since ciprofloxacin absorbs light in the UV range (Figure 6), the pure photochemical degradation, i.e., in the absence of catalyst, was also performed for comparison purposes.

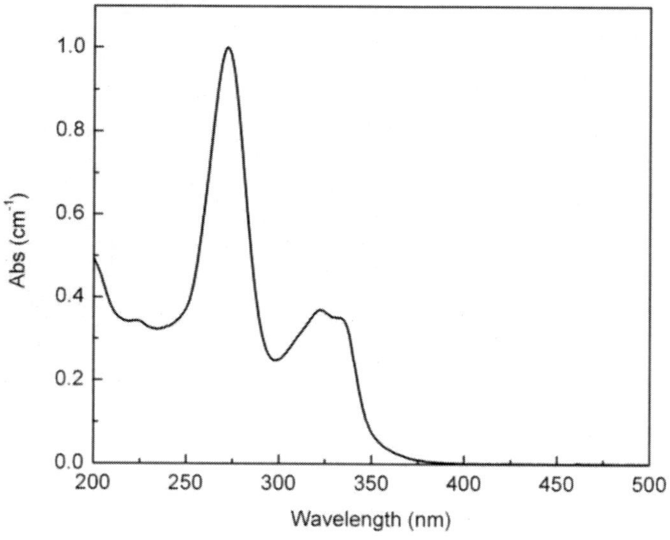

Figure 6. UV-Vis absorption spectrum of ciprofloxacin.

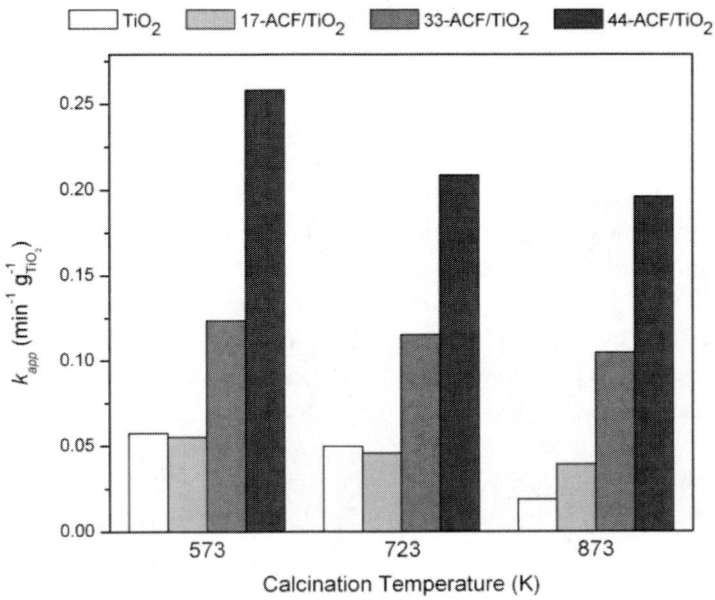

Figure 7. Apparent first order kinetic constant normalized by mass of TiO_2 for the reactions using TiO_2 and ACF/TiO_2 catalysts prepared at different calcination temperatures.

It has been observed that both photochemical and photocatalytic reactions follow pseudo-first order kinetics. The efficiency of the degradation process was evaluated in terms of the first order kinetic constant (k_{app}) normalized by the amount of TiO_2 present in each catalyst (Figure 7).

The kinetic constant for the pure photochemical reaction (k_{UV}) was 0.026 min^{-1}. It was observed that the presence of a catalyst increased the rate of the degradation reaction.

As expected, for the neat TiO_2 materials, the efficiency of ciprofloxacin removal decreased with the calcination temperature at which the catalysts were obtained. This behavior is attributed to the sintering of the TiO_2 crystallites at higher temperatures as well as to the phase transformation from anatase to rutile, which normally presents lower photocatalytic activity.

The same tendency was observed for the ACF/TiO_2 catalysts. The introduction of low amounts (17% *wt.*) of ACF did not provide any effect in the efficiency of the catalysts calcined at 573 and 723 K when compared to the respective neat TiO_2. Nevertheless, a 50% increase in k_{app} was obtained with 17-ACF/TiO_2-873 when compared with TiO_2-873, which can be attributed to the presence of a larger amount of anatase in the hybrid material. A large increase in the photoefficiency of the degradation process was observed for the reactions using ACF/TiO_2 composites with 33 and 44% of ACF. Nevertheless, it has to be noted that in those cases, adsorption of ciprofloxacin during the dark period prior to the photocatalytic reaction is very significant, as can be observed by the values of C_0 (Table 1). The best catalyst was found to be 44-ACF/TiO_2-573, which consists only of the anatase phase with very low crystallite dimensions (5 nm), and which presents the highest surface area. These results indicate that in ACF/TiO_2 catalysts, the main role of ACF is to co-adsorb ciprofloxacin, which is then more efficiently transferred to the TiO_2 surface, where it is immediately photocatalytically degraded.

3.3. Photocatalytic Degradation of Ciprofloxacin Using Au/TiO_2 Materials under Visible Light Irradiation

Gold-loaded TiO_2 materials were tested for the photocatalytic degradation of ciprofloxacin under different irradiation wavelength ranges (Figure 8). It was observed that, when irradiating with near UV-Vis light (Figure 8a), bare TiO_2 and Au-loaded TiO_2 show similar efficiencies. In this case, no profit was obtained by introducing gold nanoparticles in the metal oxide semiconductor surface.

Under visible light irradiation, which can be used to mimic solar light conditions, an increase in the efficiency of the photocatalytic degradation of ciprofloxacin was observed using gold deposited on TiO_2 (Figure 8b).

In fact, at these conditions, TiO_2 shows very little efficiency due to its almost negligible absorption at this wavelength range.

The introduction of gold leads to an increase in visible light absorption, as previously displayed in the DR UV-Vis spectra (Figure 5), due to the plasmon resonance of the Au nanoparticles.

The role of gold nanoparticles on Au/TiO_2 photocatalysts can be rationalized in terms of two different mechanisms [15]. The first assumes direct photo-excitation of TiO_2, leading to the generation of electrons in the semiconductor conduction band and electron holes in the valence band. Then, electrons in the conduction band will move to the gold nanoparticles acting as electron buffer. This is expected to be the main mechanism when using near UV-Vis radiation.

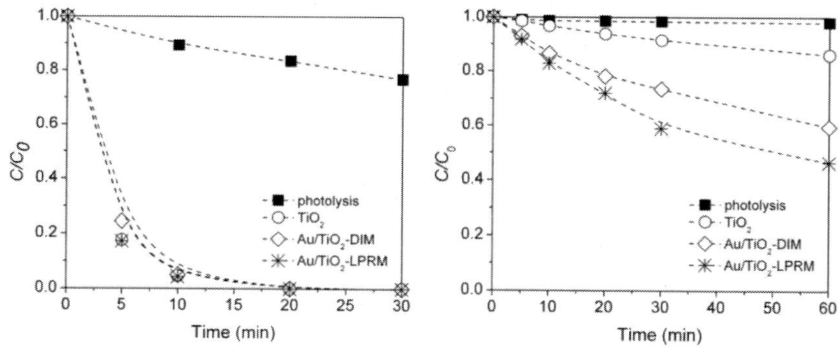

a b

Figure 8. Evolution of the normalized concentration (C/C_0) of ciprofloxacin under irradiation with: (a) near UV-Vis ($\lambda \geq 365$ nm); (b) visible light ($\lambda \geq 400$ nm).

In our case, we have observed no effect upon addition of gold to TiO_2 when irradiating with $\lambda \geq 365$ nm, which indicates that this process does not occur to a significant extent. On the other hand, when irradiating at $\lambda \geq 400$ nm (visible range), minute absorption by TiO_2 is expected and a distinct mechanism is proposed in which, upon photo-excitation of Au nanoparticles, electrons from Au are injected into the TiO_2 conduction band, leading to the generation of holes in the Au nanoparticles and electrons in the TiO_2 conduction band. This special feature is related to the very small dimensions of gold nanoparticles [50].

As in other heterogeneous catalytic systems, it is well known that the nanoparticle size of gold is related to its catalytic activity [51]. In fact, the catalyst prepared by LPRD presents the smallest Au nanoparticles, which can explain the highest photocatalytic activity under visible light irradiation.

CONCLUSION

TiO_2-mediated photocatalytic processes are efficient for the degradation of ciprofloxacin antibiotic in aqueous medium. In the case of ACF/TiO_2, ACF load and calcination temperature play an important role in the efficiency of the composite catalysts under UV irradiation. The main function of ACF in the hybrid materials is to act as co-adsorbent, rapidly transferring ciprofloxacin to the TiO_2 phase to be further degraded. Loading TiO_2 with gold nanoparticles has negligible effect when irradiating in the near-UV to visible light range ($\lambda \geq 365$ nm). Yet, under visible light ($\lambda \geq 400$ nm), an increase in the performance of the photocatalytic process is observed by using Au/TiO_2 catalysts, the most efficient being the material prepared by LPRM. This behavior can be ascribed to the lower dimensions of the gold nanoparticles, which can be photo-excited by visible light, initiating the photocatalytic degradation process.

ACKNOWLEDGMENTS

This work was supported by projects PEst-C/EQB/LA0020/2013 and PTDC/EQU-ERQ/123045/2010, financed by FEDER through COMPETE - Programa Operacional Factores de Competitividade, and by FCT - Fundação para a Ciência e a Tecnologia, and co-financed by QREN, ON2 and FEDER (Project NORTE-07-0124-FEDER-0000015).

SACC acknowledges the FCT Investigator 2013 Programme. Dr. Carlos M. Sá (CEMUP) is acknowledged for assistance with SEM/EDXS analyses.

REFERENCES

[1] Andreozzi, R., Caprio, V., Insola, A., Marotta, R. *Catal. Today* 1999, 53, 51-59.

[2] Oller, I., Malato, S., Sánchez-Pérez, J. A. *Sci. Total Environ.* 2011, 409, 4141-4166.

[3] Merayo, N., Hermosilla, D., Blanco, L., Cortijo, L., Blanco, Á. *J. Hazard. Mater.* 2013, 262, 420-427.

[4] Sui, M., Xing, S., Sheng, L., Huang, S., Guo, H. *J. Hazard. Mater.* 2012, 227–228, 227-236.

[5] Perini, J. A. D. L., Perez-Moya, M., Nogueira, R. F. P. *J. Photochem. Photobiol.,* A 2013, 259, 53-58.

[6] El-Kemary, M., El-Shamy, H., El-Mehasseb, I. J. *Lumin.* 2010, 130, 2327-2331.

[7] Van Doorslaer, X., Demeestere, K., Heynderickx, P. M., Van Langenhove, H., Dewulf, J. *Appl. Catal.,* B 2011, 101, 540-547.

[8] Meichtry, J. M., Blesa, M. A., Litter, M. I., Lin, H. J., de la Fuente, L., Levy, I. K., Gautier, E. A. *J. Sol. Energy Eng*. 2005, 129, 119-126.
[9] Bahnemann, D. *Sol. Energy* 2004, 77, 445-459.
[10] Carp, O., Huisman, C. L., Reller, A. *Prog. Solid State Chem*. 2004, 32, 33-177.
[11] Di Paola, A., García-López, E., Marcì, G., Palmisano, L. *J. Hazard. Mater.* 2012, 211–212, 3-29.
[12] Liu, G., Wang, L., Yang, H. G., Cheng, H.-M., Lu, G. Q. *J. Mater. Chem*. 2010, 20, 831-843.
[13] Silva, A. M. T., Silva, C. G., Dražić, G., Faria, J. L. *Catal. Today* 2009, 144, 13-18.
[14] Primo, A., Corma, A., García, H. *Phys. Chem. Chem. Phys*. 2011, 13, 886-910.
[15] Silva, C. G., Juárez, R., Marino, T., Molinari, R., García, H. *J. Am. Chem. Soc*. 2011, 133, 595-602.
[16] Choi, S. K., Yang, H. S., Kim, J. H., Park, H. *Appl. Catal*., B 2012, 121–122, 206-213.
[17] Wang, X., Liu, G., Chen, Z.-G., Li, F., Wang, L., Lu, G. Q., Cheng, H.-M. *Chem. Commun*. 2009, 3452-3454.
[18] Leary, R., Westwood, A. *Carbon* 2011, 49, 741-772.
[19] Pastrana-Martínez, L. M., Morales-Torres, S., Likodimos, V., Figueiredo, J. L., Faria, J. L., Falaras, P., Silva, A. M. T. *Appl. Catal.,* B 2012, 123–124, 241-256.
[20] Silva, C. G., Faria, J. L. *Appl. Catal*., B 2010, 101, 81-89.
[21] Liu, L., Ma, X., Zhu, L., Li, D. *Polym. Compos*. 2014, DOI: 10.1002/pc. 22913.
[22] Morales-Torres, S., Pastrana-Martínez, L. M., Figueiredo, J. L., Faria, J. L., Silva, A. M. T. *Appl. Surf. Sci*. 2013, 275, 361-368.
[23] Sampaio, M. J., Marques, R. R. N., Tavares, P. B., Faria, J. L., Silva, A. M. T., Silva, C. G. *J. Environ. Chem. Eng*. 2013, 1, 945-953.
[24] Cybula, A., Priebe, J. B., Pohl, M. M., Sobczak, J. W., Schneider, M., Zielińska-Jurek, A., Brückner, A., Zaleska, A. *Appl. Catal*., B 2014, 152-153, 202-211.
[25] Lee, C. W., Antoniou Kourounioti, R., Wu, J. C. S., Murchie, E., Maroto-Valer, M., Jensen, O. E., Huang, C. W., Ruban, A. *J. CO_2 Util*. 2014, 5, 33-40.
[26] Hussain, M., Ahmad, M., Nisar, A., Sun, H., Karim, S., Khan, M., Khan, S. D., Iqbal, M., Hussain, S. Z. *New J. Chem*. 2014.
[27] Guillard, C., Disdier, J., Monnet, C., Dussaud, J., Malato, S., Blanco, J., Maldonado, M. I., Herrmann, J. M. *Appl. Catal*., B 2003, 46, 319-332.
[28] Konstantinou, I. K., Albanis, T. A. *Appl. Catal*., B 2004, 49, 1-14.
[29] Molinari, R., Pirillo, F., Loddo, V., Palmisano, L. *Catal. Today* 2006, 118, 205-213.
[30] Onesios, K., Yu, J., Bouwer, E. *Biodegradation* 2009, 20, 441-466.
[31] Yu, J. T., Bouwer, E. J., Coelhan, M. *Agric. Water Manage*. 2006, 86, 72-80.
[32] Daughton Christian, G. In: *Pharmaceuticals and Care Products in the Environment*; Ed.; ACS Symposium Series; American Chemical Society: 2001, pp. 2-38.
[33] Baquero, F., Martínez, J.-L., Cantón, R. *Curr. Opin. Biotechnol*. 2008, 19, 260-265.
[34] Batt, A. L., Bruce, I. B., Aga, D. S. *Environ. Pollut*. 2006, 142, 295-302.
[35] Kolpin, D. W., Furlong, E. T., Meyer, M. T., Thurman, E. M., Zaugg, S. D., Barber, L. B., Buxton, H. T. *Environ. Sci. Technol*. 2002, 36, 1202-1211.
[36] Cek, M., Tandoğdu, Z., Wagenlehner, F, Tenke, P., Naber, K., Bjerklund-Johansen, T. *World J. Urol*. 2014, 1-8.
[37] Oliphant, C. M., Green, G. M. *Am. Fam. Physician* 2002, 65, 455-464.

[38] Harris, S., Morris, C., Morris, D., Cormican, M., Cummins, E. *Environ. Sci. Processes Impacts* 2013, 15, 617-622.
[39] Vasconcelos, T. G., Kümmerer, K., Henriques, D. M., Martins, A. F. *J. Hazard. Mater.* 2009, 169, 1154-1158.
[40] Gad-Allah, T. A., Ali, M. E. M., Badawy, M. I. *J. Hazard. Mater.* 2011, 186, 751-755.
[41] Tokumura, M., Morito, R., Hatayama, R., Kawase, Y. *Appl. Catal.*, B 2011, 106, 565-576.
[42] Vasconcelos, T. G., Henriques, D. M., König, A., Martins, A. F., Kümmerer, K. *Chemosphere* 2009, 76, 487-493.
[43] Paul, T., Miller, P. L., Strathmann, T. J. *Environ. Sci. Technol.* 2007, 41, 4720-4727.
[44] Bowker, M., Nuhu, A., Soares, J. *Catal. Today* 2007, 122, 245-247.
[45] Sunagawa, Y., Yamamoto, K., Takahashi, H., Muramatsu, A. *Catal. Today* 2008, 132, 81-87.
[46] Carabineiro, S. A. C., Silva, A. M. T., Dražić, G., Tavares, P. B., Figueiredo, J. L. *Catal. Today* 2010, 154, 293-302.
[47] Sing, K. S. W., Everett, D. H., Haul, R. A. W., Moscou, L., Pierotti, R. A., Rouquerol, J., Siemieniewska, T. *Pure Appl. Chem.* 1985, 57, 603-V.
[48] Silva, C. G., Faria, J. L. *Photochem. Photobiol. Sci.* 2009, 8, 705-711.
[49] Carabineiro, S. A. C., Machado, B. F., Dražić, G., Faria, J. L., Figueiredo, J. L. In: *Nanotechnology*, Vol. 3 - Synthesis and Characterization; J. N. Govil, S. S., N. K. Navani; Ed.; Studium Press LLC: Houston, 2012, pp. 125-140.
[50] El-Sayed, M. A. *Acc. Chem. Res.* 2001, 34, 257-264.
[51] Carabineiro, S. A. C., Thompson, D. T. In: *Gold: Science and Applications;* C. Corti, R. H.; Ed.; CRC Press, Taylor and Francis Group: Boca Raton, London, New York, 2010, pp. 89-122.

In: Titanium Dioxide
Editor: Jerri Brown

ISBN: 978-1-63321-391-3

Chapter 4

PHOTOCATALYTIC ACTIVITY OF IRON- AND SULFUR- CO-DOPED COATINGS PREPARED BY PLASMA SPRAYING

P. Ctibor*[*1], Z. Pala[1] and V. Stengl[2]
[1]Institute of Plasma Physics, ASCR, Praha, Czech Republic
[2]Institute of Inorganic Chemistry, ASCR, Husinec-Rez, Czech Republic

ABSTRACT

This study examines the photocatalytic activity (PCA) of coatings produced by atmospheric plasma spraying. The applied spraying tool was water-stabilized plasma gun. The TiO_2–Fe_2O_3 powders with and without Na_2SiO_3 additive were produced as agglomerates suitable for feeding into the plasma jet. The TiO_2–Fe_2O_3–Na_2SiO_3 coating was also post-treated in a furnace at 750°C for 2 hours. The composition of coatings was analyzed by X-ray diffraction. Photocatalytic degradation of butane under UV and under visible light (VIS) was quantified for all coatings. The coatings show a lamellar structure, as is typical for this process. Anatase titania from the feedstock powder was converted to rutile phase in all coatings but the doping with Na_2SiO_3 decreased the extent of this transformatuion. The coating FeT-VS was photocatalytically active especially under UV, whereas under visible light also FeT and FeT-VS-ann coatings were sufficiently active.

Keywords: Photocatalysis, plasma spraying, dopants

1. INTRODUCTION

When TiO_2 is illuminated by photons having energy higher than TiO_2 band gap, charge carriers are photogenerated promoting oxidation and reduction reactions. This photocatalytic activity is applied, for example, in water purification [1,2] and air cleaning [3]. The photocatalytic activity of the coatings can vary with the composition, stoichiometry, hydroxyl

* Email: ctibor@ipp.cas.cz.

and impurity concentrations, porosity, surface roughness as well as with the micro- or nano-structure [4]. Another influential factor on the photocatalytic activity is, of course, the phase composition of the material. There exists a demand for fabrication techniques of TiO_2 films, in both crystalline as well as amorphous state, and coatings applicable to large area deposition and high quality coverings achieved even at low substrate temperatures [5].

A search for ways of improvement of the photocatalytic efficiency of the TiO_2-based coatings is a desirable subject of research and development. It has been also realized that the band gap of anatase TiO_2, which is about 3.2 eV, means that the electron can only be excited by the high power UV light irradiation with a wavelength shorter than 387 nm [6]. A multiple band gap photocatalytic reaction cell was formed in composite TiO_2–Fe_3O_4 coatings [7]. Part of the material reacts in plasma according the following formula:

$$Fe_3O_4 + TiO_2 \rightarrow FeTiO_3 + Fe_2O_3 \quad (1)$$

$FeTiO_3$, as a p-type semiconductor with band gap about 2.5 eV, is advantageously combined with a sub-stoichiometric TiO_2 that is n-type semiconductor, thus, a solid p-n junction formation in the sprayed coating is possible [8]. The TiO_2–Fe_3O_4 coatings consisted of anatase TiO_2, rutile TiO_2 and pseudobrookite Fe_2TiO_5 phase which appeared when the content of Fe_3O_4 additive was over 10 wt. % [6]. Relatively low addition of Fe_3O_4 led to occurrence of ilmenite $FeTiO_3$ phase in the coatings. The content of anatase in the sprayed coatings decreased with the increasing Fe_3O_4 content. The photocatalytic activity was improved with an increase of $FeTiO_3$ content in the coating, which was explained by the good photoabsorbance and by the two-step electron transfer model [6].

During the photocatalytic reactions, when a semiconductor containing TiO_2 and $FeTiO_3$ is UV-irradiated, the electron-hole pairs possibly form in two steps [8]. In the first step, the electron is initiated from the valence band to the conduction band of $FeTiO_3$, and in the second step, the electron in the conduction band of $FeTiO_3$ is initiated to the conduction band of TiO_2. The benefit, called also inter-semiconductor hole-transfer mechanism [9], is extension of the light absorption range and possibly recombination of the excited holes and electrons.

The plasma sprayed TiO_2 is typically slightly sub-stoichiometric (TiO_{2-x}) [10,11] and we propose the use of Fe_2O_3 instead of Fe_3O_4 in order to avoid this phenomenon. In this case, part of the material will follow the reactions (2), especially when using a high-temperature jet of the water-stabilized plasma gun [12] available for our spray tests.

$$Fe_2O_3 + 2TiO_2 \rightarrow FeTiO_3 + TiO + FeO + O_2 \quad (2)$$

The free oxygen, released according to the formula (2) can replace the oxygen deficiency in TiO_{2-x} resulting from the reducing environment of the plasma jet of WSP. The FeO evaporates preferentially because of its melting point at 1370 °C which is about 180 °C lower than melting points, and simultaneously decomposition points, of both Fe_2O_3 and Fe_3O_4 at 1550 °C. TiO is created in small amount only.

The goal of the present paper is to examine the influence of plasma spray process on the composition, microstructure and photocatalytic activity of TiO_2 powder doped with Fe and with or without the simultaneous S doping, according to the above described considerations.

2. EXPERIMENTAL

2.1. Powder Synthesis

An original technique was used for preparation of the TiO_2-Fe_2O_3 composition. Ten liters of 1.6 M solution $TiOSO_4$ (99.9% purity; Sigma-Aldrich) was diluted in 60 liters of water. Then 1.6 kg of $FeSO_4$ (99.0% purity; Sigma-Aldrich) and 6 kg of urea was added. The reactive mixture was boiling at 98°C for 8 hours, decanted, filtered and dried at 105 °C. This method is based on thermal decomposition of urea at temperature higher than 60°C. The products of these reactions are mostly spherical agglomerates composed of sub-micrometric individual spheres.

After drying, disintegration of dry TiO_2-Fe_2O_3 large agglomerates and sieving, we obtained a feedstock with sizes 63-125 μm for the WSP spray process. This feedstock is further labeled "FeT feedstock" and contains TiO_2 with nominally 5 wt.% of Fe_2O_3. However, a majority of the powder disintegrated at sieving back to very fine particles, unsuitable for plasma spraying. We searched for an easy re-agglomeration process using an inorganic substance which can be released at a thermal processing. The ideal candidate would be a substance enhancing structural features that would promote the photoactivity as well. We have chosen Na_2SiO_3, because it can influence the surface state and phase transformations of TiO_2 [13,14]. The mixture was prepared mechanically with 1:1 ratio of the disintegrated FeT fine powder with so called "water glass", which contained 38 wt. % of Na_2SiO_3 and the balance was silicic acid, and dried the resulting product at room temperature. After careful crushing and further sieving, we obtained a feedstock for the WSP spray process, with the size 63-125 μm. This feedstock is further labeled "FeT-VS feedstock". Its phase composition can be seen in the results section.

2.2. Spray Process

The device used for plasma spraying was a water-stabilized plasma gun (WSP) working with a d.c.-current and water vapor as a plasma-forming gas. The parameters used were: Feeding distance (FD) 105 mm; Spray distance (SD) 350 mm; Torch current 480 A and voltage 320 V; Powder feeding gas – air (flow rate 3.25 slpm). Powder feed rate was set to 12-15 kg/h and substrates were stinless steel plates.

The resulting coatings are labeled FeT-VS and FeT (from powder with/and without Na_2SiO_3, respectively). This last type of coating was also heat post-treated in air atmosphere within a furnace at 750°C for 2 h (label FeT-VS-ann).

The reference sample "R", sprayed using a commercial powder prepared from a nanometric particles by their agglomeration, was included in our experiments. The nanometric powder was Altium™ TiNano 40 VHP (Altair Nanomaterials, Reno, NV, USA). An agglomerated fraction from 45 to 75 microns in size was utilized for WSP spraying at processing conditions represented by feeding distance 52 mm and spray distance 450 mm.

2.3. Characterization Techniques

X-ray diffraction (XRD) was performed on D8 Discover Bruker theta-theta Bragg-Brentano diffractometer using Cu-Kα radiation and 1D LynxEye detector. The obtained diffraction patterns provided information about the phase composition of feedstock powders and coatings. The influence of the instrumental broadening was taken into account by measuring calcium fluoride standard.

The reflectance measurement in this work has been implemented using ultra-violet/visible/near-infrared (UV/Vis/NIR) spectrometer (Shimadzu, Japan). Prior to the actual measurement, the calibration process was conducted using the $BaSO_4$ reference mirror in order to minimize the error from environment.

The accuracy guaranteed by the equipment manufacturer is that of wavelength ±0.3 nm and the uncertainty of the measurement less than 0.2 %. The wavelength of incident light used for the reflectance measurement was in the range from 250 to 2000 nm, and the diameter of the measured area was about 2 cm^2. The corresponding band gap energy was estimated after converting the reflectance curves to absorbance and recalculation [15] to Kubelka-Munk absorbance units. Kinetics of the photocatalytic degradation of butane was measured by using a home-made stainless steel batch photo-reactor [16] with a Narva black-light fluorescent lamp at wavelength 365 nm and warm-white fluorescent lamp at wavelength up to 400 nm (input power 8 W, light intensity 6.3 mW cm^{-2}).

Gas concentration was measured with the use of a quadrupole mass spectrometer JEOL JMS-Q100GC and a gas chromatograph Agilent 6890N.

A high-resolution gas chromatography column (19091P-QO4, J&W Scientific) was used. The sample from the reactor was taken via a sampling valve at time intervals of 2 hours.

The reactor with the total volume of 3.5 liters was filled with oxygen by 1 liter per minute flow rate. Blank tests - a layer of poly(hydroxyethyl methacrylate) without titania - were performed to establish the effect of photolysis and catalysis on the conversion of butane. The UV irradiation detects that there was no or immeasurable conversion of butane, as a testing gas, into CO and/or CO_2, and consequently butane also did not absorb on the poly(hydroxyethyl methacrylate) matrix. The injection volume of butane into the photoreactor was 30 ml.

3. Results and Discussion

The kinetics of butane decomposition is shown in Figures 1 and 2. The behavior of coatings can be confronted with band-gap energy (E_{bg}), extracted from the diffuse reflectance measurements, assuming direct VB-CB electron transfer.

The FeT coating with as high E_{bg} as 3.30 eV was, surprisingly, more efficient catalyst at 400 nm than at 365 nm. The FeT-VS coating with E_{bg} 3.08 eV was better under UV than under VIS. The FeT-VS-ann coating had E_{bg} only 2.60 eV, but composed from various unstable states signalized by a tail on the absorption coefficient characteristics. Its slow kinetics by butane decomposition under UV turned to much faster course under VIS radiation.

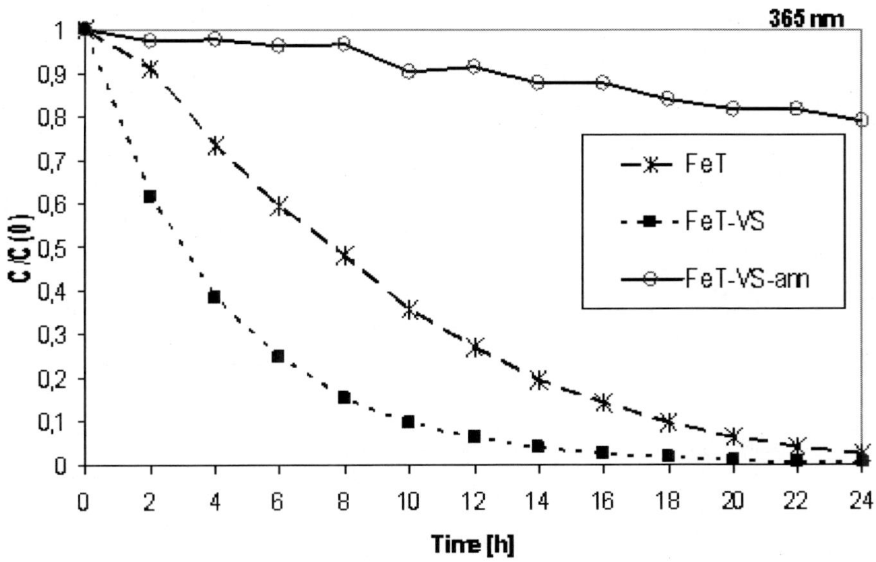

Figure 1. Kinetics of butane decomposition at 365 nm (UV).

The FeT coating contained besides rutile TiO_2 also ilmenite $FeTiO_3$ whereas the FeT-VS coating was partly amorphous. After heat-treatment (FeT-VS-ann) it crystallized to rutile – 87 %, $FeTiO_3$ – 6 %, and Fe_2TiO_5 – 7 %. This last phase is most probably that one, which blocked the UV PCA of this sample. Various other microstructural aspects as surface roughness and porosity would therefore be included in the study.

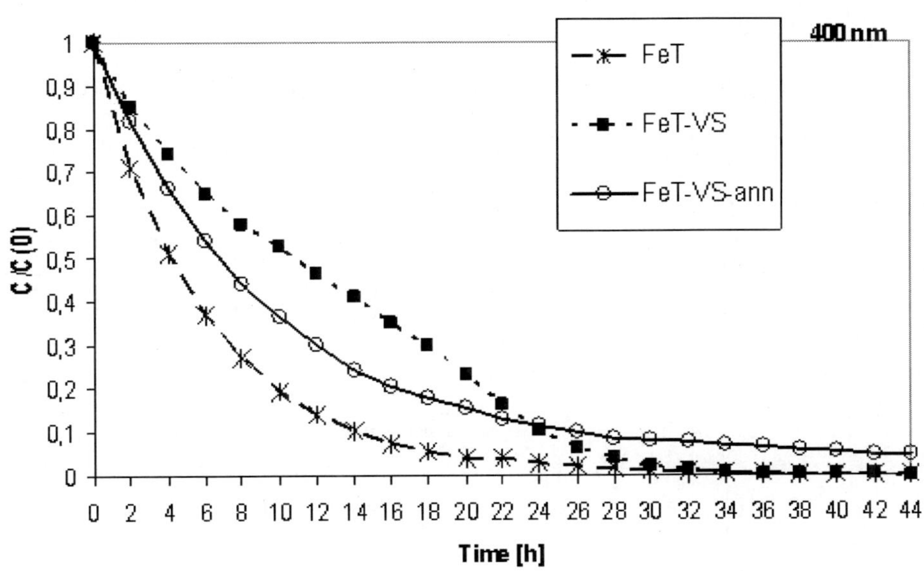

Figure 2. Kinetics of butane decomposition at 400 nm (VIS).

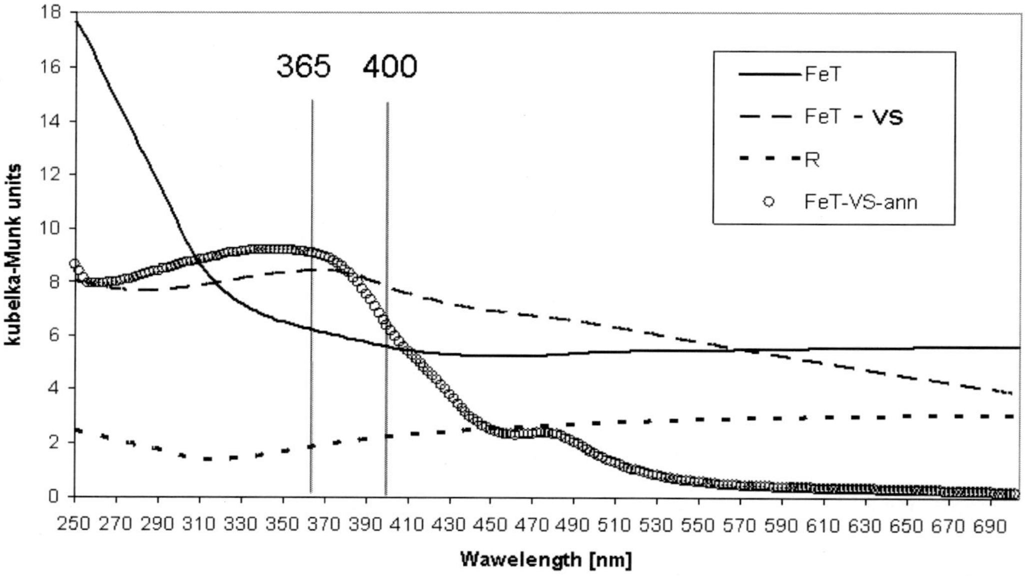

Figure 3. Absorbance of the coating expressed in the KM units.

In the Figure 3 the absorbance of the coating expressed in the Kubelka-Munk (KM) units. The annealed sample FeT-VS-ann absorbs highly the radiation at 365 nm. It is followed by the as-sprayed FeT-VS coating, FeT coating and the reference sample R. At 400 nm the highest absorbance was recorded for FeT-VS, whereas for FeT-VS-ann it drops nearly to the same level as for FeT coating. The reference sample remained the less absorptive.

Table 1. Qualitative phase composition in [%] from the crystalline material

Sample	**Anatase TiO_2**	**Rutile TiO_2**	**Hematite Fe_2O_3**	**Ilmenite $FeTiO_3$**	**$(NH_4)_2SO_4$**	**Fe_2TiO_5**
FeT feedstock	85		6		9	
FeT-VS feedstock	yes		yes		yes	
FeT- coating	3	84		13		
FeT-VS coating	22	78				
FeT- VS -ann		87		6		7

The addition of Na_2SiO_3 into the feedstock contributed to the limitation of the anatase to rutile phase transition at plasma spraying, Table 1. For the decomposition of Na_2SiO_3 and evaporation of Na in plasma, certain portion of plasma enthalpy is consumed, which at spraying of FeT feedstock is the driving force of the anatase to rutile phase transition of TiO_2.

Also microstructural aspects of the coating play a role at the photocatalysis. Table 2 summarizes the surface roughness results. Average roughness R_a as well as maximum roughness $R_{y\,max}$ were selected as representative parameters. FeT-VS-ann sample is markedly coarser than the second coarsest coating FeT-VS.

Table 2. Surface roughness of the coatings

Coating	R_a	$R_{y\ max}$
FeT	14.0 ± 0.5	105.2 ± 1.8
FeT-VS	23.8 ± 1.5	168.2 ± 6.7
FeT-VS-ann	44.4 ± 4.9	259.3 ± 25.5
R	13.3 ± 0.7	94.9 ± 9.4

It signalizes that at annealing when the amorphous material crystallized and the crystalline phases transformed, certain coarsening of the surface took place. FeT coating and R coating are the smoothest and similar together. For the photocatalysis the coarse surface is advantageous with only exception or potential problem, when the products of decomposition of the model contaminant tend to adsorb on the surface and slower the reactions. At our setup this is not the case. So the high roughness of the FeT-VS-ann sample represents large real area per a nominal area unit and by this way it is advantageous for photocatalysis.

The butane concentration C/C(0) for the reference coating R is about 0.1 after 20 hours of the test under 365 nm radiation, but about 0.6 after 20 hours of the test under 400 nm radiation.

So the R-coating lost its photoactivity when longer wavelength is applied, as predicted by the theory of undoped TiO_2 [17]. The FeT sample has similar kinetics of the butane decomposition at 365 as well as at 400 nm. The FeT-VS sample has different kinetics of the butane decomposition at 365 and at 400 nm. And finally FeT-VS-ann is markedly worse than other samples at 365 nm whereas it is very active at 400 nm.

Concerning the interpretation the role of sulfur, it is eligible to point out the following factors. Sulfur species were coordinated with TiO_2 in the network producing more acidic sites on the TiO_2 surface [18]. These acidic sites could capture more photo-induced electrons that could effectively protect the photo-induced holes from recombination with electrons. Thus, the quantum efficiency of the TiO_2 photocatalyst is improved by sulfur doping, resulting in the higher photocatalytic activity (tested on methyl orange solution in [18]). Sulfur-doped TiO_2 has enhanced activity in the presence of visible light, i.e., for acetaldehyde decomposition or for a removal of microorganisms as algae [19].

Fe-N-S-tridoped TiO_2 photocatalysts were synthesized by one step in the presence of ammonium ferrous sulfate [20]. Fe (III) and S^{6+} were incorporated into the lattice of TiO_2 through substituting titanium atoms, and N might coexist in the forms of substitutional N (O-Ti-N) and interstitial N (Ti-O-N) in tridoped TiO_2. Fe-N-S-tridoped TiO_2 catalyst exhibited a higher visible light photocatalytic activity for the degradation of phenol than that of pure TiO_2 and Degussa P25 TiO_2 [20]. Simultaneous doping of TiO_2 by Fe and S was however not mentioned in the literature.

Conclusion

The goal of this study was to examine the photocatalytic activity of coatings produced by atmospheric plasma spraying. The plasma gun used was water-stabilized one, advantageous by its high feed-rate. TiO_2–Fe_2O_3 powder with and without Na_2SiO_3 additive was synthesized and subsequently agglomerated into particles suitable for plasma spraying.

Anatase titania from the feedstock powder was converted to rutile phase whereas a presence of $FeTiO_3$ was detected in the WSP coating without Na_2SiO_3. The coating from powder with Na_2SiO_3 was partly amorphous. This last type of coating was also post-treated in air atmosphere within a furnace at 750°C for 2 hours. During this annealing new phases raised, predominantly in expense of the amorphous fraction.

All here addressed coatings were more photocatalytically active than a reference TiO_2 coating (without dopants). A charge-transfer transition between Fe and S dopant ions via the conduction band took probably place in this coatings, leading to most efficient butane conversion under UV and also mainly under visible light.

ACKNOWLEDGMENTS

This work was supported by the Grant Agency of the Academy of Sciences of the Czech Republic (Project No. IAAX00430803).

REFERENCES

[1] R. Molinari, F. Pirillo, M. Falco, V. Loddo, and L. Palmisano, *Chem. Eng. Processes* 43, 1103-1114 (2004).

[2] N. Serpone, I. Texier, A.V. Emeline, P. Pichat, H. Hidaka, and J. Zhao, *J. Photochem. Photobiol. Sci. A – Chemical*, 136, 145-155 (2000).

[3] M.C. Canela, R.M. Alberici, R.C.R. Sofia, M.N. Eberlin, and W.F. Jardim, *Environ. Scie. Tech.*, 33, 2788-2792 (1999).

[4] S. Boukrouh, R. Bensaha, S. Bourgeois, E. Finot, and M.C. Marco de Lucas, *Thin Solid Films*, 516, 6353–6358 (2008).

[5] Y. Ju, L. Li, Z. Wu, and Y. Jiang, *Energy Procedia*, 12 450 – 455 (2011).

[6] F.X. Ye, T. Tsumura, K. Nakata, and A. Ohmori, *Mat. Sci. Eng. B,* 148, 154–161 (2008).

[7] F.X. Ye, and A. Ohmori, *Surf. Coat. Technol.* 160, 62–67 (2002).

[8] F.X. Ye, A. Ohmori, and Ch. Li, *Surf. Coat. Technol.* 184, 233–238 (2004).

[9] Y.J. Kim, B. Gao, S.Y. Han, M.H. Jung, A.K. Chakraborty, T. Ko, Ch. Lee, and W.I. Lee, *J. Phys. Chemistry* C, 113 (44) 19179-19184 (2009).

[10] R.S. Lima, and B.R. Marple, *Surf. Coat. Technol.* 200, 3428– 3437 (2006).

[11] M. Bozorgtabar, M. Rahimipour, M. Salehi, and M. Jafarpour, *Surf. Coat. Technol.* 205, 229–231(2011).

[12] P. Ctibor, and M. Hrabovsky, *J. Europ. Ceram. Soc.* 30, 3131–3136 (2010).

[13] H. Zheng, Ch. Wang, A. Zhang, L. Zhang, S. Xia, and Z. Zhao, *Advanced Powder Technology*, 22, 581–586 (2011).

[14] M.A. Rahman, S. Kaneco, T. Suzuki, H. Katsumata, K. Ohta, and A. M. Shafiqul Alam, Pak. *J. Anal. Environm. Chemistry*, 8, 1 & 2, 26–35 (2007).

[15] A. A. Christy, O. M. Kvalheim, and R. A. Velapoldi, *Vibrational Spectroscopy*, 9, 19-27 (1995).
[16] P. Ctibor, H. Ageorges, V. Stengl, N. Murafa, I. Pis, T. Zahoranova, V. Nehasil, and Z. Pala, *Ceram. Intl.* 37, 2561–2567 (2011).
[17] M. Anpo, *Pure Appl. Chem.* 72 (9), 1787–1792 (2000).
[18] Y. Liu, J. Liu, Y. Lin, Y. Zhang, and Y. Wei, *Ceram. Intl.* 35, 3061–306 (2009).
[19] A. Zaleska, *Recent Patents on Engineering*, 2, 157-164 (2008).
[20] B. Li, X. Cheng, X. Yu, L. Yan, and Z. Xing, *Adv. in Mater. Sci. Eng.*, Article ID 348927, 5 pages (2012).

In: Titanium Dioxide
Editor: Jerri Brown

ISBN: 978-1-63321-391-3

Chapter 5

TITANIUM DIOXIDE: PRODUCTION, PROPERTIES AND APPLICATIONS

V. I. Lakshmanan, Ashok Bhowmick and Md Abdul Halim
Process Research ORTECH Inc., Mississauga, Ontario, Canada

ABSTRACT

The article is an overview on titanium dioxide that describes the various production processes with beneficiation and upgrading of source materials and the important properties that led to potential applications. The article discusses on significant industrial production processes through several subsections. Following through the description of different natural and synthetic source materials, a discussion is given on the beneficiation and upgrading of ores. In that, several processes of upgrading the titanium slags have been given in detail through respective flowsheets. Existing commercial processes of production of TiO_2 for example, the sulphate and the chloride processes are discussed in terms of their advantages and disadvantages.

The second part of the article discusses on various properties of TiO_2 for example, structural, electronic and photocatalytic properties and their important applications. The artificial phases of the material initiated through structural phase transitions by the application of high pressure and temperature are discussed because of their implications in creating the doped materials for the possibility of having an altered semiconducting band configuration. The fundamentals of the photocatalytic property and the advantage of using the nanoparticles have been discussed and elucidated through several applications. The synthesis of TiO_2 nanotube and the self-organized system of nanotube arrays are described comprehensively. Some properties of this amorphous material and its possible applications are indicated.

1. INTRODUCTION

The research studies on titanium dioxide (TiO_2) have accelerated tremendously in the recent past. A quick survey over a span of the last 100 years shows that on average, 910 scientific papers are published each year on this material. On that basis, it could be considered to be one of the most studied oxide materials. Since its discovery in 1791 from ilmenite ($FeTiO_3$) ore, titanium (IV) oxide, TiO_2 or Titania has been used in several applications based on its physical and chemical properties. First manufactured commercially in 1916, it has acquired wide acclaim in the painting industry where it is known as the pigment white 6 and subsequently, in the food industry as colouring agent E171.

Three different phases of this oxide are found naturally; anatase, discovered in 1801; rutile, discovered in 1803; and the brookite, found in 1825. The brookite phase is comparatively rare and is also difficult to synthesize compared to the anatase and rutile ones which are photocatalytically active. The rutile phase is the most stable while both the anatase and brookite phases can be converted irreversibly into the rutile phase through heat treatment. While the very stable rutile phase is widely used in the paint industry, in many recent applications, the anatase phase has gained more importance because many interesting nanostructures can be synthesized from the anatase phase as the starting material.

Commercial TiO_2 is produced by either the sulphate process or the thermochemical chloride process from ores and concentrates. The major disadvantages in sulphate process are large scale generation of acidic and solid waste, toxic emissions, higher energy consumption and a variable quality of product due to batch processing. In thermochemical chloride process, though the amount of waste is much reduced, the environmental issues including considerable CO_2 and other toxic emissions still persists. A recently developed innovative hydrometallurgical process using the concepts of solution chemistry that involves direct chloride leaching followed through solid-liquid separation, has been proven to be superior in terms of minimum CO_2 emissions, significantly less energy consumption and the production of lesser environmental waste due to the closed loop operation.

56% of the global production of TiO_2 is utilized as paint whereas plastic and paper industries have 23% and 9% utilization respectively. Besides these large scale industrial utilizations, several new and innovative applications have been developed recently; especially TiO_2 synthesized as nanoparticles, nanotube and nanotube arrays. These properties are extremely promising for new technological applications, most notably for sustaining continual development in the area of renewable energy and creating a better environment for tomorrow.

In this chapter, the various production processes coupled with beneficiation and upgrading of source materials are discussed along with the important properties and their potential applications. The first part of the article concentrates on the production processes of titanium dioxide (TiO_2), which are commercially available and under development as well. In the second part, the fundamental properties and applications are discussed in the context of both the existing and recently developed technologies.

2. Production of Titanium Dioxide

2.1. Raw Materials for Titanium Dioxide Production

Titanium (Ti) is the 9th most abundant element in the earth's crust and is the 4th most abundant metallic element [1]. Pigment grade titanium dioxide is manufactured from a variety of ores including ilmenite ($FeTiO_3$), rutile, anatase and leucoxene ($TiO_2.xFeO.yH_2O$), which are mined from deposits located throughout the world. The most common titanium minerals and their chemical compositions are listed in Table 1. Titanium dioxide may also be recovered from slag produced during iron smelting and from synthetic rutile produced from ilmenite. Figure 1 illustrates the various processing routes to produce high grade TiO_2 from different titanium bearing minerals.

2.1.1. Ilmenite and Leucoxene

Ilmenite is the most wide spread form of titanium dioxide-bearing ore around the world. It is a common accessory mineral in igneous rocks such as gabbro, diorite and anorthosite. It also forms in veins and pegmatites. Ilmenite contains 40–65% TiO_2, depending on its geological history. Ilmenite is also found in beach sand in existing or fossil coastlines. Surface processes alter the ilmenite in these deposits to produce submicroscopic mixtures of minerals that include anatase, rutile and amorphous phases.

Leucoxene ($Fe_2O_3 \cdot nTiO_2$) is a natural alteration product of ilmenite, typically containing more than 65% TiO_2. Leucoxene is recovered from some deposits and treated separately from ilmenite; however, the quantities produced are small in comparison with those of ilmenite. The concentrates obtained from ilmenite sand, which are depleted of iron, are generally richer in titanium dioxide than those from the massive deposits. Other elements in these concentrates include magnesium, manganese, vanadium, aluminum, calcium, chromium and silicon. The chemical composition of ilmenite ores found in different locations is shown in Table 2 [2].

Major igneous deposits are found in Brazil, Canada, Norway, the Russian Federation and the Ukraine [3]. Important heavy mineral sands are found along the eastern and western coasts of Australia, the eastern coast of South Africa, the southeastern coast of the USA, the west coast of South Island, New Zealand, the eastern coast of China, the northeastern coast of Sri Lanka, at various locations along the southern coast of India, in coastal Malaysia and in alluvial deposits in Sierra Leone [3]. World reserves of ilmenite are estimated at 650 million tonnes in terms of TiO_2 content (Table 3). Major reserves are in China (31%), Australia (15%), India (13%), South Africa (10%), Brazil (7%), Madagascar and Norway (6% each) and Mozambique (2%).

Table 1. Titanium minerals and their chemical compositions with contents of TiO_2

Mineral	Composition	TiO_2 content (%)
Rutile	TiO_2 (tetragonal, twinned)	94-99
Anatase	TiO_2 (tetragonal, near octahedral)	90-95
Brookite	TiO_2 (orthorhombic)	90-100
Brookite	TiO_2 (orthorhombic)	90-100
Ilmenite	$FeTiO_3$	40-65
Leucoxene	$Fe_2O_3TiO_2$	60-90
Perovskite	$CaTiO_3$	40-60
Loparite	$(CeNaCa)(TiNb)O_3$	38
Titanite	$CaTiSiO_3$	30-42
Titaniferous magnetite	$(Fe\cdot Ti)_2O_3$	2-20

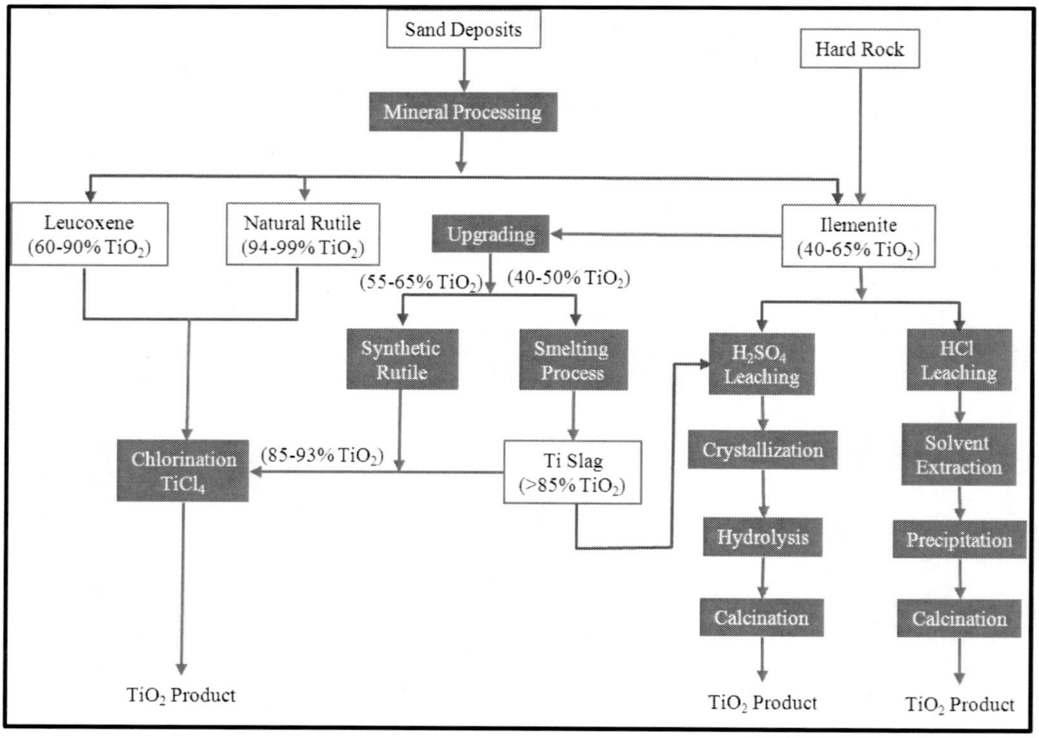

Figure 1. Overview of the titanium mineral processing industry from the raw material to the final product.

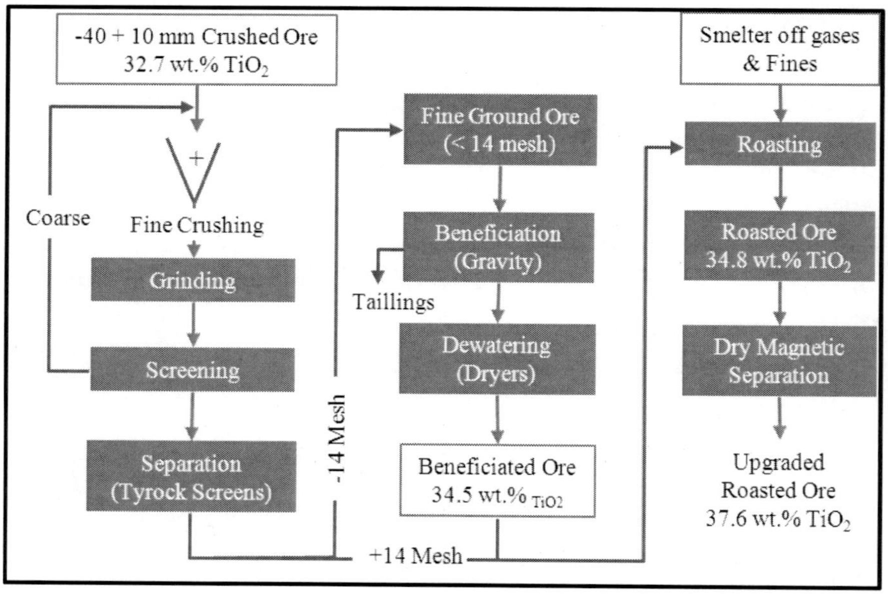

Figure 2. A flowsheet of the beneficiation process for hemo-ilmenite ore [4].

Table 2. Chemical composition of ilmenite ores (in %) [2]

(%)	North and South America			Europe			Asia			Africa		Australia	
	U.S.A.	Canada	Brazil	Norway	Russia	Portugal	India	Malaysia	Sri Lanka	Senegal	Sierra Leone	West Coast	East Coast
TiO_2	43–50	35-73	48-61	37-44	44	52.2	52.2	51-54	53.61	54-56	42.3	52-55	45-55
FeO	35–39	31-33	26-27	32-36	32.4	42.1	9-26	35-38	20.67	14-Jul	28	20-23	28-33
Fe_2O_3	1.6–13.8	20	14-15	11–13	16.9		12-27	3-6	20.96	28-30	25	15-19	1.5-5
SiO_2	1.4–3.0	0.8-4.0	1.4	0.6-3.0	1.84	0.27	0.9-1.4	0.5-1.11	0.38	0.9-1.3		1.41	1.2
Al_2O_3	0.2–1.21	1.05-1.7	0.25	0.85-1.8	0.2	0.29	1.0-1.11	1.09	0.54	0.5		0.25	1
P_2O_5	0.07–1.01			0.01-0.2	0.15	0.03	0.17-0.26	0.09		0.14-0.15		0.15	
ZrO_2	0.05–0.55		0.25-0.3	0.3-1.0			0.60-2.1			2.37		0.08	
MgO	0.6–2.35	1.0-2.0	0.3-0.35	1.6-3.0	2.76	0.03	0.65-1.01	0.21	0.92	1.9		0.01	
MnO	0.1–0.52	0.03-0.04	0.1-0.2	0.2-0.3	0.72	5	0.40-0.48	3.5	0.95	1.32		1.44	1.25-1.5
V_2O_5	0.05–0.27	0.2-0.35	0.06-0.2	0.2-0.3	0.09	0.09	0.03-0.26	0.04		0.27	0.38	0.13	0.23-1.0
Cr_2O_3	0.02–0.28	0.15	0.1-0.5	0.03-0.07	0.02	0.02	0.03-0.26	0.02	0.05	0.23		0.03	1.84-4.9
REE Oxide	0.08	0.07	0.08	0.068	0.07		0.08	0.079		0.06		0.089	0.12

Table 3. World reserves and production of ilmenite and rutile in 2012 (In '000 tonnes)

Country	Ilmenite		Rutile	
	Reserves	Production	Reserves	Production
United States*	2000	6300	–	–
Australia	100000	940	480	18000
Brazil	43000	45	5	1200
Canada	31000	700	–	–
China	200000	700	–	–
India	85000	550	25	7400
Madagascar	40000	280	–	–
Mozambique	16000	380	8	480
Norway	37000	350	–	–
Sierra Leone	–	–	100	3800
South Africa	63000	1030	131	8300
Sri Lanka	–	60	–	–
Ukraine	5900	300	60	2500
Vietnam	1600	500	–	–
Other countries	26000	40	17	400
World total (rounded)	650000	6200	830	42000

* United States rutile production and reserve data are included with ilmenite.
Source: Mineral Commodity Summaries, 2013.

World production of ilmenite concentrate was 10.9 million tonnes in 2011. During *2012*, *world production of ilmenite* increased by 2.9% to 11.3 million tonnes. Canada contributed 23% of ilmenite production, followed by South Africa (13%) and Australia (12%). World reserves and production of titanium minerals are listed in Table 3.

2.1.2. Rutile and Anatase

The second most commonly available ore is the buff-coloured mineral rutile, which contains about 95% titanium dioxide with smaller amounts of iron and other impurities. Its deposits are often found in coastal areas such as beach sands, and are the simplest to mine and concentrate in a form adaptable for titanium dioxide pigment production. Rutile sands are mostly found in Australia, Sierra Leone and South Africa. The importance of mineral rutile to the titanium dioxide industry is waning. In the 1970s, it accounted for 20% of the feedstock, but now accounts for less than 10% due to diminishing reserves. The world reserves of rutile are 42 million tonnes in terms of TiO_2 content. Major rutile reserves are located in Australia (43%), followed by South Africa (20%), India (18%), Sierra Leone (9%) and Ukraine (6%). World production of rutile concentrates was 0.83 million tonnes in 2012. Australia produced 57% of world rutile output, followed by South Africa with 16% and Ukraine 12%. World reserves and production of rutile are also shown in Table 3. Anatase, like rutile, is a modification of titanium dioxide. The largest reserves of this mineral are found in carboniferous intrusions in Brazil.

2.2. Beneficiation and Upgrading of Ore

2.2.1. Beneficiation and Upgrading of Ilmenite Ore

Beneficiation of titanium ore, particularly ilmenite is generally carried out by crushing, grinding, sorting, sizing, gravity concentration and magnetic separation. Most beneficiation operations result in the production of three materials: a concentrate; a middling or very low-grade concentrate, which is either reprocessed or stockpiled; and a tailing (waste), which is discarded. A flow sheet of the complete beneficiation process for hemo-ilmenite ore is presented in Figure 2 [4]. The average mass fractions of ilmenite and hematite in the hemo-ilmenite grains are 70 and 30 wt.%, respectively. Other obiquitous gangue minerals coming from anorthosite host rock, such as silicates, oxides, sulphides, and sulfosalts, are also present along with hemo-ilmenite grains. The major part of the silica, alumina, and calcia present in the ore comes from gangue minerals such as plagioclase feldspar and andesine.

Primary crushing and screening of ore take place at the mine site. Primary crushing is accomplished by using gyratory and conecrushers. The ore is then crushed and sized at a secondary milling facility. Secondary milling (comminution) further reduces particle size and prepares the ore for beneficiation processes that require finely ground ore feed. Subsequent fine grinding further reduces the ore particles to the consistency of fine powder. Most operations employ rod and/or mill grinding. Milled ore is then transferred to the next beneficiation step, gravity concentration. Gravity concentration is used to suspend and transport lighter gangue (nonmetallic or nonvaluable rock) away from the heavier valuable mineral. This separation process is based primarily on differences in the specific gravities of the materials and the size of the particles being separated, therefore, particle sizes must be kept uniform with the use of gravity classifiers. Waste tailings from gravity concentration are made up of gangue in the form of coarse and fine-grained particles and process water. Several variations of rakes are commonly used in thickeners to reduce moisture content and reclaim water from both the concentrate and tailings streams. When concentrates are thickened, the solid material resulting from these operations is collected for further upgrading by magnetic separation.

The intricate comminution techniques cannot liberate the ilmenite and hematite phases due to the mixture of these phases being fine grained. Therefore, the gravity concentrated ore is then subjected to an oxidative roasting to allow:

(i) The formation of ferromagnetic domains within hemo-ilmenite grains, to allow their magnetization in order to separate them from residual gangue minerals by magnetic separation.
(ii) The oxidation of sulphide and sulfosalt minerals for the removal of most of the sulphur content.

Hematite and ilmenite are initially antiferromagnetic minerals and the roasting step induces atomic reordering within the ilmenite and hematite crystal lattices that modify their magnetic behavior. During roasting conducted in air, the ore is treated at a temperature ranging between 900 °C and 1000 °C. The hematite reacts with the Fe_2O_3 from the ilmenite to produce magnetite (Fe_3O_4) and pseudobrookite (Fe_2TiO_5) according to the following reaction [5]:

$$FeTiO_3 + 2Fe_2O_3 = Fe_3O_4 + Fe_2TiO_5 \quad (1)$$

After roasting, the roasted ore is subjected to a magnetic separation step using rare earths magnets that remove most ore from the nonmagnetic gangue materials, specifically plagioclases, spinel, and phlogopite mica that form the black tailings. The resulting upgraded roasted ilmenite ore is then used for production of titanium slag and synthetic rutile.

2.2.2. Beneficiation and Upgrading of Natural Rutile

Prior to processing the basic approach in treatment of natural rutile ore is concentration by conventional mineral beneficiation techniques. The ore is crushed and ground. The iron content (magnetic) is removed by magnetic separation. Gravity separation methods including Hymphrey spiral, tabling, filtering and drying; and flotation are used to separate silica, silicate and aluminate. Electrostatic separation is used extensively to remove zircon, etc. to obtain high-grade titanium dioxide from beach sand. A flow sheet of the beneficiation process for natural rutile is shown in Figure 3.

2.2.3. Production of Titanium Slag from Ilmenite Ore

Ilmenite can be smelted at high temperatures (approximately 1650 to 1700 ^{0}C) with a suitable carbonaceous reductant in electric furnaces to produce high titanium slag as the major product and liquid iron is produced as a by-product. The iron product can subsequently be modified by secondary treatment (de-sulphurisation, carburization, alloy additions) to optimize the value of the product. The TiO_2 content (total Ti expressed as TiO_2) of the slag after reduction can range between 75% and 86%. Not all the iron can be reduced, as some FeO in the slag is required to act as a fluxing agent. The final quality of the titanium slag is dependent on both the quality of the ilmenite and the reductant. The reduction process can broadly be described by the following reactions [6]:

$$Fe_2O_{3(s)} + C_{(s)} = 2FeO_{(l)} + CO_{(g)} \quad (2)$$

$$FeO_{(s)} + C(s) = Fe_{(l)} + CO_{(g)} \quad (3)$$

$$2TiO_{2(s)} + C_{(s)} = Ti_2O_{3(s)} + CO_{(g)} \quad (4)$$

In the smelting process, contact between the molten slag and the furnace refractories is prevented by the presence of a solid freeze lining. This freeze lining is required to prevent attack on the refractories by the corrosive molten slag. Due to the high electrical conductivity of the slags the electrodes are above the slag level in an "open arc" position. The high titanium slag is tapped into ladles, water-cooled and crushed to produce the final products. Titanium slag is used for both the sulphate process and the chloride process routes for production of high grade TiO_2 (Figure 1).

A schematic process flowsheet for production of titanium-rich slag from upgraded roasted ore as explained in section 1.2.1 is presented in Figure 4. In this process, anthracite coal is used as carbonaceous reductant in the electric furnace. Due to the presence of trace moisture and volatile matter in the anthracite coal, some hydrogen forms and combines with the carbon monoxide to yield smelter gas (i.e., $CO + H_2$). At the high temperatures existing in the furnace, iron forms dense liquid droplets($\rho = 6800$ kg/m^3) that sink by gravity settling to the bottom of the furnace, forming a pool of a molten iron-carbon alloy, while a thick layer of titanium-rich slag floats due to its lower density above the molten iron bath.

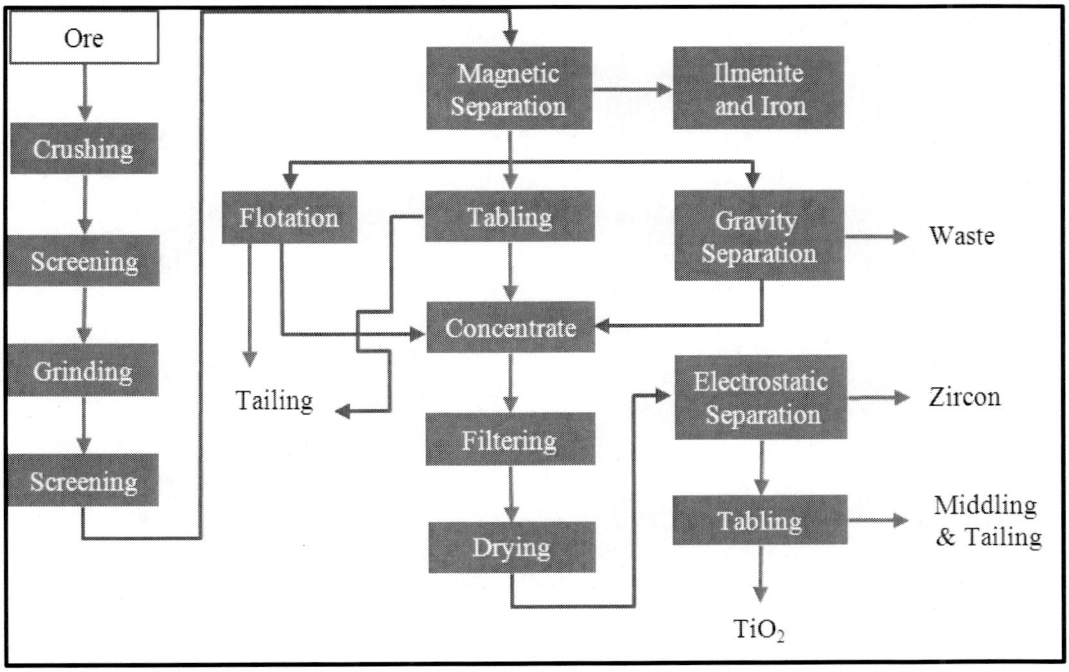

Figure 3. A flowsheet for processing natural rutile [2].

During the slagging process, Ti(IV) is partially reduced to Ti(III), as indicated by the high Ti_2O_3 content of the bulk titanium slag (Reaction (4)), while reducible minor metal cations such as V(III), Cr(III), and Mn(III) along with some Si(IV) are reduced partially and are distributed between the underlying molten iron layer and the titanium-rich slag. Regarding the traces of sulphur coming from the anthracite coal, they combine with the volatile matter and the traces of moisture to give hydrogen sulphide gas (H_2S) and to a lesser extent carbonyl sulphide (COS) entrained with the smelter gas and are later trapped and removed by scrubbing. The molten iron metal and titanium-rich slag are tapped at regular intervals from the electric furnace through separate tap holes. The titanium-rich slag is poured directly into steel wagons lined with crushed slag bedding. The hot solidified slag blocks are subsequently cooled under water sprays for several hours in order to build a thick protective skull. They are then demoulded and stored in the field until air cooling ensures the complete solidification of the inner core. Some remaining traces of sulphur also distribute either into the titanium-rich slag or into the molten iron. Therefore, the molten iron has to be further purified to remove all these impurities prior to being converted into a high-purity iron product.

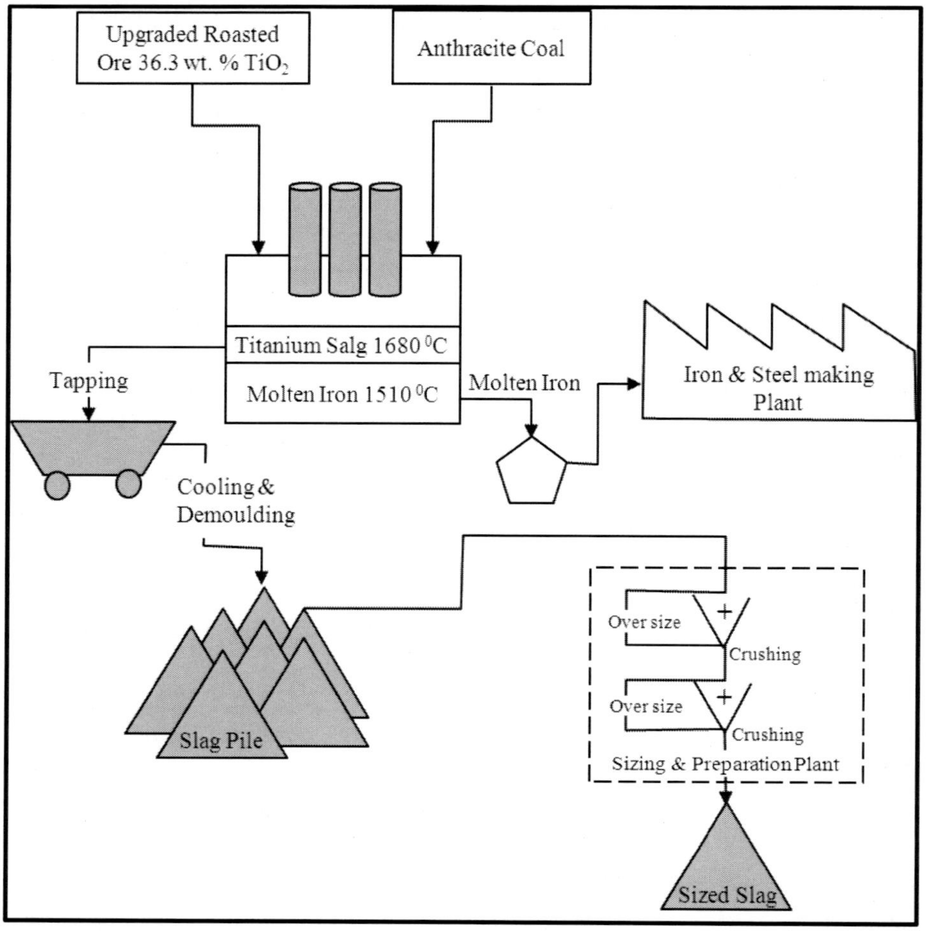

Figure 4. A process flowsheet for production of titanium slag from upgraded roasted ilmenite ore [4].

The low-iron and titanium-rich feedstock is used in the sulphate process as a substitute or in combination with ilmenite for production of TiO_2 pigment. In this process, the titanium slag is digested with sulphuric acid and precipitated, washed, dried, and calcined to yield anatase or rutile pigment. Since 1997 however, a part of the production of the titanium-rich slag is also further enriched to upgraded titanium rich slag (UGS) to be used in the chloride route. The process for upgrading of titanium slag is described in the following section.

2.2.4. Upgrading of Titanium Slag

The process for upgrading of titanium slag is based on the carbochlorination of the titanium for producing a titanium tetrachloride called "Ticle", which is then oxidized to yield titanium dioxide pigment. The scope of the upgraded titanium slag (USG) process is to increase the titanium dioxide content and consequently to lower the level of deleterious elements such as magnesium, calcium, and iron and to a lesser extent aluminum. The flowsheet of the UGS process is shown in Figure 5.

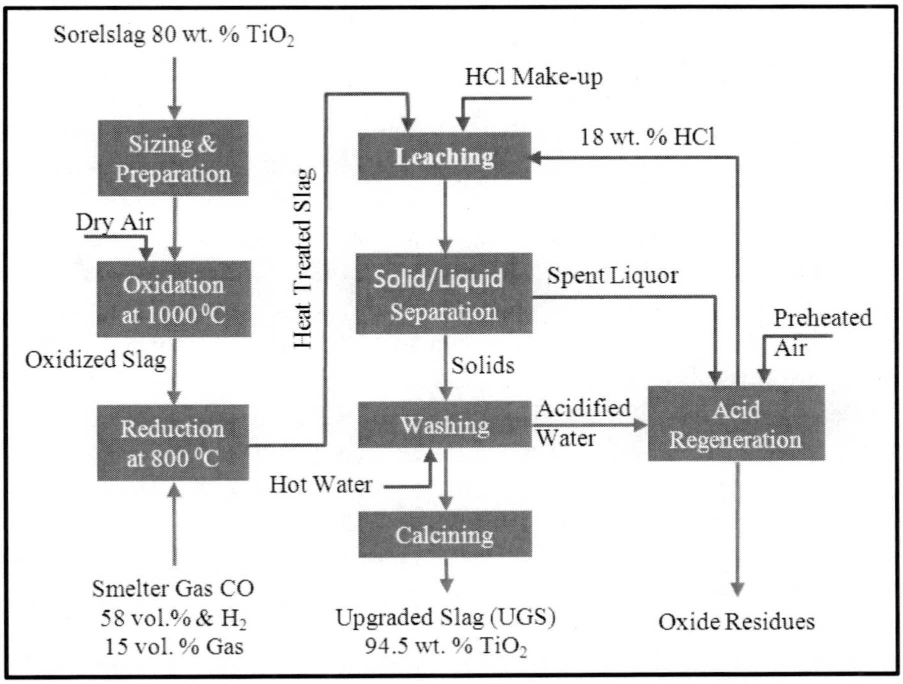

Figure 5. A flowsheet for upgraded titanium slag (USG) process [4].

2.2.4.1. Oxidation of Titanium Slag

The crude slag is first transferred to the sizing and preparation plant (SPP), where it undergoes crushing and screening. The sized and ground slag is then directed to the oxidation-reduction plant (ORP). The slag undergoes thermal oxidation which is performed in the oxidizer at 1000 ^{0}C. The aim of this step is to oxidize all the Ti(III) into Ti(IV) to render the titanium insoluble as titanium dioxide (TiO_2) during the subsequent high-pressure hydrochloric acid leaching. During this step all other reducible cations are oxidized to their higher oxidation state such as Fe(III), V(V) and Mn(III) while the Mg(II), Ca(II), Al(III), and Cr(III) remain unchanged.

2.2.4.2. Reduction of Oxidized Slag

The oxidized titanium-rich slag undergoes a thermal reduction at 800 ^{0}C ensured by smelter gas (i.e., 85 vol. % CO–15 vol. % H_2). The reducing conditions allow the reduction of the iron oxides back into ferrous oxide while titanium dioxide remains unaffected. In addition, during the reduction process the iron cations initially contained in the iron-rich silicates tend to concentrate around pores, forming a thin micrometer-size layer and also migrate to the edges of the titanium phase forming a secondary ilmenite, which is not at all related to the ilmenite originally found in the ore. This dual thermal oxidation/reduction process is mandatory because it greatly enhances the acid leaching of impurities contained in the original slag with minimal consumption of hydrochloric acid values, minimal loss of titanium dioxide and without degradation of the original particle size.

2.2.4.3. High-Pressure Acid Leaching

The heat-treated titanium slag exiting the reducer after cooling is sent to the acid leaching plant (ALP), where it is acid leached under high pressure with an azeotropic solution of hydrochloric acid. During this step, a major part of the leachable cations such as iron, magnesium, aluminum, manganese, calcium, vanadium and chromium are removed from the product and dissolved into the acid as metal chlorides or oxychlorides (eg., $FeCl_2$, $MgCl_2$, $AlCl_3$, $MnCl_2$, $VOCl_2$ and $CrCl_3$). The leached titanium-rich slag is then washed, dried, and calcined to yield the so-called upgraded titanium slag (UGS) with at least 94.5 wt. % TiO_2 and less than 1.1 wt. % MgO and 0.1 wt.% CaO [4].

In this process, the spent acid liquor that contains all the metal impurities as chlorides is regenerated by pyrohydrolysis in the acid regeneration plant (ARP). During this process, the pyrohydrolyser is continuously fed with the spent acid and hot gases. The existing bed of oxides acts as seed material to promote the nucleation and growth of new crystals. The process generates micrometer-size beads of metal oxides that crystallize around the previous seeds and hydrochloric acid (HCl) leaves the reactor and is absorbed fresh hydrochloric acid. The fresh concentrated hydrochloric acid is reused upstream in the process with some make-up acid to balance the process losses while the solid oxide residues are continuously discarded.

Table 4. Metallurgical process options for beneficiation of ilmenite to titanium slag [7]

Process	Description	Plant status
QIT – Electrosmelting	Carbothermic smelting of hard rock ilmenite at 1700 °C to pig iron and sulphatable titanium slag (86–87% TiO_2).	Sorel Quebec 1000000 tonnes/year slag.
RTZ iron and titanium Electrosmelting	Same as above: but customized to suit beach sand ilmenite.	Richards Bay, South Africa, 1000000 tonnes/year of slag and 500000 tonnes/year pig iron.
Process	**Description**	**Plant status**
Submerged arc smelting process	Pre-reduction of ilmenite (50% TiO_2), followed by smelting in arc furnace to produce pig iron and titania slag (87% TiO_2).	TINFOS, Norway, 200000 tonnes/year of slag and 100000 tonnes/year of pig iron.
Plasma DC arc smelting technology (South Africa Anglo American Corp.)	Carbothermic smelting of ilmenite in DC arcplasma furnace to pig iron and titanium slag.	NAMAKWA Sands Ltd., South Africa, 1100000 tonnes/year of slag and 450000 tonnes/year of pig iron ISCOR, South Africa 220000 tonnes/year of slag from 1999.
Upgraded slag technology	Conditioning of the Sorel slag and its leaching followed by calcination to UGS (up-graded slag) (95% TiO_2).	Sorel, Quebec 200000 tonnes/year of slag to be expanded to 600000 tonnes/year.

2.2.4.4. Comparison between Different Smelting Technologies for Titanium Slag Production

There are four major companies in the world (excluding Russia) using various smelting technologies to produce high titanium slag (Table 4). The producers are Richards Bay Minerals (RBM) and Namakwa Sands in South Africa, QIT-Feret Titanate Inc. in Canada, a subsidiary of the RTZ (Rio Tinto-Zinc) and Tinfos in Norway. RBM and QIT use 6-in-line alternating current furnaces, while Namakwa Sands use two circular furnaces with direct current configurations. The chemical compositions of titanium slag produced in different smelting processes are presented in Table 5 [8].

2.2.5. Upgrading Ilmenite to Synthetic Rutile

Synthetic rutile (TiO_2) is produced through the upgrading of ilmenite ore to remove impurities (mostly iron) and yield a feedstock for production of titanium tetrachloride through the chloride process. The titanium tetrachloride is then converted to produce titanium dioxide pigment. The chemical composition of synthetic rutile is similar to that of natural rutile, but differs in physical form. Synthetic rutile concentrates are composed of very fine crystals and are porous, whereas natural rutile grains are composed of single crystals. There are a number of commercial and processes under development to produce synthetic rutile from ilmenite. These processes involve a combination of thermal oxidation and reduction by roasting, leaching and physical separation steps. Iron is converted to soluble ferrous or elemental forms by reduction at high temperatures followed by acid leaching to obtain synthetic rutile (TiO_2) products. The process features are summarized in Table 6 [9] and some of commercialized processes are subsequently discussed.

Table 5. Chemical compositions (%) of titanium slag produced worldwide [8]

	Content (%)				
	Sulfate slag		Chloride slag		
	Sorel slag	Tinfos slag	RBM slag	Namakwa slag	UGS
TiO_2	80	77	85.5	86	95
Fe_2O_3	9	–	–	–	–
FeO	–	8	10.8	9	1.17
SiO_2	2.4	4.5	2.1	1.8	1.95
MgO	5	6	1.1	0.7	0.6
Al_2O_3	2.9	1.7	1.3	1.4	0.6
Cr_2O_3	0.17	0.12	0.17	0.09	0.03
(U + Th) (mg/L)	<5	–	15-30	20	<5

2.2.5.1. The Becher Sulphate Process

An industrial process for upgrading ilmenite to synthetic rutile is typically represented by the Becher process. Ilmenite containing 40–65% titanium as TiO_2, with the rest being iron oxide is used as raw material in the Becher process. The Becher process removes the iron oxide, leaving a residue of synthetic rutile, which contains more than 90% TiO_2. A schematic flowsheet of the improved Becher process is shown in Figure 6. This process is essentially a pyrometallurgical operation followed by a number of wet chemical process steps. It consists of the following four main process steps [10]:

Table 6. Summary of upgrading ilmenite to synthetic rutile [9]

Process	Pyro-treatment	Leaching	Advantage	Disadvantages
The Becher sulphate process	Iron oxidized to Fe_2O_3 and reduced to metallic Fe by at 1200 °C.	(a) NH_4Cl/O_2 (b) 0.5 M H_2SO_4	Allowing diverse ilmenite ores feed.	Multi step iron conversions and leaching High energy consumption Emission of CO_2.
The Murso process	Similar to the Becher process, but fluidized beds for the conversion.	20% HCl	Improved efficiency by using fluidised beds Easier HCl recycle than sulphate system.	Similar to the Becher processes.
The Laporte process	Lower temperatures for iron conversion to FeO with controlled CO_2 pressure.	18% HCl with a bed contactor	Free of formation of fine TiO_2 particles Ease for leaching FeO.	Similar to the Becher processes in spite of lower temperature used.
The Benelite process	Iron conversion to Fe(II) forms by carbon thermo-reduction.	18–20% HCl	Simple one step conversion of iron.	Limited ilmenite types as the feed.
The Austpac process	Magnetization of the ilmenite at 800–1000 °C.	25% (w/w) HCl	Magnetic separation for higher N97% TiO_2.	Higher acidity need for leaching remaining magnetic iron form.
The Dunn process	Selective chlorination of iron inilmenite with Cl_2	–	Cl_2 recycle by oxidation of $FeCl_2$ to Fe_2O_3.	Handling highly corrosive Cl_2.
The Kataoka process (in Japan)	Conversion to ferrous form	H_2SO_4	Less corrosive using H_2SO_4 than HCl, low leaching temperature.	Produce large iron sulphate wastes.

(i) Oxidation involves heating the ilmenite in a rotary kiln with air to convert the iron in ilmenite to iron oxides. This allows for the use of a wide range of ilmenite materials with various Fe(II) and Fe(III) forms for subsequent steps.

$$4FeTiO_{3(s)} + O_{2(g)} = 2Fe_2O_3{\cdot}TiO_{2(s)} + 2TiO_{2(s)} \quad (5)$$

(ii) Reduction is performed in a rotary kiln with pseudo brookite ($Fe_2O_3{\cdot}TiO_2$), coal and sulfur mixture at 1200 ^{0}C to reduce iron oxide to metallic iron. Sulphur is added to assist with the removal of manganese. Manganese sulfide is produced and then separated from the final product by a subsequent leaching step. The reduced ilmenite is separated from the gangue materials by screening and magnetic separation. Coarse char recovered from the process is recycled to the kiln.

$$Fe_2O_3{\cdot}TiO_{2(s)} + 3CO_{(g)} = 2Fe_{(s)} + 2TiO_{2(s)} + 3CO_{2(g)} \quad (6)$$

(iii) The metallic iron is oxidized and precipitated from the solution as a form of slime in the aeration "rusting" step in large tanks with 1.0% to 1.5% ammonium chloride solution at 80 ^{0}C. The mixture of synthetic rutile and iron oxide from the aeration step is separated in a multi-stage hydrocycloning system. At this stage, the aerated product contains approximately 90% TiO_2.

$$4Fe_{(s)} + 3O_{2(g)} = 2Fe_2O_{3(s)} \quad (7)$$

(iv) The aerated product is separated from the iron oxide by leaching with 10% sulphuric acid to obtain a high quality synthetic rutile product. The leached product is separated from the acidic liquor in a spiral classifier. The remaining acid is removed from the synthetic rutile by counter-current washing on a belt filter. The synthetic rutile can then be dried in a flash or fluidized bed dryer. The final product contains more than 93% TiO_2.

2.2.5.2. The Murso Process

The Murso process can be applied successfully to all commercial ilmenites whether they are in the form of beach sands or rock deposits. The process is such that where ferruginous impurities such as chromite occur in association with ilmenite they can be removed readily without any significant loss of TiO_2 values. The flowsheet of the Murso process is shown in Figure 7. It consists of the following major process steps:

(i) Pre-oxidation of ilmenite is carried out in a fluidized bed at temperatures ranging from 900–950 °C in order to oxidize all ferrous to the ferric state. Oxidation is an exothermic reaction as shown below:

$$FeTiO_2 + 1/2O_2 = FeO.TiO_2 \quad \Delta H = -52.9 \text{ Kcal} \quad (8)$$

$$12FeO.TiO_2 + 3O_2 = Fe_2O_3 + 5Fe_2O_3.TiO_2 + 7TiO_2 \quad \Delta H = -324.7 \text{ Kcal} \quad (9)$$

(ii) The hot oxidized ilmenite ore is transferred into another fluidized bed where ferric ion is reduced by a reducing agent. A typical temperature of reduction is 850 ^{0}C. High purity hydrogen is not necessary, a typical reductant gas composition being 70% H_2, 13% CO_2 and 4% water vapour. The reduction of oxidized ilmenite with hydrogenmay be represented by the following reactions:

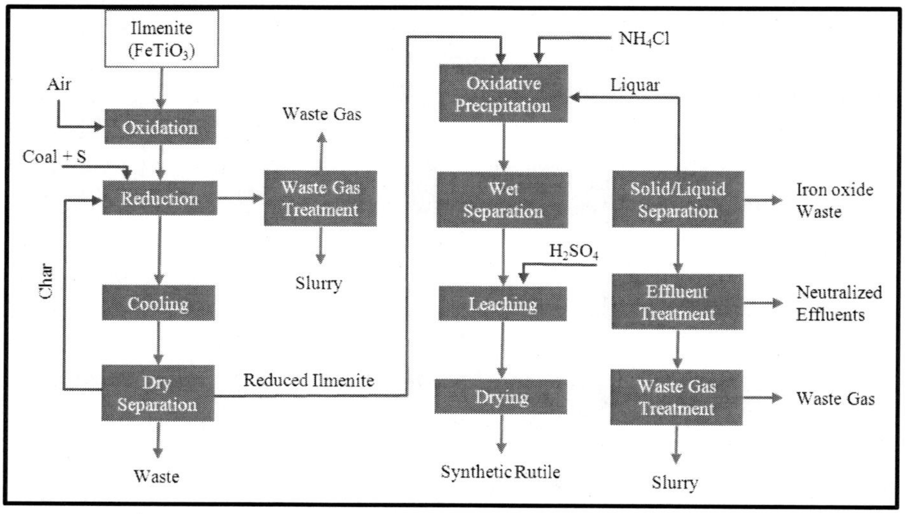

Figure 6. A flowsheet of the improved Becher process for upgrading ilmenite to produce synthetic rutile [10, 11].

$$Fe_2O_3.TiO_2 + H_2 = 2FeO.TiO_2 + H_2O \qquad \Delta H = -5.55 \text{ Kcal} \qquad (10)$$

$$Fe_2O_3 + 5Fe_2O_3.TiO_2 + 7TiO_2 + 6H_2O = 12FeO.TiO_2 + 6H_2O \qquad \Delta H = -32 \text{ Kcal} \qquad (11)$$

$$FeO.TiO_2 + H_2 = Fe + TiO_2 + H_2O \qquad \Delta H = 9.78 \text{ Kcal} \qquad (12)$$

(iii) The product from the reduction-fluidised bed is then leached with about 20% HCl at 108–110 °C for 3-4 hours. Oxides of manganese, magnesium and vanadium, which are structurally associated with TiO_2 in the ilmenite lattice, are effectively leached out with iron and partial leaching of aluminum oxide occurs also. The leaching of iron with hydrochloric acid can be shown by the following reaction:

$$FeO.TiO_2 + 2HCl = FeCl_2 + TiO_2 + H_2O \qquad (13)$$

(iv) The leached solids are classified into a coarse and fine fraction using conventional equipment such as classifier, filter and thickener. The coarse fraction has a grain size similar to the feed ilmenite. It is washed, filtered and calcined at 450 °C to remove residual traces of moisture, iron chloride and hydrochloric acid. A typical analysis of raw ilmenite, synthetic rutile produced by the Murso process and commercial rutile is shown Table 7. The fine fraction is also washed and, preferably, spray dried to the solid state.

(v) Around 19-20% of the hydrochloric acid is regenerated from the leach solution by hydrolysis of ferrous chloride. The regenerated hydrochloric acid is recycled to the leaching stage. The predominant hydrolysis reaction is:

$$2FeCl_2 + 1/2O_2 + 2H_2O = Fe_2O_3 + 4HCl \quad (14)$$

2.2.5.3. The Benelite Cyclic Process

The Benelite process was developed in order to overcome operational drawbacks associated with the direct acid leaching of ilmenite. In India there are plants located in Kerala, Sahuparum and Orissa. The Malaysian Titanium Corporation operates a plant in Ipoh, Malaysia and Kerr McGee operates a plant in Alabama, United States of America. A flowsheet for production of synthetic rutile from ilmenite by the Benetile process is shown in Figure 8 [13]. It consists of the following main process steps [14]:

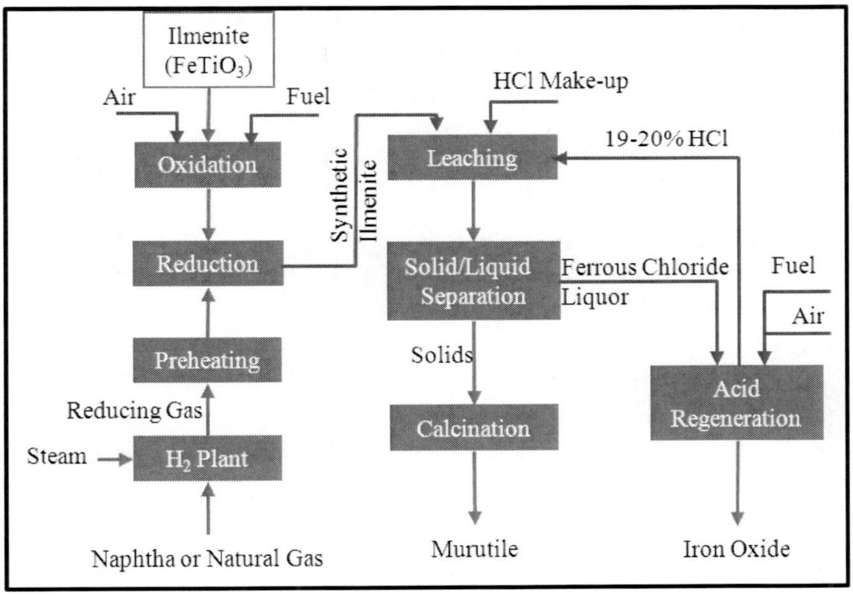

Figure 7. A flowsheet of the Murso process for production of synthetic rutile from ilmenite [12].

Table 7. Analyses of raw ilmenite, the Murso rutile and commercial synthetic rutile [12]

Oxides	Content (wt. %)		
	Ilmenite	The Murso rutile	Commercial rutile
TiO_2	52.9	96.2	96.6
FeO	30.8	–	–
Fe_2O_3	11.9	1.5	0.5
MnO	1.77	0.05	0.02
MgO	0.54	0.04	0.15
Cr_2O_3	0.14	0.15	0.23
Al_2O_3	0.25	0.23	0.4
V_2O_3	0.16	0.04	0.6

(i) Raw ilmenite sand containing 54 - 65% TiO_2 is roasted with heavy fuel oil in a rotary kiln at 870^0C for the partial reduction of ilmenite. The fuel oil functions as a reducing agent, converting ferric iron (Fe^{3+}) in the ilmenite to the ferrous (Fe^{2+}) state. The fuel oil is burnt at the discharge end of the kiln, and the resulting gases are passed through a cyclone and an incinerator to remove solids and unreacted hydrocarbons.

(ii) The reduced ilmenite is then batch-digested in rotary-ball digesters with 18-20% HCl at 140 °C. Ferrous oxide in the ilmenite is converted to soluble ferrous chloride, and the TiO_2 portion of the ilmenite is left as a solid. Spent acid liquor, which contains excess HCl and ferrous chloride, is sent to an acid regeneration circuit.

(iii) The TiO_2 solids are washed with water filtered and calcined at 870 °C, yielding synthetic rutile with approximately 94% TiO_2. Exhaust gases from the calciner are treated to remove solids and acidic gases before they are released to the atmosphere.

(iv) In the acid regeneration circuit, the spent acid liquor is sent to a pre-concentrator where one-fourth of the water in the liquor is evaporated. The concentrated liquor is sprayed through atomizers, causing the droplets to dryout, yielding HCl gas and ferric oxide powder. The gas is cycloned and then sent to an absorber to remove HCl forreuse. The ferric oxide powder is slurried with water to create the waste stream iron oxide slurry.

2.3. Process of Titanium Dioxide Production

Production of high grade TiO_2 by different routes is schematically presented in Figure 1. There are two processes to commercially produce pigment grade titanium dioxide: the sulphate process and the thermochemical chloride process. These two processes differ in both their chemistry and raw material requirements. In the sulfate process, the feed material, high grade ilmenite or titanium slag is digested with concentrated sulfuric acid to produce titanium sulfate followed by the bulk removal of iron through crystallization, then hydrolyze and precipitation of titanium dioxide. In contrast, the thermochemical chloride process involves reacting the high grade feed materials mainly titanium slag, natural rutile, or synthetic rutile with petroleum coke and chlorine gas at high temperatures, forming a titanium tetrachloride ($TiCl_4$) vapor to produce TiO_2 pigment. The upgrading of ilmenite to titanium slag and synthetic rutile is an important part of these processes. The world production capacity of TiO_2 pigment by the chloride process is an estimated 3.4 million metric tons per year.

Utilizing the knowledge of chloride chemistry, an innovative hydrometallurgical process (CTL Process) has recently been developed and is likely to be commercialized to produce TiO_2 pigment. The process involves direct chloride leaching at 65-80 °C of ilmenite material, solid-liquid separation and metal separation by solvent extraction, HCl recovery, precipitation and calcination to produce TiO_2. Using the concepts of solution chemistry, titanium is selectively extracted from the leach liquor at controlled pH and Eh conditions. In addition to these processes, other innovative processes are subsequently discussed.

2.3.1. Production of Titanium Dioxide by the Sulfate Process

As noted in the previous section, titanium dioxide pigments are commercially produced by two main processes: the sulfate process and the chloride process. The sulfate process is the older of the two methods. This process was first used for production of high purity TiO_2

pigment in Norway in 1915. The ilmenite ore or synthetic titanium slag is used as the raw material in this process. A flowsheet of the sulfate process for production of TiO_2 pigment is shown in Figure 9. The main process steps are as follows [15, 16].

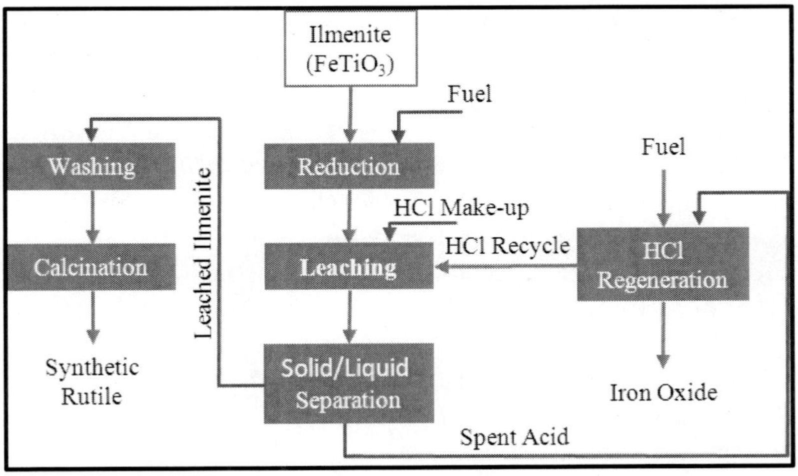

Figure 8. A flowsheet of the Benetile process for production of synthetic rutile from ilmenite [13].

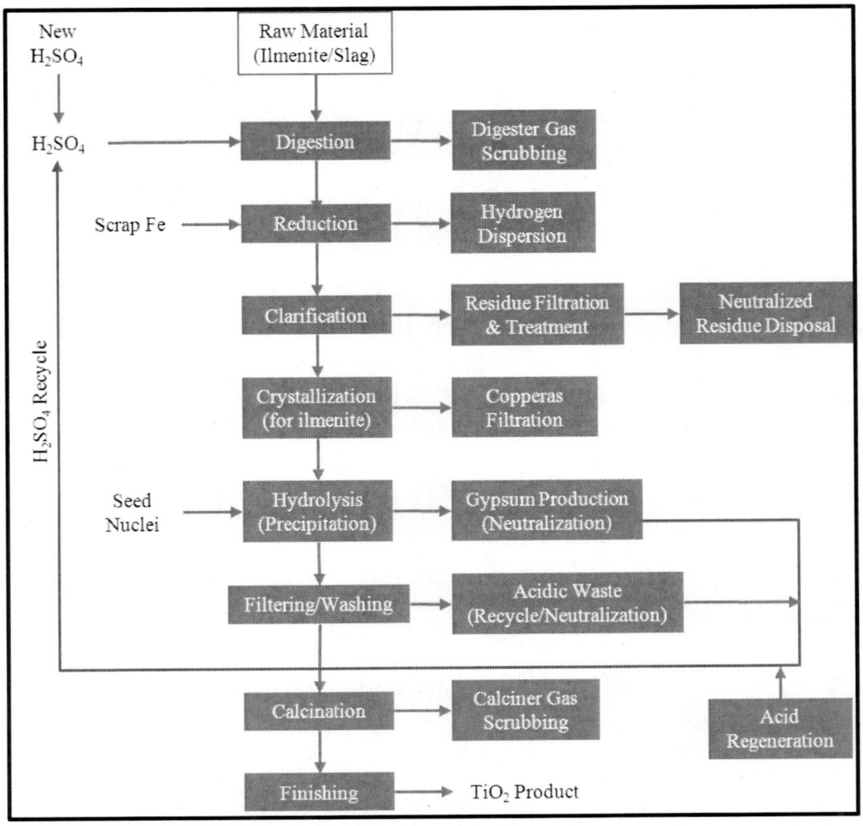

Figure 9. A flowsheet for production of TiO_2 pigment from ilmenite or slag by the sulphate process [17].

2.3.1.1. Acid Digestion and Clarification

(i) The raw materials are dried (<0.1% moisture content) then finely ground (<40 μn) in ball mills before sulfuric acid digestion.
(ii) The ground raw material ilmenite or titanium slag or mixtures of the two is mixed with concentrated sulfuric acid (80-98%) in a digestion vessel. The reaction can be started by adding water to a suspension of the raw material in concentrated sulfuric acid. Excess of free acid, based on TiO_2 content, is important to increase digestion efficiency. The temperature initially increases to 50-70 °C due to the heat of hydration of the sulfuric acid. The exothermic sulfate formation then increases the temperature to 170-220 °C to form solid fusion cake.
(iii) After completion of the reaction, the cake is leached with cold water or diluted sulfuric acid recycled from the process. Scrap iron is added to reduce any remaining Fe^{3+} ($Fe + Fe^{3+} = Fe^{2+}$). Due to the addition of scrap iron, some of the Ti^{4+} is reduced to Ti^{3+} which prevents the re-oxidation of iron during subsequent process steps. Titanyl sulfate ($TiOSO_4$) is produced in the leach solution. For ilmenite as raw material, the leaching and reduction reactions can be written as follows:

$$FeTiO_3 + 2H_2SO_4 = TiOSO_4 + FeSO_4 + 2H_2O \tag{15}$$

$$Fe_2(SO_4)_3 + Fe^0 = 3FeSO_4 \tag{16}$$

(iv) Following the leaching process, pregnant leach liquor is completely separated from un-dissolved solid material by flocculation, sedimenta-tion and processing through filter presses to remove fines.

2.3.1.2. Crystallization

(i) The concentration of ferrous sulfate ($FeSO_4$) in pregnant leach solution is dependent on the starting material. The $FeSO_4$ concentration will be at least 3-4 times higher for ilmenite digestion compare to titanium slag digestion. If ilmenite is used as the starting material, pregnant leach solution is cooled to less than 15 °C under vacuum to crystallize Fe^{2+} sulphate heptahydrate ($FeSO_4.7H_2O$) which is also known as copperas.
(ii) The concentrated titanyl sulfate solution is separated from copperas by filtration or centrifugation. Copperas ($FeSO_4.7H_2O$) has several commercial applications including water purification, Fe_2O_3 pigment production, agriculture, and chromate reduction for cement production.

2.3.1.3. Hydrolysis Precipitation and Purification

(i) The titanyl sulfate solution is heated to 95-110 °C to precipitate titanium oxide hydrate. The hydrolysis is accelerated by adding some seeds to the feed solution. The seeds are either produced by neutralization with sodium hydroxide (Mecklenburg method), or by heating of a small part of the titanyl sulfate solution with water (Blumenfeld method). The overall reaction is shown below:

$$TiOSO_4 + (n+1)H_2O = TiO_2.nH_2O + H_2SO_4 \qquad (17)$$

(ii) After hydrolysis, the precipitate is filtrated and re-pulped with Ti^{3+} solution and water. Re-pulping of titanium oxide hydrate with Ti^{3+} solution is facilitated to remove impurities including Fe, Cr, Cu, Mn and V by more soluble oxidation states. Re-pulping can also be performed with dilute sulfuric acid (3-10%) at 50-90 °C.

(iii) The filtrate of titanium dioxide hydrate contains 20-28% sulfuric acid and different amounts of dissolved sulfate depending on raw materials. Sulfuric acid is regenerated from this filtrate and washing solutions. Regenerated H_2SO_4 is recycled to digestion vessel.

2.3.1.4. Doping, Calcination and Grinding

(i) The purified titanium dioxide hydrate is treated with alkali metal compounds (for rutile) and phosphoric acid (for anatase) in order to produce specific pigment grades. Rutile seeds are used to produce rutile pigments. Rutile seeds are produced either by converting the purified titanium oxide hydrate to sodium titanate or by precipitating from titanium tetrachloride solutions with sodium hydroxide solutions.

(ii) The doped titanium oxide hydrate is filtered with pressure rotary filters or automatic filter presses to remove water.

(iii) After filtration, the hydrate is directly treated with gas or oil in counter-current flow in rotary kilns. Moisture is removed between 200°C and 300 °C. At around480 °C, TiO_2 crystals begin to form, growing as the temperature rises. A final temperature of 800-850 °C results in the anatase form as the final product. For the rutile form, a final temperature of 900-930 °C is required. Excessive temperatures will, in both cases, result in larger crystals and poorer physical properties. The overall reaction is shown below:

$$TiO_2.nH_2O = TiO_2 + nH_2O \qquad (18)$$

(iv) After cooling, the rutile/anatase TiO_2 product can be milled and surface treated to form the TiO_2 base pigment specific to its intended application.

Typical waste generated by the sulphate process includes digestion scrubber water from the scrubbing of gasses generated during the digestion step. Digestion sludge is generated after the filtering of the bottom solids from the settled titanyl sulfate liquor generated during digestion. A waste acid is generated as a result of the filtering of the titanium dioxide hydrate. This waste acid is used in the production of commercial gypsum. Other wastewaters are generated during the calcination and finishing steps of the process. Product milling sands is also generated during the finishing process.

2.3.2. Production of Titanium Dioxide by the Chloride Process

The chloride process was commercialised in the United States in 1958 [18] and more than 60% of the world's production of titanium dioxide (TiO_2)pigment is now produced by this process. The newer chloride process generates up to 3 times less waste than the sulphate process [19]. The commercial chloride process is continuous and natural rutile is the preferred raw material for this process. However, due to dwindling reserves other titanium feed stocks including synthetic rutile, upgraded titanium slag (UGS) and high grade QIT slag are used as feed material for TiO_2pigment production. A flowsheet of a typical chloride process for production of TiO_2 pigment is shown in **Figure 10**. The main process steps are as follows [18, 20]:

2.3.2.1. Chlorination

After thorough drying, the feed stock and coke as reductant are mixed and continuously fed into the fluidized-bed chlorinator, are reacted with chlorine gas introduced through the bottom of the chlorinator. The chlorinator is a long, brick lined steel reactor, with a gas chamber and gas distributor at the bottom and a feed port and product discharge line at the top. The feedstock and coke are charged from the top of the reactor. As the chlorination reactions proceed, carbon monoxide (CO) is produced by the endothermic reaction of carbon with carbon dioxide (CO_2) and thus air or oxygen is added to the reactor to maintain temperatures between 800 and 1000 °C [21]. The gas liberated from the chlorinator contains titanium tetrachloride ($TiCl_4$) vapour and carbon monoxide. Impurity metals contained in the feed stocks, such as iron and aluminum, are simultaneously chlorinated, and are carried by the $TiCl_4$ gas into the cooler in the next step. Ignoring intermediate steps the most important chlorination reactions are [22]:

$$TiO_2 + 2Cl_2 + C = TiCl_4 + CO_2 \quad (19)$$

$$TiO_2 + 2Cl_2 + 2C = TiCl_4 + 2CO \quad (20)$$

$$C + CO_2 = 2CO \quad (21)$$

Reactions 19 and 20 are exothermic and reaction 21 is endothermic. The overall reaction is exothermic so an external heat supply is not required; however air or oxygen is supplied to chlorinators to maintain reaction temperature [23]. The non-volatile chlorides and the unreacted coke and feedstock solids are removed from the gas stream and from the bottom of the chlorinator.

Most industrial scale chlorinators are limited to 1000 °C because of bed sintering and refractory damage that occurs at higher temperatures [24]. Static bed chlorinators were used in the past but today all pigment producers use fluidised bed chlorinators [24]. The excellent heat transfer and absence of channelling in the fluidized reactor permit a three to fourfold production rate per unit area as compared to the fixed bed reactor [25].

2.3.2.2. Condensation of Gas and Purification of $TiCl_4$

The gas stream from the chlorinator is directly/indirectly cooled with liquid $TiCl_4$ to less than 300°C. At this low temperature, the chlorides are separated from $TiCl_4$ by condensation or sublimation. The cooled gaseous product stream is transferred through a series of separation steps to recover the $TiCl_4$ from the other chlorides. The small amounts of $TiCl_4$ and Cl_2 remaining in the exhaust gases (CO, CO_2 and N_2) are scrubbed using water and caustic solutions prior to venting to the atmosphere. The conversion of TiO_2 to $TiCl_4$ is 95 – 100% and is dependent on reactor design and chlorine velocity.

Most impurities are removed by fractional distillation, however, separation of vanadium tetrachloride (VCl_4) and vanadyl chloride ($VOCl_3$) from the $TiCl_4$ by distillation is more challenging owing to the similar boiling points. The removal of these vanadium chlorides is carried out by reducing the valence of vanadium from V^{4+} and V^{5+} to V^{3+} followed by distillation. A number of reducing reagents including gaseous hydrogen sulfide, titanium tetrachloride, amines, fatty acids, soaps and hydrocarbons are used.

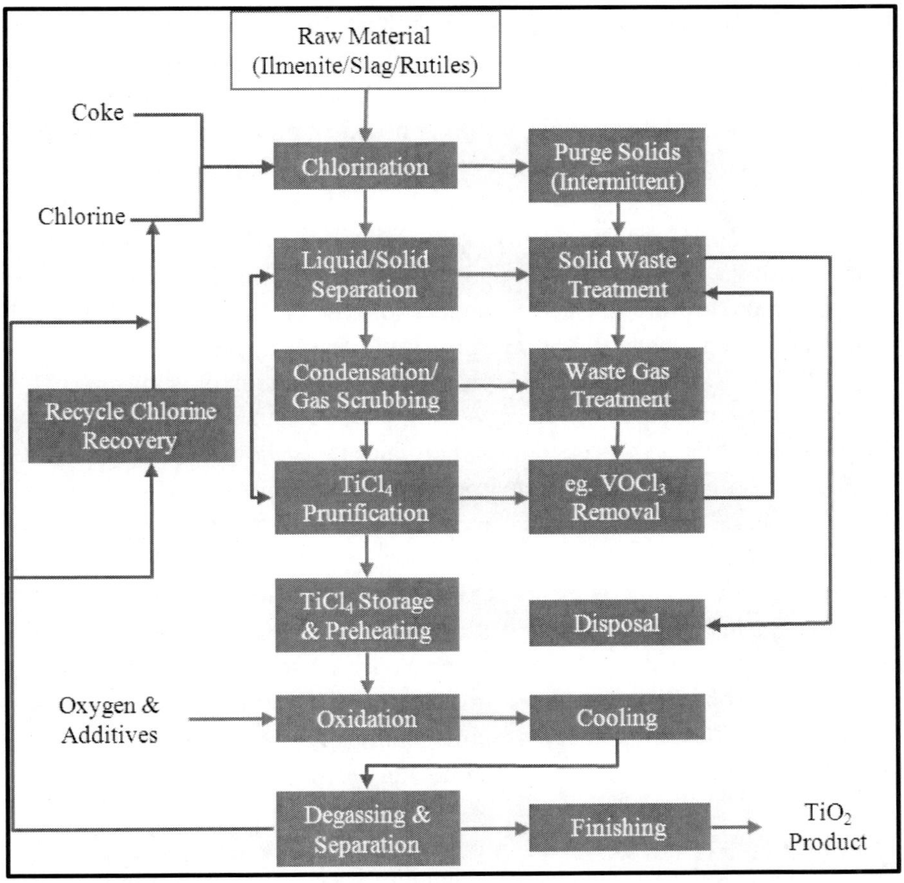

Figure 10. A flowsheet for production of TiO_2 from natural rutile or synthetic rutile or slag or high grade ilmenite by the chloride process [17].

2.3.2.3. Oxidation of $TiCl_4$ and Recovery of Titanium Dioxide

The purified $TiCl_4$ is burnt with oxygen gas at 900-1400 °C to convert to TiO_2. Aluminium chloride ($AlCl_3$) is added in this step to promote/catalyse the formation of titanium dioxide in the rutile form. In contrast, the addition of phosphorus trichloride (PCl_3) or silicon tetrachloride ($SiCl_4$), suppresses rutile formation and forms a pure anatase. The resulting TiO_2 particles of this stage are known as raw pigment. The overall reaction is shown below:

$$TiCl_4 + O_2 = TiO_2 + 2Cl_2 \qquad (22)$$

Chlorine gas is liberated during this oxidation reaction. The product is very rapidly cooled to a temperature lower than 600 °C and is separated from the chlorine gas by quenching with nitrogen or air. The regenerated chlorine gas is recycled to the fluidised bed reactor. In the final stage of the process, the raw pigment is milled to the required particle size, treated with chemicals (i.e., sodium aluminate and sodium silicate) to enhance surface properties, dry milled and then packaged.

Typical wastes generated by the chloride process includes wastewaters from chlorinator coke and ore solids recovery, reaction scrubbers, chemical tank storage scrubbers, product finishing operations and wastewater treatment solids decantation. Solid wastes are also generated during the production of titanium tetrachloride. Waste sands from finishing (milling) of the titanium dioxide product, scouring of oxidation process units, and blasting of reactor internal surfaces prior to replacement of refractory are also generated.

2.3.3. Production of Titanium Dioxide by Hydrometallurgical Chloride Processes

As mentioned in the previous section (2.3), commercial pigment grade TiO_2 is produced by either the sulfate process or the thermochemical chloride process from high grade ilmenite ($FeTiO_3$), titanium slag, naturally occurring rutile and synthetic rutile. The major disadvantages of the sulfate process include an abundant generation of acidic and solid waste, SO_2 emission, a large consumption of energy, and a variable quality of product due to batch processing. While the chloride process does not produce the same magnitude of wastes as the sulfate method, it still has several environmental issues including considerable CO_2 emissions. Utilizing the knowledge of chloride chemistry, some innovative hydrometallurgical processes have recently been developed and one of them is likely to be commercialized to produce TiO_2 pigment [26-36]. The process involves atmospheric chloride leaching of ilmenite material, solid-liquid separation, titanium recovery either by solvent extraction or direct precipitation. Using the concepts of solution chemistry, titanium is selectively extracted from the leach liquor at controlled pH and Eh conditions. The process promises to minimize CO_2 emissions, consume significantly less energy, and produce less environmental waste due to closed loop operation with recycle.

2.3.3.1. Titanium Speciation in Hydrochloric Acid Solution

Understanding the solution chemistry of titanium species is one of the most important tasks for its separation and purification by solvent extraction to produce pigment grade TiO_2. The Pourbaix diagram of the titanium–water system is shown in Figure 11 [37]. Based on this diagram, TiO^{2+} could be the dominant species in the aqueous solution with the acid activity of greater than 1 M and Ti(IV) forms $TiO_2 \cdot H_2O$ precipitate in the pH range of greater than 5. Kislik and Eyal [38] summarized seven ranges of Ti(IV) species in hydrochloric acid (Table 8). In high acidity solution ($[H^+]>4$ M), the proposed Ti(IV) species in Table 8 is quite different from those reported in the Pourbaix diagram in Figure 11 due to the presence of chloride ions. It is evident that both in Figure 11 and Table 8 the TiO^{2+} species is stable and predominant in the acidity range of $pH < 0.3$ under a normal solution potential range of about 0.3 V–1.3 V.

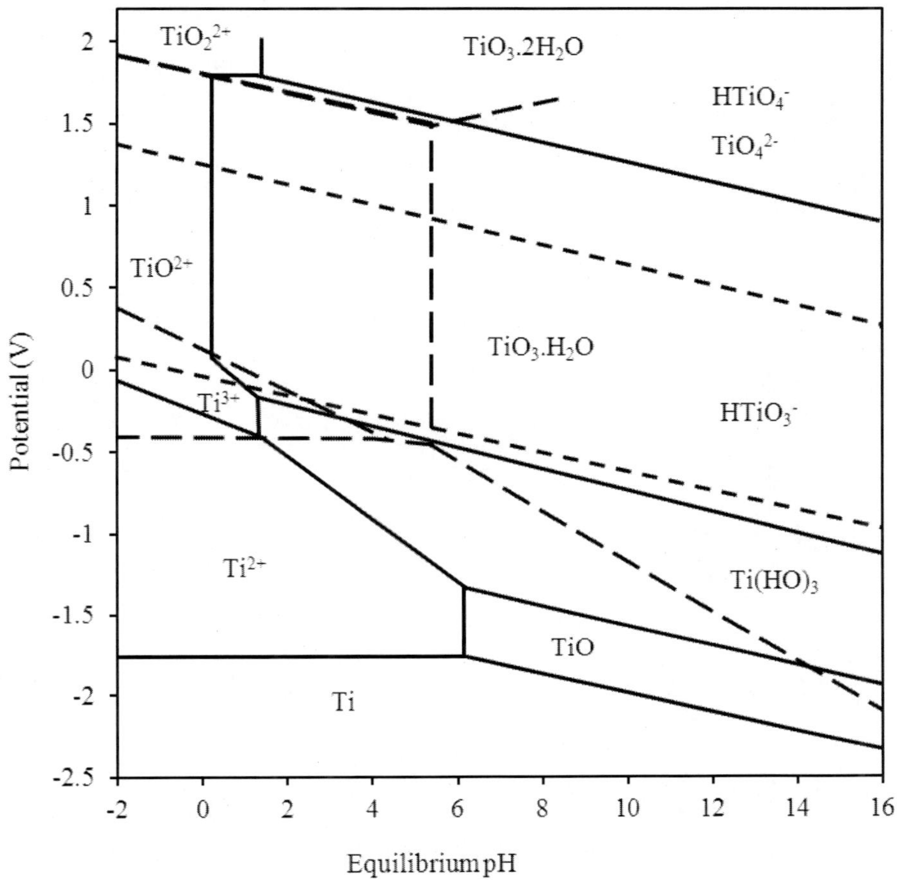

Figure 11. The Eh-pH diagram for Ti-H_2O system at 25 °C [37].

Table 8. Supposed species of Ti(IV) at various pH/acidity of hydrochloride solutions [38]

pH/ acidity	pH > 3.3	pH = 3.3–2.0	pH = 2.0–0.3	pH = 0.3– $[H^+]$ = 4M
Supposed species	$(Ti(OH)_4{\cdot}4H_2O)^0$ $(Ti(OH)_5{\cdot}3H_2O)^-$ $(Ti(OH)_6{\cdot}2H_2O)^{2-}$	(TiOOH·2H2O)+ $(Ti(OH)_3{\cdot}H_2O)^+$ $(Ti(OH)_4{\cdot}2H_2O)^0$	$(TiOH)^{3+}$ $(Ti(OH)_2)^{2+}$ $(TiOOH)^+$	TiO^{2+}
Acidity Supposed species	$[H^+]$ = 4 M–8 M $(TiCl)^{3+}$; $(TiCl_2)^{2+}$ $(TiCl_3)^+$; $TiOCl)^+$	$[H^+]$ = 8 M–11 M $(TiCl_3{\cdot}2HCl)^+$ $(TiOCl{\cdot}HCl)^+$ $(TiCl_4{\cdot}2HCl)^0$	$[H^+]$ > 11 M $(TiCl_4{\cdot}4HCl)^0$ $(TiCl_5{\cdot}3HCl)^-$ $(TiCl_6{\cdot}2HCl)^{2-}$	

Figure 12 represents the calculated titanium speciation with 0.1 M Ti(IV) activity in 1 M HCl solution at 298 K [39]. In chloride solutions, TiO^{2+} tends to combine with Cl^- to form different complexes. At about 4.5 M Cl^- concentration, the neutral charged ion of $TiOCl_2$ reaches a maximum which may explain the solvation mechanism of Ti(IV) extraction with cation ion exchange extractant in high chloride concentration solutions.

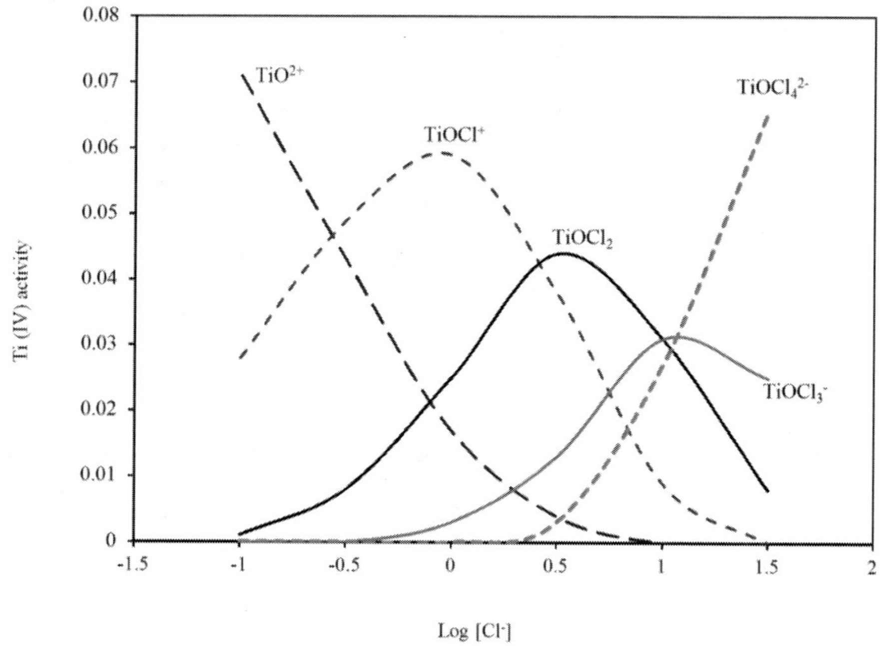

Figure 12. Distribution diagram of Ti (IV) chloride species with 0.1 M Ti(IV) activity at 298 K [39].

2.3.3.2. The Mixed Chloride Leaching Process

Process Research ORTECH Inc. (PRO) has developed a proprietary mixed chloride technology (CTL Process) for production of high purity titanium dioxide from ilmenite ore or concentrates [28-32]. The flowsheet for the process is shown in Figure 13. It consists of the following main process steps:

(i) *Leaching:* Leaching of the ore or concentrate at a mesh size of –100 is carried out using a mixed chloride lixiviant consisting of HCl and $MgCl_2$at temperature less than 80 °C. The presence of $MgCl_2$ in the lixiviant enhances the activity of the hydrogen ion by orders of magnitude [40]. With an increase in magnesium chloride concentration, titanium extraction reached >95% at elevated amount of HCl. The lixiviant also contained an oxidant, e.g., chlorate salts or chlorine for oxidation of Fe^{2+} to Fe^{3+}. The redox potential (Eh) of the solution is controlled to at least 300-400 mV. The following important reactions occur during the leaching process:

$$FeTiO_3 + 4HCl = FeCl_2 + TiOCl_2 + 2H_2O \qquad (23)$$

$$Fe_2O_3 + 6HCl = 2FeCl_3 + 3H_2O \qquad (24)$$

(ii) *Liquid-solid separation*: The leach slurry is subjected to liquid solid separation. The resultant leach solution is used for subsequent solvent extraction steps.

(iii) *Iron solvent extraction:* After the liquid-solid separation, the resultant pregnant leach solution is subjected to solvent extraction of iron with a suitable extractant (e.g., high molecular wt. of ketones and amines) in tridecanol and kerosene. Iron is stripped from the loaded organic and regenerated organic is recycled to the extraction stage. The pregnant strip solution is used in pyrohydrolysis to produce Fe_2O_3 and HCl which is further recycled to leaching stage. Results of Fe extraction with an organic mixture is shown in Figure 13.

(iv) *Titanium solvent extraction:* The raffinate of iron extraction is subjected to solvent extraction of Ti(IV)with a suitable extractant in tridecanol (modifier) and kerosene (diluent). The main organophosphorus extractants, neutral extractant and amine extractants used for Ti(IV) solvent extraction study are listed in Table 9. The reaction time to reach extraction equilibrium of Ti(IV) are summarized in Table 10. Extraction mechanisms of Ti(IV) with D2EHPA ($(HA)_2$) and Cyanex 923 (TRPO) are shown in reactions (25) and (26), respectively [42, 43].

$$TiOCl_2 + 2(HA)_2 = TiO(HA_2)_2 + 2H^+ + 2Cl^- \quad (25)$$

$$TiOCl_2 + 2TRPO = TiOCl_2 2TRPO \quad (26)$$

The concentration of Ti(IV) in the organic phase is obtained by mass balance. The distribution ratio, D is defined as the ratio of the concentration of Ti(IV) in organic phase to that in the aqueous phase. Since the organic to aqueous phase ratio is 1(O/A=1), the extraction percentage of Ti(IV) is calculated using the following equation:

$$\%\,Extraction = \frac{D_e \times 100}{D_e + V_{aq}/V_{org}} \quad (27)$$

where D_e = $[M]_{org}/[M]_{aq}$; M = Ti(IV); V_{aq} = volume of the aqueous phase; V_{org}= volume of the organic phase. After scrubbing the impurities from the loaded organic with pure titanium chloride solution, the scrubbed and loaded organic is contacted with HCl to strip titanium. In addition to HCl, results of titanium stripping from different organics with different acid solutions containing H_2O_2 are summarized in Table 11. The stripping percentage of Ti(IV) is calculated using the following equation:

$$\%\,Stripping = \frac{D_s \times 100}{D_s + V_{aq}/V_{org}} \quad (28)$$

where D_s = $[M]_{aq}/[M]_{org}$; M = Ti(IV); V_{aq}= volume of the aqueous phase; V_{org} = volume of the organic phase. Titanium stripped organic is recycled to extraction stage. The pregnant strip is used to thermally precipitate of titanium dioxide (TiO_2) at 95-100 °C. After solid-liquid separation, TiO_2 is dried and calcined. In this process, 100% rutile is obtained.

(v) *Impurity removal:* The raffinate of titanium extraction is subjected to impurity removal by pH adjustment with MgO followed by liquid-solid separation.

(vi) *MgO and HCl regeneration:* After removal of impurities, the resultant solution is subjected to pyrohydrolysis for regeneration of HCl and MgO. This MgO is used for impurity removal from Ti-raffinate while HCl is recycled to the leaching stage.

This process aims to minimize CO_2 emissions, consume significantly less energy, and produce less environmental waste due to closed loop operation with recycling of reagents. It is noted that solvent extraction has been incorporated into processes to separate titanium from iron and other impurities to enable the use of low grade ilmenite as feed materials, the elimination of the chlorination for the production of high purity TiO_2.

Table 9. Commonly used acidic organophosphorus, neutral organophosphorus and amine extractants in solvent extraction study [41]

Type/Extractant	Chemical name	Formula
Acidic organophosphorus extractants		
Alkyl phosphoric acids	Di(2-ethylhexyl)phosphoric acid (D2EHPA)	$(RO)_2(OH)P = O$ R = $CH_3(CH_2)_3CH(C_2H_5)CH_2$
Alkyl phosphonic acids	2-ethylhexyl phosphonic acid mono-2-ethylhexyl phosphoric ester (EHEHPA)	$R(RO)(OH)P = O$ R = $CH_3(CH_2)_3CH(C_2H_5)CH_2$
Alkyl phosphinic acids	di-2,4,4-trimethylpentyl phosphinic acid (Cyanex 272)	$R_2(OH)P = O$ R = $CH_3CH(CH_3)_2CH_2CH(CH_3)CH_2$
Thio-alkyl phosphinic acids	di-2,4,4-trimethylpentyl monothiophosphinic acid (Cyanex 302)	$R_2(OH)P = S$ R = $CH_3CH(CH_3)_2CH_2CH(CH_3)CH_2$
	di-2,4,4-trimethylpentyl dithiophosphinic acid (Cyanex301	$R_2(SH)P = S$ R = $CH_3CH(CH_3)_2CH_2CH(CH_3)CH_2$
Neutral organophosphorus extractants		
TBP	Tri-n-butyl phosphate	$(RO)_3P=O$, R=butyl
TOPO	Tri-octyl phosphine oxide	$R_3P=O$, R=octyl
Cyanex 923 (TRPO)	Tri-alkyl phosphine oxide	A mixture of $R_3P=O$, $R_2R'P=O$ and $RR'_2P=O$, R=octyl, R′=hexyl
Amine extractants		
Alamine 336	Tri-octylamine	$(C_8H_{17})_3N$
Aliquat 336	Tri-octyl methylammonium chloride	$(C_8H_{17})_3CH_3N^+Cl^-$

Table 10. Contact time for Ti(IV) extraction to get equilibrium [41]

Extractant	Contact time (min)	Primary conditions		
		T (°C)	HCl (M)	Diluent
D2EHPA	40	20	–	Kerosene
	120	30	1.0 and 0.5	Kerosene
EHEHPA	<5	30	0.2–1.5	Xylene
	300	20	–	Kerosene
	90	30	1–2	Kerosene
	60	Room	0.5	Kerosene
Cyanex 272	120	30	0.14	Xylene
Cyanex 302	<5	30	0.1	Xylene
	<2	25±2	0.01–5	Toluene
Cyanex 301	<2	25±2	0.01–5	Tolulene
TBP	3	Room	10	16% Decanol + Dodecane
TOPO	5	Room	8	7% Decanol + Dodecane
Cyanex 923	<3	30	1 − 7	Xylene
	<2	30	1 − 4	Kerosene

Table 11. Stripping of Ti(IV) from loaded acidic organophosphorus extractants by strip solution containing H_2O_2 at room temperature and an A/O ratio of 1[41]

Extractant	Conc. (M)	Stripping solution	Stripping (%)
D2EHPA	0.01	3% H_2O_2 + 1 M H_2SO_4	99.7
EHEHPA	0.01	3% H_2O_2 + 1 M H_2SO_4	99.4
	0.2	1% H_2O_2 + 2 M HCl	61.2
		2% H_2O_2 + 2 M HCl	95.5
		1% H_2O_2 + 2 M H_2SO_4	68.2
		2% H_2O_2 + 2 M H_2SO_4	>99.9
	0.5	1% H_2O_2 + 3 M H_2SO_4	50
		3% H_2O_2 + 3 M H_2SO_4	80
		2% H_2O_2 + 2 M H_2SO_4	Complete (2 stages)
Cyanex 272	0.2	1% H_2O_2 + 2 M H_2SO_4	Complete
Cyanex 302	0.1	3% H_2O_2 + 0.5 M H_2SO_4	99.1
Cyanex 301	0.1	3% H_2O_2 + 0.5 M H_2SO_4	98.2

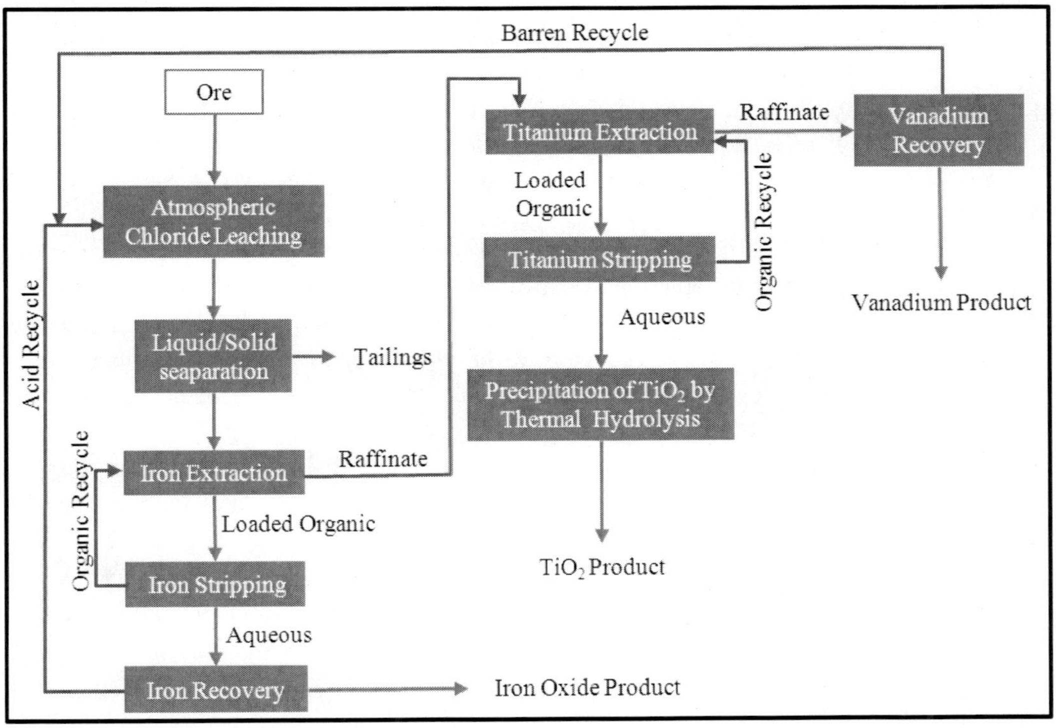

Figure 13. A flowsheet for production of TiO2 from titanium containing raw materials by the mixed chloride leaching process [33].

The discussions in section 2 present the development of different synthesis processes of TiO_2 and its commercialization and highlight the benefits of the hydrometallurgical chloride process with respect to the others. In the following sections, the important properties of TiO_2 including that of the nanoscale materials and the major applications will be discussed.

3. PROPERTIES OF TiO_2

3.1. Structural Property

In general, one titanium atom together with six oxygen atoms forms the TiO_6 ployhedra in TiO_2 lattice that gives the Ti atom a coordination number of six. Both the anatase and rutile phases have tetragonal unit cells while brookite settles in an orthorhombic structure [44-47]. Brookite has the largest unit cell volume while anatase is 9.1% more voluminous than the rutile; thereby, anatase is 3.2 times taller than rutile along the *c*-axis [45]. Table 12 summarizes the important crystallographic parameters of all different varieties of TiO_2. The building block of the crystal structure for both anantase and rutile are the distorted TiO_6 octahedra which are corner shared in rutile while edge shared in anatase. Both the phases have D_{2h} symmetry, and each Ti atom is bonded to six O atoms which makes each O atoms coordinated to three Ti atoms. The distortion of the octahedra happens through the fact that two Ti—O bonds are slightly larger than the four others which makes the O—Ti—O bond angle to deviate from 90°. The anatase phase is more distorted than the rutile. The third phase brookite has 8 formula units per unit cell unlike the anatase and rutile which have 12 and 6 units respectively. The TiO_6 octahedra are edge shared in brookite as well, however it is has a larger cell volume and it is the least dense among all the three stable varieties. Brookite has six different Ti—O bonds giving rise to twelve different O—Ti—O bond angles, thereby the TiO_6 octahedra are more distorted as compared to that in the other two phases. Figure 14 shows the crystal structures of the three stable phases of TiO_2 [47].

There are eight more structural polymorphs of TiO_2, discovered subsequently over the years, of which three are metastable having different space group symmetries than the stable ones. The remaining five are the high pressure forms synthesized artificially at high pressure and temperature; in some cases heating locally within the diamond anvil cell by using a laser beam. Recently, the baddeleyite type monoclinic high pressure variety has been discovered in the shocked gneisses in Ries Crater, Germany.

The three metastable TiO_2 varieties, as given in Table 12, have in fact been derived through complete oxidation of the titanium bronze structures e.g., sodium titanium bronze ($Na_xTi_4O_8$, x = 0.8), titanium bronze (K_xTiO_2, x=0.25), potassium tetratitanate ($K_2Ti_4O_9$), lithium titanate bronze (Li_xTiO_2, x = 0.5) etc. [48-50]. The metastable TiO_2(B) is a white compound having monoclinic unit cell and is synthesized through the complete chemical and thermal oxidation of potassium tetratitanate which has been synthesized as the parent by heating the anatase TiO_2 with KNO_3 in a 2:1 molar ratio at 1000 °C. Hydrolysis of the parent compound is done using nitric acid for complete removal of potassium and then followed by thermal treatment at 500 °C resulting into the desired TiO_2(B) variety [48]. The tetragonal hollandite type metastable oxide TiO_2(H) is obtained through a complete oxidation of the parent titanate having potassium in the range $0.13 \leq x \leq 0.25$; the parent compound is synthesized by reacting potassium metal with TiO_2 at 1250 °C for 48 hrs. in a vacuum sealed atmosphere. The oxidation of the slurry is then undertaken in an acidic solution of hydrogen peroxide. The hydrated light yellowish TiO_2(H) is heated thereafter at 300 °C in air atmosphere to remove the water content. It has tetragonal unit cell which is more than double that of rutile in its length along the *a*-axis. It is interesting to note that TiO_2(H) converts back to the anatase phase instead of rutile, on heating [49]. The third Ramsdellite-type metastable

TiO_2(R) is derived by topotactic oxidation of lithium titanate bronze $Li_{0.5}TiO_2$ which, as the parent, is synthesized by reacting lithium metal with TiO_2 at 1200 °C. Atmospheric exposure for topotactic oxidation might be as long as 1 year resulting into the most oxidized crystal species $Li_{0.14}TiO_2$ which is then treated in HCl for approximately 10 days for complete removal of Li and obtaining the colourless crystal of TiO_2(R) variety. This phase converts to brookite TiO_2 on heating above 350 °C [50].

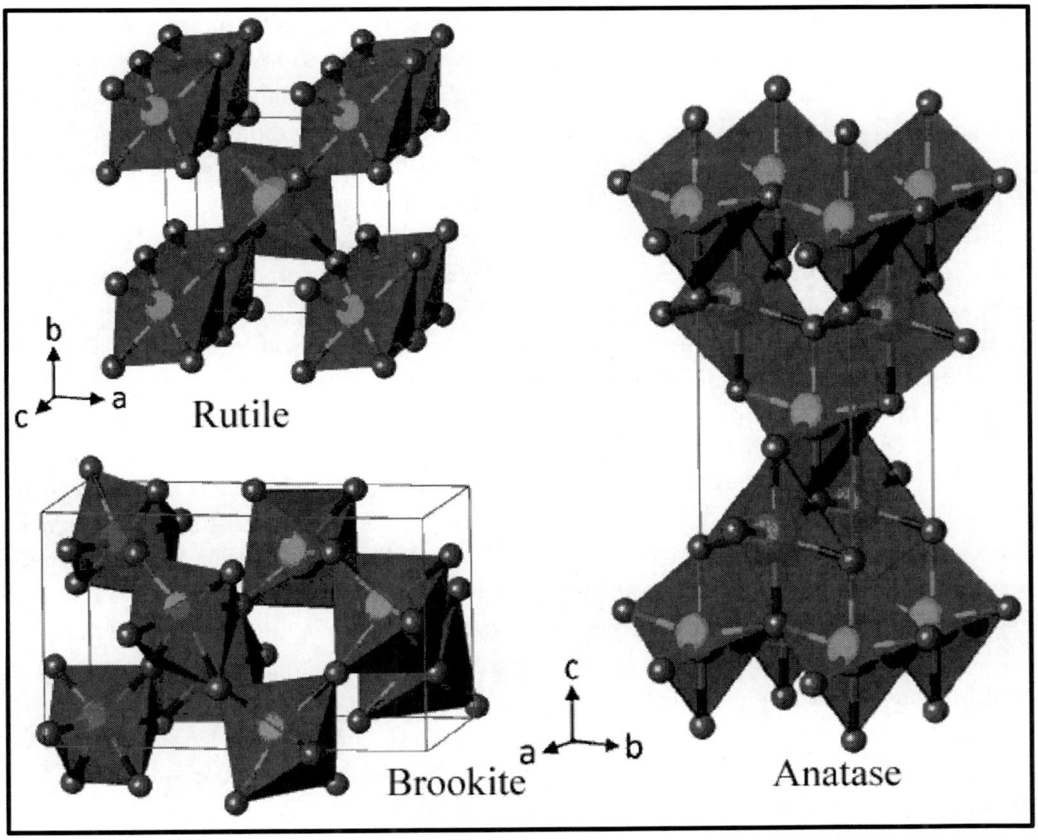

Figure 14. Crystal structure of the three stable phases of TiO_2; Titanium atoms are seen inside the oxygen octahedra, the building blocks of the material. Respective unit cells are superimposed on the structures [47]. Cell parameters are given in Table 12.

Due to the fact that the unit cells of TiO_2 have corner or edge shared distorted octahedra, it is a good sample to study the structural phase transitions under pressure and if needed, aided by application of heat. The objective behind such a drive has partly been in seeking the possibility of tuning the band-gap of this extraordinary oxide semiconductor so that a stable variety could be synthesized physically which would have the photocatalytic response in the visible range of the electromagnetic spectrum. The yield of the pressure treatment has indeed shown many interesting new polymorphs of TiO_2creating substantial fundamental as well as other application interests [51-56]. The development in this regard actually has a span of over 40 years resulting into five new structural varieties. In this, invariably, the technology of the diamond anvil cell to create high pressure within a confined small area and intense local heating by applying suitable laser pulse, has made the studies possible. The cubic phase is the most recently reported in 2004. As given in Table12, the orthorhombic TiO_2 II having the space group symmetry *Pbcn* has been the first to be reported in this class of compounds. This α-PbO_2 type variety has been synthesized from brookite by applying 4 GPa pressure at 450 °C and the cell volume has shrunk by over 52% though remains orthorhombic [51]. Figure 15 shows a variation of cell volume of different high pressure phases of TiO_2 with respect to applied pressure. Space group *Pbcn* has four formula units per unit cell and the cell volume turns out to be in between that of the anatase and the rutile. The Ti—Ti distance for the shared octahedra has been measured to be 3.07 Å. It is noteworthy that TiO_2 II phase was reported way back in 1967 after the crystal structure of the base material was revealed for the first time in 1961. Thereby, the work opened up a chapter of pressure induced structural phase transition in TiO_2. The baddeleyite type stable ZrO_2 phase having a monoclinic unit cell (MI) has been achieved in TiO_2 at a pressure above 14 GPa applied to powdered crystals of 99.99% purity rutile material and treated at 770 °C [52]. It is pertinent to mention here that X-ray diffraction measurements of these high pressure varieties requires intense synchrotron radiation and often the samples need to be packed with binder materials which in the present case is Co metal. Baddeleyite is the stable ambient phase of ZrO_2 and it happens to be a distorted fluorite structure where the metal ion has coordination with seven oxygen atoms. HfO_2 is also found to have this structure phase at ambient pressure. This monoclinic seven coordinated phase of TiO_2 is an important variety which shows that despite having a much smaller ionic radii, Ti can settle into other oxide structures containing larger ionic radii metals. The cell volume of the MI phase is further reduced compared to TiO_2 II, but it is more than that of the starting rutile material which has the least among all different TiO_2 phases discovered so far. There are however questions on the stability of the MI phase as to when the pressure is released out of the system. Decompression below 7 GPa is actually found to transform this baddeleyite type phase over to the α-PbO_2 type TiO_2 II orthorhombic one. Ten years later with the synthesis of the MI phase, such a polymorph has been discovered as a natural formation in Ries crater in Germany [53]. Though there is a nominal variation in the cell parameters, the structure is found to be the same and the material is 11% denser than the rutile phase. This finding adds additional importance to the high pressure studies. In fact, several post-rutile polymorphs were discovered earlier in the Russian craters, but none corresponded to this variety [53] which emphasizes the role of high temperature (and possibly melting) in the transformation process since the discovery happened in a shocked gneisses. An intermediate orthorhombic OI phase has been reported to form on prolonged local (laser) heating of either the anatase or rutile material at 1100-1200 °C under 28-32 GPa pressure. This phase persists until 38 GPa, however transforms back to the MI phase on decompression below 20 GPa [54]. OI phase has a structure similar to that of the ambient PbO_2, HfO_2 or ZrO_2 and a cell volume much larger compared to other high pressure varieties (Table 12). It has the same space group as the brookite phase but slightly shorter *b* and *c* values of the unit cell. It can be considered as an intermediate phase between MI and the OII type, where the Ti atom is seven coordinated. The OII phase has a cotunnite-type structure similar to that of ambient $PbCl_2$ and has been obtained at a much higher pressure above 55 GPa and approximately 850 °C. It is considered to be the hardest phase ever synthesized [55]. The Ti atom in this *Pnma* orthorhombic phase is coordinated with 9 oxygen atoms and it has the smallest cell volume among all the high pressure varieties. The extrapolated bulk hardness of this phase is estimated to be as high as diamond. At ambient temperature, this phase exists over a large pressure range of 30—80 GPa. However, decompression below 25 GPa transforms it back to the MI phase that ultimately goes back to the TiO_2 II phase below 7—8 GPa. The latest structure is the identification of a CaF_2-type cubic phase at 48 GPa and a very high temperature of above 1600 °C, applied locally using intense laser beam [56]. The formation of this phase is somewhat unusual because normally, group IVa dioxides do not follow such structural metamorphosis. Compression-decompression studies reveal that it usually follows the path of, rutile →TiO_2 II→ MI → OI → OII. On that account, as this phase has been identified as mixed together with the OII cotunnite-structured phase, the cubic phase might have resulted as an offshoot of in-situ synthesis under high pressure and temperature. It is interesting to note that the cubic phase follows through a decompression down to 9 GPa at ambient temperature. Heat capacity studies under the adiabatic conditions of a diamond anvil cell might enlighten the understanding on its formation. Nevertheless, the pressure dependent structural phase transition studies reveal the possibility to explore the doping of the base material by various higher size ions to create similar internal stress within the lattice leading to high-pressure type stable varieties that could have a smaller band gap.

3.2. Electronic and Optical Properties

TiO_2 is a well-known semiconductor. The first determination of whether it is *p* or *n* type, has actually been conducted back in 1941 in an elaborate measurement of conductivity under the variation of oxygen pressure [57]. This study concluded that the charge carriers to be electrons and measured the activation energy to be 1.7 eV that corresponds to the band gap of the rutile material reported in early calculations as 1.78 eV [46]. The same first principles OLCAO calculations also reported the band gaps of the anatase and brookite to be 2.04 eV and 2.20 eV respectively. This calculation is based on the density functional theory (DFT) in its local approximation and is applicable only at the ground state configuration. Over the years, there have been many sophisticated DFT calculations to determine the band structure of stable TiO_2 phases (mostly anantase and rutile) more and more accurately [58-63]. There is however variations in the values calculated in different methods that spans over 1.88—3.73 eV in rutile, 1.94—4.05 eV in anatase and 1.86—3.68 eV in brookite. Very recent frequency-dependent G_0W_0 calculation that takes into account the quasi particle (QP) corrections due to self-energy effects within the GW approximation, shows the band gaps as: 3.46 eV for rutile, 3.73 eV for anatase and 3.45 eV for the brookite phases [64] which presumably might be considered as the latest update.

Photoabsorption, photoluminescence and Raman spectroscopy measure-ments have revealed the band gaps in rutile and anatase phases. Such measurements near the fundamental band gap of rutile TiO_2 single crystals at low temperature estimated the band gap to be 3.031 eV which corresponds well with the DFT calculation results [65]. Time integrated photoluminescence studies reveal a peak at $\hbar\omega$ = 3.031 eV which is attributed to the $2p_{xy}$ excitons. Temperature dependent polarized optical transmission measured at the absorption edge of anatase TiO_2 single crystals estimated the optical gap to be 3.42 eV [66], a value quite reasonably greater than that of the rutile phase, as has also been shown in DFT calculations. It has been concluded that the excitons in anatase are self-trapped but remains free in the rutile structure. Such opposite nature of the excitons in the two types of phases is also confirmed in the photoluminescence studies [65]. For the two phases of a single material having the same symmetry in their structures (unit cell), such a behaviour of the excitons is unique. This variation in the excitonic states is explained to be a consequence of the difference in the network of the TiO_6 octahedra which are the essential building blocks of the material. The excitons in such structures are actually termed as charge-transfer excitons which result due to the iteration between conduction band electrons in the titanium *3d* states and the holes created in oxygen *2p* states. It is to be noted that in rutile, each octahedral are connected to ten neighbouring ones through corner sharing or through single edge while in anatase, each octahedra shares with eight neighbouring ones making it comparatively less compact. Thereby, the transfer of excitons from one to the other inter sites is weaker in anatase which favours the localization through the coupling with phonons [66]. Revealing such fundamental features in other interesting structural polymorphs of TiO_2 would be of profound interest, however, the experimental studies on them are still lacking in this regard.

3.3. Photo-Activity

From the discussion on band gap of TiO_2, it follows that this semiconductor should directly respond to the UV band of the electromagnetic spectrum corresponding to 3.10—4.43 eV energy. The absorption edge that has been referred to describe the band gap, are different for different phases of TiO_2; it is 416 nm in rutile while 386 nm in anantase. This suggests that UV light having a wavelength in the range 280—400 nm should be able to release photoelectrons from the valence band to the conduction band and thus create an electron-hole $(e^-—h^+)$ pair. This photochemistry can be represented as: $TiO_2 + h\nu \rightarrow e^- + h^+$. Both rutile and anatase are found to exhibit such UV photon incited activity whereas brookite is found to be photo inactive. The *(e—h)* pair generated may recombine within the bulk or migrate to the surface which might be used directly for photovoltaic application to produce electricity or indirectly to drive a chemical reaction, a phenomenon usually referred as photocatalysis. The reactants adsorbed on the surface of the material can get oxidized or reduced coming in contact with the electrons or holes respectively in the process of photocatalysis. Within the energy content of UV photons, TiO_2 has been found to be a very effective photocatalyst ushering in various applications, especially in self-cleaning the atmosphere and surfaces. Table 13 shows some typical examples of such photosensitive applications of TiO_2. On the other hand, there being no absorption of visible light in TiO_2, the application as highly reflective white paint has become prominent since the early years of its discovery. Rutile material has been used in this regard since 1940 for its high opacity and durability more than the anatase phase which has been in the industry since 1916. As nanoparticles, TiO_2 exhibits the highest reflectance among all commercial pigments of high opacity [67]. This happens because the sticking coefficients throughout the unevenness of a surface is enhanced as the particle size is reduced i.e., the smaller the particles are, the easier it is to fill in the micrometer scale troughs and steps on a normal surface.

Table 12. Crystallographic data of various TiO_2 polymorphs

Forms	Normal						High pressure and Temperature				
Phases	Stable			Metastable			4 GPa ~ 723K	>14 GPa ~ 1043K	28-32 GPa 1300–1500K	48 GPa 1900-2100K	>55 GPa 1100K
	Anatase TiO_2(A)	Rutile TiO_2(R)	Brookite TiO_2(B)	TiO_2 (B)	TiO_2 (H) Hollandite type	TiO_2 (R) Rams-dellite oxide	TiO_2 II α-PbO_2	Baddeleyite ZrO_2 type (MI): ultra-dense polymorph	OI PbO_2 / HfO_2 type	CaF_2 type phase	Cotunnite OII $PbCl_2$ type
Unit cell	Tetra-gonal	Tetra-gonal	Ortho-rhombic	Mono-clinic	Tetra-gonal	Ortho-rhombic	Ortho-rhombic	Monoclinic	Ortho-rhombic	Cubic	Ortho-rhombic
Space group	*I4₁/amd*	*P42/mnm*	*Pbca*	*C2/m*	*I4/m*	*Pbnm*	*Pbcn*	*P2₁/c*	*Pbca*	*Fm3m* / *Pa3*	*Pnma*
Density (gms/cc)	3.79	4.13	3.99	3.64	3.46	3.80	4.329	4.72			
Cell-para-meters (Å)	a=3.784 c=9.515	a=4.5936 c=2.9587	a =9.184 b =5.447 c =5.145	a=12.163 b=3.735 c=6.513 β=107.29°	a = 10.161 c =2.97	a=4.901 b=9.453 c=2.958	a=4.515 b=5.497 c=4.939	a=4.64/4.61 b=4.76/4.99 c=4.81/4.93 β=99.2°	a=9.052 b=4.836 c=4.617	a =4.516	a=5.163 b=2.989 c=5.966
Nominal cell volume (Å)^3	136.24	62.43	257.38	282.51	306.64	137.04	122.58	104.87 111.95	202.11	92.100	92.07
Ti Co-ordina tion no.	6	6	6	6	6	6	6	7	6	7	9
Refer-ences	JACS, 77, 1995, p.4708	JACS, 77, 1995, p.4708	Acta. Crys, 14, 1961, 214	Mat.Res. Bull, 15(8), 1980, p.1129	J. Solid. State. Chem, 81, 1989, p.78	J. Solid. State. Chem, 113, 1994, p.27	Acta Cryst., 23, 1967, p.334	Science, 251, 1991, p.786 Science, 293, 2001, p.1467	Phys. Rev. Lett., 87, 2001, p.275501	Phys. Rev. B, 70, 2004, p.212101	NATURE, 410, 2001, p.653

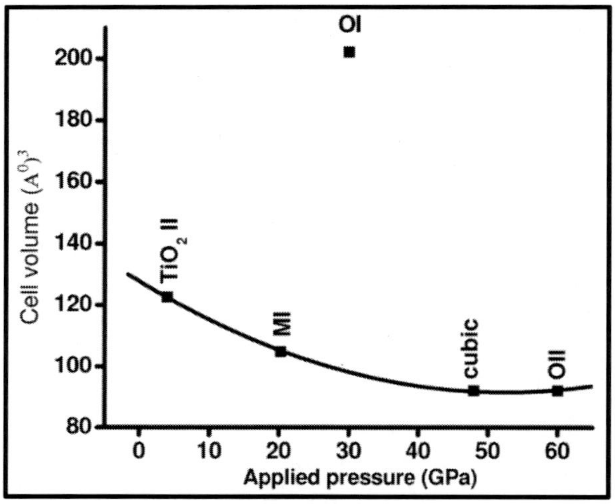

Figure 15. Cell volumes of various high pressure phases with respect to applied pressure. The solid line represents a 2nd order polynomial fit. The OI phase does not accommodate onto it. Cell parameters are given in Table 12.

The photovoltaic application of TiO_2 was first reported by Fujishima and Honda in 1972 in their study of electrolysis of water using the single crystal of rutile TiO_2 exposed to an UV light source [68]. This was the first photo-electrochemical cell of TiO_2 where the following reaction scheme has been proposed: $TiO_2 + 2hv \rightarrow \frac{1}{2} O_2 + H_2$. Figure 16 shows their cell design (included here for its historical importance) in which they observed the anodic current proportional to the intensity of the exposed light for wavelengths shorter than 415 nm which is equivalent to 3 eV.

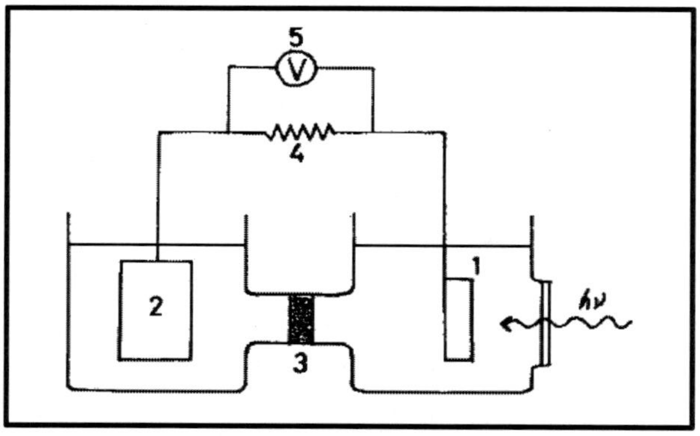

Figure 16. The electrochemical cell used by Fujishima and Honda in their photolysis experiment [68].

This study has actually initiated a new domain on the possible application of TiO_2 as a photocatalyst. The much higher oxidation potential (3.0 eV) makes the *(e—h)* pair produced in photolysis to act as a strong redox system which creates highly reactive radicals. The following reactions are probable in splitting water on top of TiO_2 surface:

$$OH^- + h^+ \rightarrow {}^{\bullet}OH;\ H_2O + h^+ \rightarrow {}^{\bullet}OH + H^+;\ \text{and}\ O_2 + e^- \rightarrow {}^{\bullet}O_2^-$$

Table 13. Photon induced applications of TiO_2

<table>
<tr><th colspan="6">Photo activation of TiO₂</th></tr>
<tr><td>Photovoltaics</td><td colspan="4">Photocatalysis</td><td rowspan="5">Hydrophilicity</td></tr>
<tr><td>Fujishima's Electrolytic cell</td><td colspan="4">Water splitting</td></tr>
<tr><td rowspan="3">Dye Sensitized cells or DSSC or Grätzel cell</td><td colspan="2">Hydrogen production</td><td colspan="2">Photosynthesis</td></tr>
<tr><td colspan="4">Decontamination</td></tr>
<tr><td>Air purification</td><td>Water purification</td><td>Self-Cleaning glass</td><td>Breathing cement</td></tr>
</table>

These produce highly reactive hydroxyl and peroxyl radicals which are capable to degrade the organic pollutants and break them as follows:

$$\text{(organic)} + {}^{\bullet}OH \rightarrow CO_2 + H_2O + \text{(residue) and (organic)} + {}^{\bullet}O_2^{-} \rightarrow CO_2 + H_2O + \text{(residue)}$$

Figure 17 depicts the situation schematically. At the level of titanium atoms, the redox reaction would mean a conversion of the Ti^{4+} ions into Ti^{3+} and back, as follows:

$$OH^{-} + Ti^{4+} \rightarrow {}^{\bullet}OH + Ti^{3+} \text{ and } Ti^{3+} + O_2 \rightarrow Ti^{4+} + {}^{\bullet}O_2^{-}$$

In short, the reaction initiated by photons on TiO_2 may thereby be expressed as,

$$O_2 + H_2O + h\nu \rightarrow {}^{\bullet}OH + {}^{\bullet}O_2^{-}.$$

The production of superoxide (or peroxyl) radical in the photon induced process makes TiO_2 the most efficient photocatalyst. This feature, in addition to the fact that TiO_2 is safe, stable and available at low cost has made it popular for industrial applications [69-71].

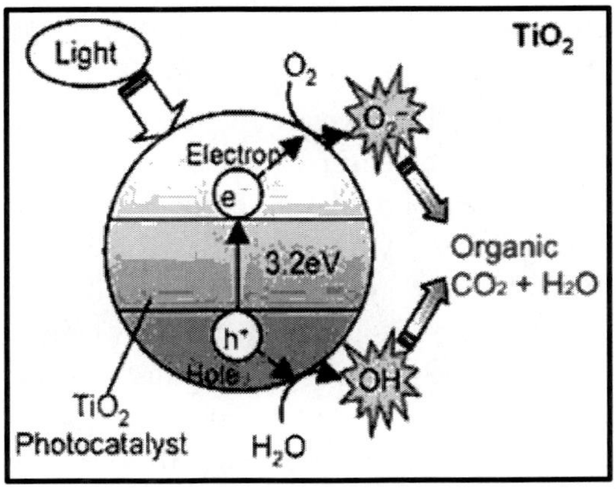

Figure 17. Schematic representation of photocatalytic action on TiO_2; an $(e^{-} - h^{+})$ pair is created by a UV photon. On moving over to the surface, the charge carriers make hydroxyl and peroxyl radicals that break the organics.

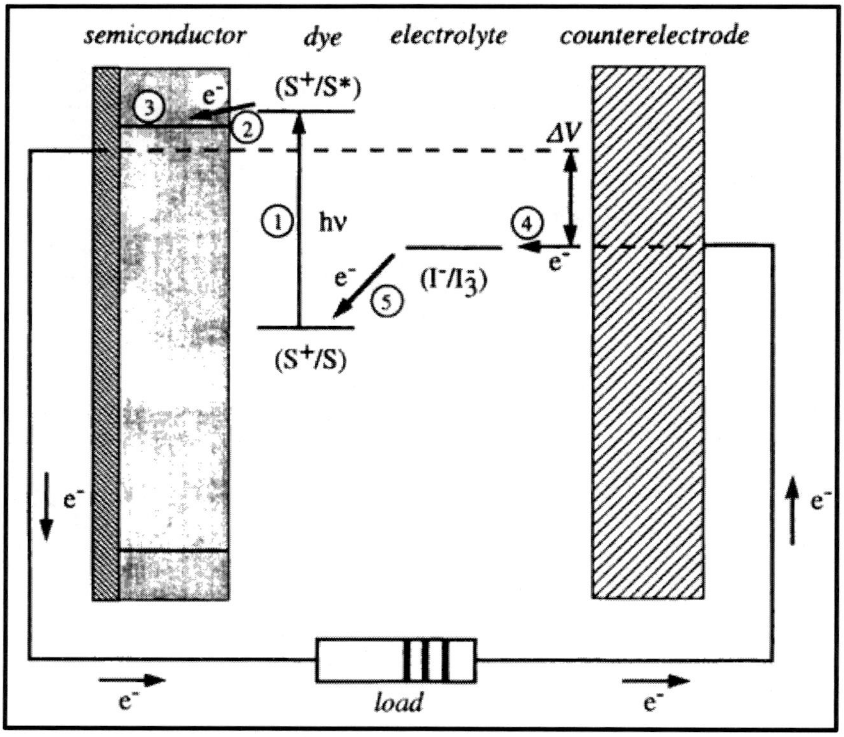

Figure 18. The dye sensitized photovoltaic cell of O'Regan and Grätzel. Details of the same could be found in [72].

The most profound application of this property came about twenty years later in devising a new type of solar cell by O'Regan and Grätzel that became synonymous to the dye sensitized solar cell or DSSC [72]. A film made out of a paste of anatase TiO_2 nanoparticles having the average diameter of 15 nm has been used. Reduction of particle size has effectively increased the active surface area of TiO_2 multi-fold which by their estimate has been a 2000-fold increase in surface area in a 10 mm thick film. This has made the packing of a very large amount of (visible) light sensitive charge-transfer dye molecules possible; thereby the light harvesting capacity has increased tremendously which had been the main bottleneck in the development of dye sensitized cells. Using a monolayer of the charge-transfer dye, trimeric ruthenium complex, the absorption onset shifted to 750 nm and it achieved 100% light harvesting efficiency for the whole visible region below 550 nm. As is known, DSSC differs in design from the conventional solar cells in that the function of light absorption and charge carrier transport happen in two different materials. The use of nanostructured TiO_2 film as the charge carrier together with charge-transfer dye as the carrier producer, has been successful to enhance the conversion efficiency of visible solar light into electricity to as large as 12% from its earlier value of less than 1% in which sintered ZnO [73-74] and SnS_2[75] were used. Figure 18 shows the original schematic of the photochemical cell designed by Grätzel [72].

The role of TiO_2 as a superior oxide in bringing a success to the power conversion efficiency of the DSSC might be appreciated by deciphering the photocatalytic mechanism involving the dye molecule. It has been shown that the photocurrent originates due to electron injection from the (visible) light sensitive dye molecules into the conduction band of the *n*-

type semiconductor TiO_2 [72]. In the process, the dye molecules are converted into the cationic radicals. The photochemistry that follows through is:

$$h\upsilon + (dye) \rightarrow (dye)^*$$
$$(dye)^* + (TiO_2) \rightarrow {}^\bullet(dye)^+ + (TiO_2)^-$$
$$(TiO_2)^- + O_2 \rightarrow (TiO_2) + {}^\bullet O_2^-$$
$$2{}^\bullet O_2^- + 2H^+ \rightarrow O_2 + H_2O_2$$
$$H_2O_2 + (TiO_2)^- \rightarrow OH^- + (TiO_2) + {}^\bullet OH$$

The complete reaction mechanism reveals that the transport actually happens through the catalytic conversion which an UV photon could initiate directly in TiO_2 by promoting an electron from the valence band [76]. The difference here is that the electron is injected from the charge-transfer dye molecule which quickly releases it into the conduction band of TiO_2. However, the same mechanism, by the very act itself, also degrades the dye and the organic electrolyte over a period of functioning, similar to what it does on the external organic dirt as the result of UV photocatalytic action [77]. The hole being not created within the semiconductor itself, the recombination factor as is experienced in UV-photocatalysis, should be less probable here, which in turn contributes to the stability exhibited in repeat performance. The improvement in the performance of the DSSC cells is a subject of constant research and various alternate measures are being proposed in order to enhance the conversion efficiency which however is not within the scope of this article.

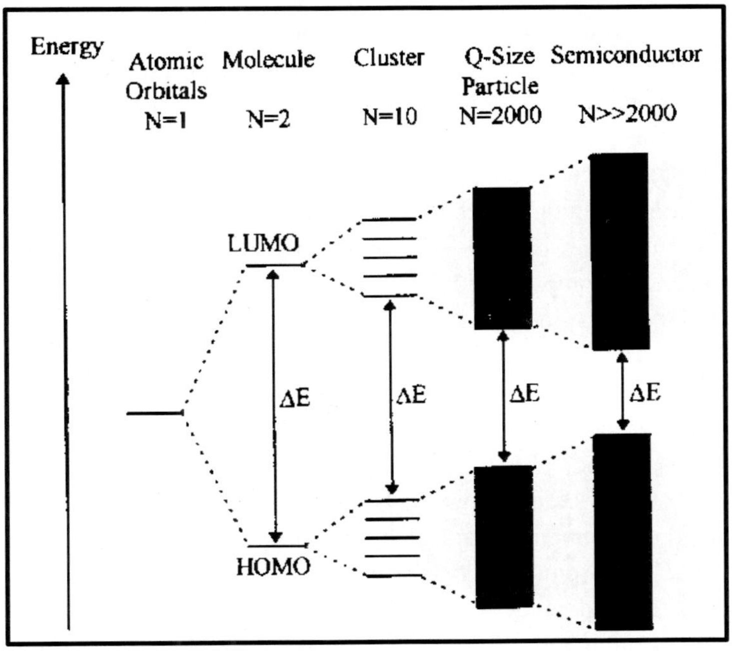

Figure 19. A model to depict the evolution of the electronic structure from the superimposition of atomic energy levels as the number of atoms increases to reach the typical band configuration of the bulk. In this respect, nanoparticles would typically contain about 2000 atoms [81].

All the applications listed in Table 13 basically follow through the same fundamental mechanism of photocatalysis. They are just the various manifestations in terms of the target of the end use. However, recombination of the *(e^-—h^+)* pair in direct UV photocatalysis is a serious drawback that does not allow exploiting the benefit unto the expected extent. Moreover, UV light is only 3% of the total sunlight that falls on earth surface. Thereby, any reasonable profiteering through large scale utilization of direct photocatalysis e.g., self-cleaning glass incited through super-hydrophilicity, breathing cement enabling cleaning of the local air-mass etc. that proposes the modern concrete landscapes to be engaged in air cleaning, has to exert extra fundamental research effort in preventing the loss due to carrier recombination. In this regard, tuning TiO_2 to be sensitive to visible light is a more pragmatic approach. In fact, most of the current research efforts on TiO_2 have an objective in this direction. First is, the studies on creating new high pressure ploymorphs discussed earlier. This has implications in finding the suitable doping agent that might create similar internal pressure at the molecular level and narrow down the band gap eventually by settling into an optimum structural configuration to be sensitive to visible light. Second, is the surface modification of TiO_2 particles or nanoparticles that might tune the hybrid system to be sensitive to (visible) light. Third, is band gap nanoengineering by coupling TiO_2 nanoparticles with a narrow band gap semiconductor [78, 79]. Fourth is, devising newer forms of TiO_2 nanostructures and coupling that with semiconductor nanoparticles or quantum dots to produce light sensitive new hybrid materials.

4. Nano TiO_2

As mentioned in the previous section, TiO_2 nanoparticle having an average dimension of 15 nm has been used in DSSC. The motive behind the use of nanoparticles instead of micron size bulk anatase powder material is appreciated in terms of enhancement of the total amount of surface area that made room to accommodate much larger quantity of dye molecules. Thereby, it predicts that further reduction will add a greater amount of effective surface area. However, the electronic structure of TiO_2 is also modified as the particle size is reduced. Therefore the question is; does it impose any limit to the particular application or whether the photo-response of $TiO_2$is compromised due to the reduction in particle size? Additionally, why nanometer size should be important?

4.1. Nanoparticles

Figure 19 schematically depicts how the band structure evolves right from a single atom or molecule as we go on adding more and more atoms to finally attain a finite size that might correspond to the bulk electronic structure. The typical size between that of the clusters and the bulk is normally referred as the nanoparticle domain. Though on average they contain a couple of hundreds to thousands of atoms per particle, the lower and upper limits are not restrictive and do vary depending on what material it is. While clusters have countable number of atoms with discrete energy levels as a result of size confinement, nanoparticles could be seen to have a preformed energy band due to many closer overlaps of energy levels that produce a gap somewhat larger than what is there in the bulk. However, the lower limit to be considered a replica of the bulk in terms of number of atoms or molecules, does depend on what particular property is being addressed in the study.

Various studies showed that TiO_2 is a more effective photocatalyst as nanoparticles than as the bulk powder [80]. Particles with a size less than 10 nm have an enlarged band gap which blue-shifts the band-edge yielding larger redox potential that in turn increases the photoactivity [81-84]. As the conduction band is effectively pushed further upward with respect to the reversible hydrogen potential, injecting an electron from the charge-transfer dye molecule into the conduction band of the nanoparticle becomes more likely which is a gain at the fundamental level over and above the advantage of having a larger effective surface area in the film made out of the nanoparticles. Thereby, use of further smaller particles up to the limit of self-coagulation is advantageous for the application. On the other hand, an increase in the redox potential makes the nanomaterial more aggressive in producing highly active radicals so that the same percentage of UV radiation could be better utilized in exploiting the benefits of photocatalysis. Therefore, a paint made out of TiO_2 nanoparticles will be more effective in terms of air-purification and self-cleaning functions over and above the enhanced (visible) light reflectivity because of multifold increase in packing density. A comparison with other oxides shows that ZnO and WO_3 have the same band gap (3.2 eV) [85]. However, the conduction band of WO_3 lies below the standard hydrogen potential which does not function without additional biasing. ZnO though has a comparative energy level, it reacts with H_2O forming a coat of $Zn(OH)_2$ on top of the surface of the particles making them inactive over a period of exposure.

4.2. Nanotube and Nanotube Array

Recently, TiO_2 has been synthesized into nanotube, nanowire and nanotube array materials. While nanotubes and nanowires are synthesized from the anantase as the starting material, arrays are grown in electrochemical anodization directly on top of titanium metal surface. These are one dimensional (1-D) nanostructures having very large surface-to-volume ratio which gives rise to certain unique properties like, enhanced light absorption capability, larger specific surface area and pore volume, electron transport over comparatively longer atomic distances and ion-exchange ability etc. [86-98].

Titania nanotubes were first reported by hydrothermal synthesis in 1998 [99] two years later than the first report of TiO_2 based nanotubes by electrochemical route in a porous aluminum oxide mould [100]. Nanotubes are also grown via template which produced materials with regular and controlled morphology. Hydrothermal synthesis gives isolated longer multiwall tubes where anatase TiO_2 powder is treated in highly concentrated (15M) NaOH solution continuously for 72 hours at 150 °C and acid washed before drying and treating for about 5-6 hrs. at 400 °C. This process is very similar to that used to make the nanowires. Even though the synthesis is not difficult, the size and morphology of the nanotubes depend on maintaining the parameters critically e.g., (i) the purity of the starting material and the precursor, (ii) concentration of the alkali solution, (iii) temperature and treatment duration and (iv) post treatment calcination etc. Long isolated nanotubes are grown as shown in Figure 20 [101]. The average outer diameter of these multiwall tubes is 10-20 nm, smaller than that of the nanowires (20-50 nm), with 5-8 nm pores inside. The length of these tubes may be as large as 1000 nm. Even though the starting material is anatase that has tetragonal structure, nanotubes produced from it have the monoclinic TiO_2-B structure however, the TiO_6 octahedra are shared through one edge like that in layered trititanic acid ($H_2Ti_3O_7$) [102-103]. A more detailed account regarding this might be found in a recent review [104].

Figure 20. Hydrothermally grown TiO_2 nanotube, viewed perpendicular to its axis [101].

The monoclinic TiO_2 (B) phase of the nanotubes has a wider semiconductor band gap of 3.87 eV, larger than that of the anatase or rutile material [105]. Like in particle size reduction that causes a widening of the band gap in nanoparticles, here a reduction in dimensionality (1D material) due to the structure enhances the band gap. Likewise, TiO_2 nanosheets (2D material) also exhibit a wider band gap of 3.84 eV [106]. This makes them photocatalytically more active, however, because of a different structure from that of the particles, nanotubes have a larger specific surface area that in turn reduces the $(e^-—h^+)$ pair recombination factor and opens up the possibility of tuning the band gap down to the visible limit. Several strategies are being actively pursued in this regard, e.g., controlling the morphology and surface property, modifying the crystals by neutralizing the hydroxy and peroxy radicals with other functional groups (e.g., from polymers or aromatic organics) on the surface of the tube etc. but the most widely practiced recent approach has been doping of suitable foreign entities into the system. These are, non-metal and metal doping, low band-gap semiconductor doping like CuS and CdS (2.4 eV) and integrating with the new graphitic materials like, graphene [107—113]. The incorporation of CdS has been found to extend the absorption edge up to 540 nm, well within the limit of visible light. Doping of CuS quantum dots is found to narrow down the nanotube band gap to 2.35 eV and is found to be more effective in producing the peroxy radical which reduces the $(e^-—h^+)$ recombination factor. Initial results on integration with graphene oxide have also exhibited enhanced photocatalytic activity. All such endeavour have a common ultimate objective of trimming the semiconducting gap in the form of a hybrid TiO_2 based system which will be photoactive in the domain of visible light. Nevertheless, they are still to be recorded as promising possibilities and yet to bring in any expected breakthrough in terms of photovoltaic applications.

Synthesis of TiO_2 nanotube array via direct anodization of titanium metal has added a new chapter in this regard. Gong and co-workers have first reported in 2001 on the synthesis of self-organized highly uniform TiO_2 nanotube array achieved through the anodization of titanium metal foil in aqueous dilute HF electrolyte [114]. The fundamental difference with all other forms of TiO_2 described so far is that here the synthesis involves a slow growth mechanism and self-organization. The tubes obtained in hydrothermal process are random and the formation happens during thermal treatment; products are sometimes quite lumped up requiring separation through post-formation treatments and it is possible to isolate single tubes using microscopic resolution. While, the anodization growth is radically different where it forms an array of tubes closely spaced together as standing on top of the titanium plate that mostly happen under room temperature conditions. Isolating single tubes is impossible here because their structure sustains through the lateral interaction between the neighbouring and next-neighbour tubes. As such, this highly self-organized array should be viewed as a self-sustained low-dimensional molecular-system in which the structure of an individual tube is unimportant. In this regard, for fundamental reasons, the whole system of a particular array synthesized under particular physico-chemical condition should itself be viewed as a specially grown nanostructured thin-film and is a 2D entity that could be compared to the self-assembled alkane-thiol films.

The size of the array, length and diameter of the tubes and the overall dimension depends critically of the synthesis condition e.g., the electrolyte, anodization voltage, period of anodization, temperature etc. [115-119]. In Table 14, a comprehensive summary of all different synthesis parameters is given with references as a quick guide. It shows that ethylene glycol with 2 vol% water and 0.3 wt% ammonium fluoride produces the best quality array on a 500 micron thick pure titanium plate, for a moderate 72 hours of anodization at 60V. Figure 21 shows a typical picture of the nanotube array [120-121].

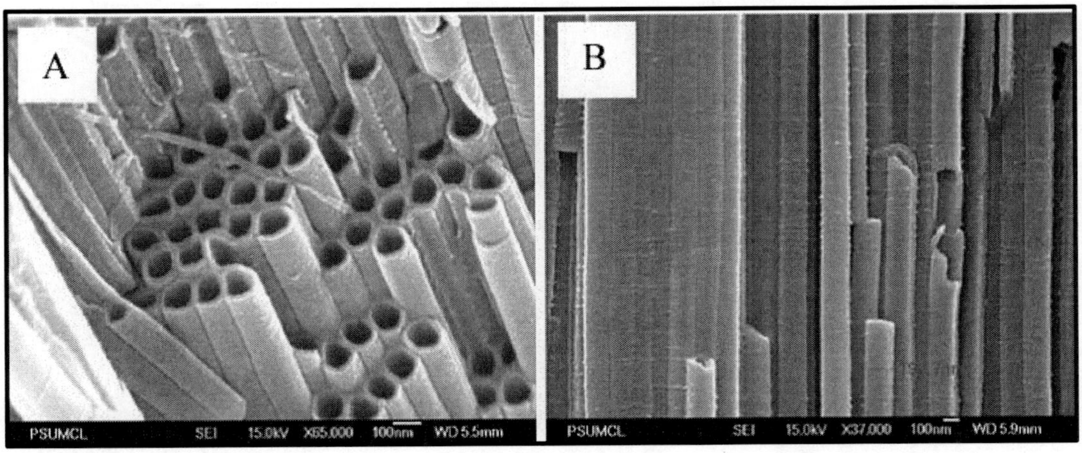

Figure 21. Typical examples of self-assembled TiO_2 nanotube array grown in electrochemical anodization of titanium foil; A. Top view and B. View from one side [120].

Table 14. TiO_2 Nanotube Synthesis Parameters

Electrolyte		Components		Voltage (V)	Time (hour)	Temp (°C)	Electrode (anode then cathode)	Tube Length (nm)	Inner Diameter (nm)	Wall Thickness (nm)	Physical Properties	References
Ammonium Fluoride (NH_4F)	*Ethylene Glycol*	0.1 wt% NH_4F	1 vol% H_2O	60	17	Room	Ti foil (0.5 mm)	Fresh: 54000 Used: 70000			Length increases with increasing NH_4F concentrations up to 0.3 wt% and increasing H_2O concentrations up to 2 vol%. Between 0.1-0.3 wt% NH_4F and 1-2 vol% H_2O, the used solution exhibited an increase ranging from 15 to 70 μm for the same applied potential and duration.	J Phys Chem B 110:16179–16184 (2006) J Phys Chem C 111:7235–7241 (2007)
			2 vol% H_2O					Fresh: 85000 Used: 70000				
		0.3 wt% NH_4F	2 vol% H_2O	20				5000	45	65		
				40				30000	70	115		
				50				45000	90	140		
				60				165000	105	155		
				65				106000	135	185		
				60	4			58000				
					17			160000				
					21			188000				
					48			289000				
					96			360000				
		0.5 wt% NH_4F	1 vol% H_2O		17			Fresh: 47000 Used: 115000				
			2 vol% H_2O					Fresh: 10000 Used: 45000				
			3 vol% H_2O					Fresh: 8000 Used: 66000				
	CH_3COOH (nearly anhydrous)	0.5 wt% NH_4F	0.2 vol% H_2O	30		Room	99.6% impurity free 0.1 mm thick Ti foil	200	10			Electrochem Commun 7:576–580 (2005)
	H_2SO_4 (0.24 mol/L)	0.5 wt% NH_4F		20		Room	Ti foil	300	120			J Mater Sci 42:5539–5543 (2007)
	H_3PO_4 (1M)	0.1-0.4 wt% [F^-]		1-25	2	Room	Ti foil	20-1100	15-120			
		0.5 wt% NH_4F	1M NaOH	20 at 0.5 V/sec from 0 to 20	2 + 200 sec			1000 ± 100	100 ± 10	10 ± 2	Electrolyte has pH of 4.5.	Electrochem Commun 7:505–509 (2006)
				3-20			See physical properties	560	80	20	Electrode made of sputter-deposited Ti films (1.9 μm thick) on p-type Si(100) substrates. Has pH of 4.5.	Nanotechnol 17:808–814 (2006)
	Diethylene Glycol (99.7%)	0.3 wt% NH_4F	2 vol% H_2O	80	48	Room	Ti foil	3500	120		Pyramidal structure.	J *Mater* Chem 18:3332–3336 (2008)
					18	Room	Ti foil	6000-7000	40-50		Have smooth walls.	Angew Chem Int Ed

						to 40						44.7463–7465 (2005)
Ammonium Fluoride (NH_4F)	*Glycerol*	0.5 wt% NH_4F	0.1M C_2H_3-NaO_2	20	8	Room	Ti foil	Without: 950 With: 4160			Electrolyte consists of 75% Glycerol, 25% water. pH adjusted to 5.6 with H_2SO_4. $C_2H_3NaO_2$ added to increase tube length.	Electrochim Acta 52:1258–1264 (2006)
	$(NH_4)_2SO_4$ (1M)	0.5 wt% NH_4F		Current Density				Layer Thickness	Tube Diameter		Tubular structure, porous.	J Electrochem Soc 153:B137–B143 (2006)
				0.5 mA/cm^2				500 ± 100 nm	40 ± 10 nm			
				0.75 mA/cm^2				900 nm	60 ± 10 nm			
				1-1.1 mA/cm^2				950 nm	75 ± 15 nm			
Hydrofluoric Acid (HF)	*HNO_3*	1 wt% HF	2.5 wt% HNO_3	20	4	Room	Ti foil	560	10-120	20	0.5M H_3BO_3 also added. Had precipitate layer 400 nm thick.	Sol Energy Mater Sol Cells 90:1283–1295 (2006)
				10		5		224	About 100	34		Nano Lett 5:191–195 (2005)
						25		176		24		
						35		156		13.5		
						50		120		9		
	H_2SO_4 (1M)	0.15 wt% HF		50 mV/sec from 0 to 20, then 20	400 sec then 2	Room	Ti foil	500				Electrochem Commun 8:544–548 (2006) Nanotechnol 16:2696–2702 (2005)
	$H_2Cr_2O_7$ (0.5 mol/L)	0.095 mol/L HF		See physical properties		Room	Ti foil	Film Thickness 150 ± 20	At 10V for 20 min: 75 ± 20		Procedure for potential consisted of increasing from 0 to 5/10 V by 5 equal steps every minute, then holding at final for 1-55 min.	Electrochim Acta 45:921–929 (1999) Surf Interface Anal 27:629–637 (1999)
	HF (0.5 wt%)			10	2	Room	Ti foil, platinum foil for cathode	200	22	13	Electrolyte is stirred magnetically. Adding acetic acid in 1:7 ratio results in more robust nanotubes.	7, 41, 57-63 from references of Fabrication of TiO_2 Nanotube Arrays by Electrochemical Anodization.
				14				260	53	17		
				20				400	76	27		
	Dimethyl Sulfoxide (DMSO)	Ethanol	1-6% HF	10-70	2-90	Room	Ti foil	2500	120	8-17 (Avg 15)	Addition of ethanol inhibited nanotube lengths. Use of previously used electrolytes eliminates debris and improves adhesion.	Mater Chem Phys 87:394–401 (2004) Chem Commun 26:3328–3330 (2005) Electrochem Commun 9:1519–1524 (2007)
									With crystallization			
									105	30		
									With "used" electrolytes			
									70-90	100		
	Diethylene	2 wt%		40				3000	140		Increasing voltage	J Phys Chem B

Table 14. (Continued)

Hydrofluoric Acid (HF)	*Glycol (99.7%)*	HF		60 100	48	Room	Ti foil		7500	150 190	100	results in greater tube separation.	110:16179–16184 (2006)
	Methanol	1 wt% HF		Current Density 10-30 mA/cm^2	30 min	Room	Ti foil		400	30-145			Scri Mater 56:373–376 (2007) Miner Eng 17:261–268 (2004) J Mater Res 20:230–236 (2005) Sol Energy Mater Sol Cells 90:1283–1295 (2006)
Hydrochloric Acid (HCl)	*HCl (0.22-1.1 wt%)*	0.06M		9-13	30 min, then 2	Room	Ti foil, copper		300	80		Voltage was increased manually and then cycled twice between a fixed voltage and zero before being left at fixed voltage.	Thin Solid Films 515:8511–8514 (2007)
		0.15M								10-20			J Phys Chem C 111:13028–13032 (2007)
		0.3M								30			J Mat Chem 18:2341–2348 (1972)
Formamide	*Dimethyl Formamide*			35	48	Room	Ti foil		70000	156	24		J Phys Chem B 110:16179–16184 (2006)
Potassium Fluoride (KF)	*KF*			25	17	Room	0.25 mm, 99.7% purity Ti foil, platinum		6000			KF bath has pH of 6.0.	J Electrochem Soc 152:B405–B410 (2005)
EFFECTS OF DIFFERENT CATHODE METALS	*Aqueous Electrolyte*	0.2M NH_4F	0.1M H_3PO_4	20	10	Room	Ti foil	Ni	1200	143		Overpotential of the cathode is a critical factor that affects dissolution, the activity of the electrolyte, and the morphology of the architectures, which is why there are different values for different cathodes.	Sol Energy Mater Sol Cells 92:1468–1475 (2008)
								Pd	1435	134			
								Pt	1520	105			
								Fe	2470	99			
								Co	1900	135			
								Cu	1265	81			
								Ta	1175	140			
								W	690	91			
								C	1300	143			
								Al	570	96			
								Sn	1200	147			
	Ethylene Glycol + 2 vol% H_2O							Ni	1510	65			
								Pd	2500	61			
								Pt	1725	65			
								Fe	2000	68			
								Co	2100	143			
								Cu	1130	83			
								Ta	1300	70			
								W	2400	114			
								C	1220	81			
								Al	1600	85			
								Sn	1060	90			

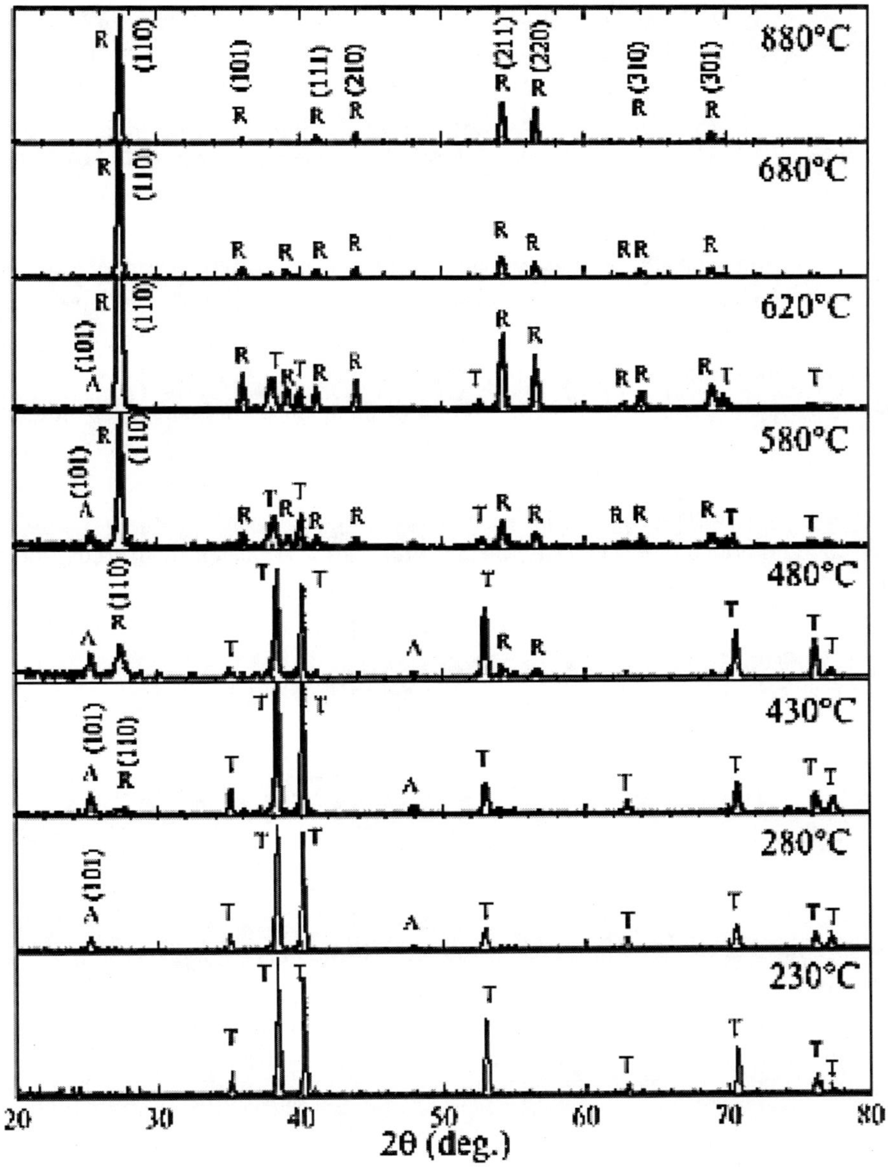

Figure 22. X-ray diffraction data of TiO_2 nanotube array revealing the changeover to the rutile phase from its amorphous structure on post-synthesis thermal annealing. A, R, and T represents for the anatase, rutile and titanium metal [121].

As grown arrays have amorphous structure which transforms over to the anatase phase on heating beyond 250 °C and then converts completely into the rutile on heating beyond 600 °C. Varghese et al. have studied this structural transformation through X-ray diffraction at different annealing temperatures as shown adapted in Figure 22 [121-123]. Peaks identified as A, R and T represents that from the anatase, rutile and the titanium metal itself. This work is important as a guideline to set the annealing condition in order to settle the system into a specific phase as might be necessary for any particular application. However, an array-film that retains the metal titanium surface underneath is also found to have anatase-phase walls with a rutile-phase barrier layer [124]. On the other hand, the as grown amorphous array is speculated to have utilization as a new bone-implant material because of its long term structural stability. This is an important factor for artificial bone implants as mismatch in mechanical properties would cause implant loosening and failure. A systematic investigation of the mechanical behavior of such films on the titanium metal substrate has been carried out applying the technique of nanoindentation and the Young's modulus has been determined to be in the range of 36—43 GPa [125].

Light absorption and the photocatalytic property of nanotube arrays are found to extensively vary on the size and morphology of the tubes e.g., the length, pore-size, wall thickness etc. [123] thereby, the conversion efficiency is actually specific to a size and the internal structure (or the shape) of the film, which is likely for a nanostructured entity because at nanoscale, the uniqueness is qualified in terms of physical parameters. Dependence of the property on size and morphology essentially pertains to allowing several handlers to physically tune the system for the desired properties. The arrays exhibit similar optical behavior as the mesostructured TiO_2 [126] having an average refractive index in the visible range (1.66) and an optical band gap of 3.34 eV with 2.4 eV band-tail measured over a 12 nm thick film [123]. It is speculated that higher lattice distortion in the array films might instigate an aggregation of vacancies acting as trap-states along the seams of the walls lowering the band-to-band transition energy [123].

5. Applications of TiO_2

5.1. Breathing Cement and Air Purification

Itlacementi, the largest Italian cement company has bagged the 'Times Best Invention Award 2008' in this regard for their product *TX-Active*. Nanophase TiO_2 has been incorporated into the cement, mortar, paint and plaster. As discussed earlier, because of the photocatalytic action, this product engages silently in breaking down the exposed polluted air from industrial activities, car emissions and smoke from heating systems etc., using the UV content in natural daylight. They reported 45—60% reduction in nitric oxide content in an atmosphere over an area of 2 acres.

5.2. Self-Cleaning Glass

Usual practice of making the self-cleaning glass is to use a hydrophobic coat that drives water away. There are several drawbacks however in any or all of the hydrophobic coatings, such as the compromise in transparency etc. that prevented their widespread use. Whereas, TiO_2 achieves the same effect through quite contrary phenomenon and instead uses the hydrophilic property that also washes off the dirt form the surface. When light falls on a thin titania coated glass, photocatalysis causes the chemical breakdown of adsorbed organic dirt as outlined above through photochemical reactions. The water produced through the neutralization of the hydroxyl and peroxy radicals in reactions with the organics present in the dirt makes the glass surface hydrophilic which forms a thin sheet of water that eventually washes the dirt off. The conversion of Ti^{4+} sites to Ti^{3+} as well favours this dissociative water adsorption.

By way of the fact, the same phenomenon also creates the glass or plastic surface anti-fogging which commonly is practiced by using the surfactants that minimize the surface tension of water e.g., soap, detergent, shampoo, shaving cream, gelatine, hydro-gels, some colloids and nanoparticles. Essentially, it is the wetting phenomenon that helps create a non-scattering film of water instead of droplets which are translucent and reduces the visibility. The hydrophilicity under exposure of light in titania makes it an excellent agent for anti-fogging glass or plastics. The advantage of using TiO_2 in such commercial production is that it is non-toxic, chemically inert in the absence of light, in addition to being inexpensive. In today's modern glass industry, use of such glass and/or plastics should be indispensable to keep them ever shining. When it removes the dirt deposited from atmosphere in a very large scale through huge areas of exposed glass, it definitely contributes in making the air around cleaner as well. Sodium diffusion from glass however has a negative effect that becomes counterproductive and a challenging issue. Pre-coating prevents the issue to some extent which demands an additional coating and/or calcination. Prior acid treatment of the surface has been tried by some researchers however with limited success. The effect of silver, cobalt, copper, gallium, molybdenum, and tantalum doping has been examined on sol–gel TiO_2 films and compared to SiO_2 pre-coating. Sodium poisoning degrades the photocatalytic activity over time by 70% which has been reduced to 10% in Mo- and Ta-doped films [127]. Such technical issues are still the hindrance in the way of commercialization of such super glass materials.

5.3. Waste-Water Treatment

Utilization of TiO_2 for waste water treatment had continued over the last two decades. The photocatalytic degradation, as explained in section 1.3, has been applied to wastewater pollutants, like aliphatic, aromatic organics, dyes, surfactants, herbicides and pesticides etc. to convert them into carbon-di-oxide, water and mineral acids. Recently, TiO_2 nanoparticles have been introduced in this regard, both as paints on large surface area constructs or coatings on top of beads of silica-gel, in order to increase the reaction area. Neutralization of a very diverse category of toxic pollutants, like different nitro-phenols from agricultural and industrial waste has been attempted through photocatalytic water treatment studies [81,128]. The dye Acid Blue 25 as pollutant from the textile industry has been degraded in pilot scale using TiO_2 nanoparticles [129]. Several insecticides, herbicides and pesticides in agricultural waste have also been degraded using aqueous TiO_2 suspensions [130]. Photodegradation of many such compounds under simulated UV radiation have been studied in aqueous suspensions of TiO_2 having a concentration of 100 mgL^{-1}. Gas chromatographic mass detection of intermediate formations shows a quite complex photodegradation pathway. However, in many cases, the initial pollutant degrades to a different form giving reactive inorganic residues, like Cl^-, $(NH_4)^+$, $(SO_4)^{2-}$ etc., [130] which need to be separated through further treatment in sequence before the crude water might be drained back to the natural resources. As such, waste water treatment is a global problem and is acute also, because modern civilization cannot be devoid of massive agriculture and industry. Therefore, cost-effective remediation is absolutely important, for all nations to implement it in order that the natural water reserves can remain safer.

5.4. Hydrogen Production as Fuel

Ever since Fujishima's discovery of photocatalysis in 1971, hydrogen production from water using TiO_2 has gained serious attention because of the demand of hydrogen in alternate energy. Several research efforts have tried to utilize TiO_2 in combination with other solar active materials to devise interesting photoelectrochemical cells to produce hydrogen. Polymer membrane electrode assembly with TiO_2 photoanode, platinum cathode and a proton exchange Nafion membrane using acidified methanol as the electrolyte have been recently devised to generate hydrogen continuously under UV excitation [131]. This reverse solar cell has generated up to 340 mA/cm^2 current producing hydrogen in the cathode.

On the other hand the TiO_2 nanotube arrays are found to be quite effective hydrogen sensor [132]. High-temperature hydrogen sensing has been reported using arrays about a few hundred nm in length which shows chemisorption of molecular hydrogen that changes the electrical resistance of the array by orders of magnitude and the effect is fully reversible. For example, 76 nm pore size tubes exhibited a variation in measured resistance by a factor of 1000 upon exposure to 500 ppm hydrogen at 290 °C. The sensitivity however is highly dependent on the physical parameters of the tubes, like pore size and the diameter so that a 22 nm pore tube is found to be 200 times more sensitive in detection than the 76 nm pore tube though the surface area increases by only 30%. The chemisorption thus differs with the morphology and wall thickness. This could be, as the wall thickness becomes comparable to the Debye length, the space charge is strongly modified and charge transfer is associated with the chemisorption [132].

5.5. Data-Storage and Memory Device

Tokyo University has developed a new material based on 5-20 nm size titanium oxide nanoparticles that can be used to make data storage disc having capacity 1000 times more than a blue-ray DVD. It is in fact a new nanocrystalline form that shows a reversible photoinduced metal-semiconductor phase transition between λ-Ti_3O_5 and β-Ti_3O_5 at room temperature so that it transforms from a metallic black-colour to a brown-colour semiconducting state when light shines on it. This unusual type of light dependent electron transport characteristic makes it a very promising material for next-generation optical storage device because λ-Ti_3O_5 nanocrystal is good for optical storage using near-field light and the memory density could be as high as 1 terabit per square inch. It might also be possible to store different types of data in different colours codes using such a storage media. Such super-discs might be commercially viable because of the use of a base material like titania. Commercial production of such devices is underway through private sector participation with academia [133].

5.6. Food Safety and Quality Control

Recently TiO_2 has been demonstrated to have potential application in food safety and quality control in various ways. The gas sensing property of the nanoparticles has been shown to be useful for possible detection of damaged packaging leading to contamination and bacterial infection especially in raw animal products like, fish and meat. UV active TiO_2 nanoparticles are placed as a visible dot at the inner surface of a transparent plastic packaging and then flushed with CO_2 before immediate sealing. Another same size dot of methylene blue indicator dye is placed at the outer surface of the same packaging so that they both appear as two juxtaposed blue dots on the package containing the raw food item inside. A localized UV exposure thereafter turns both the dots white and invisible. However, after sometime, the outer dot regains the colour as it is left exposed in air while the inner dot remain white because of the absence of oxygen inside the package until the package is opened in air [134, 135]. This method is a very good way to noninvasively detect the presence of any leak in the packaging both immediately after the production and at the retail site which also test the integrity of the seal over the period of indicated shelf-life. Such techniques do provide the retailer and the consumer an easy, visual and reliable means to choose the required food items.

Another innovative application in the detection of microorganism has been demonstrated using the TiO_2 nanowires through electrochemical measurement [135, 136]. In this, conductive TiO_2 nanowire bundles have been fabricated and then coated with the antibodies selective to the target bacterium. Such nanowires are connected between two gold connectors completing an electrochemical circuitry and the impedance is recorded with respect to a variation in frequency. On exposure to a particular contaminated sample, containing even a matrix of several types of bacteria, the particular one that responds to the coated antibody are attached onto the nanowire bundles and a modification in the impedance is recorded due to the bacterium-antibody binding events. Thereby, the nanowire bundle works as the detector. The bacteria Listeria Monocytogenes could be detected at a concentration of 4.65×10^3 cfu/ml in 1 hour without significant interference from other food-borne pathogens. The merit of the technique is that the detection limit is greatly superior to the traditional Immuno-Dot Blot analysis that only has a lower limit of 2.2×10^5 cfu/ml only.

To conclude, it might be pertinent to note that better understanding in modern science and the higher technological capability to make and manoeuvre materials, could reveal the old and familiar TiO_2's many hidden potentials for our benefits. Practical realization of these immense possibilities should be emphasized at all levels of future development, be it materials science or biotechnology or nanotechnology.

REFERENCES

[1] Sahu, KK; Alex, TC; Mishra, D; Agrawal, A. *Waste Manage. Res.*, 2006, 24, 74–79.

[2] Kothari, NC. *Inter. J. Min. Proc.*, 1974, 1, 287-305.

[3] Chang, LLY. Industrial Mineralogy: Materials, Processes, and Uses; Upper Saddle River, New Jersey, Prentice Hall, 2002.

[4] Michel, G ; Francois, C. *Min. Proc. Extra. Metal. Rev.*, 2007, 28, 1-58.

[5] Briggs, RA; Sacco, A; Jr. Metal. Trans. A: *Phys. Metal. Mat. Sci.*, 1993, 24A, 1257-1264.

[6] Chase, MW; Jr. NIST-JANAF Thermochemical Tables, Part. I and II, Physical and Chemical Reference Data, Monograph No. 9, 4th Ed.; Springer: New York, 1998.

[7] Mukherjee, TK; Patra, RN. Environ. *Waste Manage*, 1998, 52-66.

[8] Gilman, SK; Taylor, RKA. The future of ilmenite beneficiation technologies. Proceedings of the International Heavy Minerals Conference. 18-19 June 2001, Freemantle, Western Australia. Australasian Institute of Mining and Metallurgy, 2001, 3, 177-185.

[9] Zhang, W; Zhu, Z; Cheng, CY. Hydrometallurgy, 2011, 108, 177–188.

[10] Formanek, L; Lommert, H; Beyzavi, A. Synthetic rutile manufacture by the SL/RN-Becher process. Robinson, R. E. (Eds.). Heavy Minerals 1997, Johannesburg. South African Institute of Mining and Metallurgy, 1997, *Symposia Series* S, 17, 161-168.

[11] Cassidy, PW; Clements, RJ; Ellis, BA; Rolfe, PR. The AMC Narngulu synthetic rutile plant. Australia: a world source of ilmenite, rutile, monazite and zircon. September-October 1986, Victoria. Australasian Institute of Mining and Metallurgy. 1986, *Symposia Series*, 46, 61 – 67.

[12] Sinha, HN. Ilmenite upgrading by the Murso process. Light Metals, AIME-TMS, 1972, 261–274.

[13] Benelite Corporation of America. *United States Patent*, 1974, No. 3825419.

[14] United Sates. Environmental Protection Agency, 1989, Op. Cit., B-40.

[15] Buchel, KH; Moretto, HH; Woditsh, P. Industrial Inorganic Chemistry, Wiley-VCH; Weinheim: 2000.

[16] Thompson, R. The Modern Inorganic Chemicals Industry, 1977, 354-374.

[17] McNulty, GS. Production of titanium dioxide. Proceedings of the 5th International Symposium on Naturally Occurring Radioactive Material (NORM V), Seville, Spain, 19-22 March, 2007; 169-188.

[18] Fisher, JR. Developments in the TiO_2 pigments industry which will drive demand for TiO_2 mineral feedstocks, Robinson, R. E. (Eds.). Heavy Minerals 1997, Johannesburg. South African Institute of Mining and Metallurgy, 1997; *Symposia Series S*, 17, 207 – 218.

[19] DuPont. (2007) DuPont Ti-Pure titanium dioxide- Titanium dioxide for coatings http://www2.dupont.com/Titanium_Technologies/en_US/tech_ info/literature/Coatings/CO_B_H_65969_Coatings_Brochure.pdf, viewed 20/11/2010.

[20] Stanaway, KJ. Mining Engineer, 1994, 46, 1367 – 1370.

[21] Habashi, F. *Handbook of Extractive Metallurgy*, Wiley VCH: New York. 1997, 1142 – 1144.

[22] Bergholm, Transact. *Metallurgic. Soc. AIME*, 1961, 221, 1121– 1127.

[23] Youn, I; Park, KY. *Metallurgic. Transact. B.*, 1989, 20B, 959 – 966.

[24] Reeves, JW; Reeves, RG. Misconceptions about titanium ore chlorination. Robinson, R. E. (Eds.). Heavy Minerals 1997, Johannesburg. South African Institute of Mining and Metallurgy, 1997; Symposia Series S17, 203 – 206.

[25] Rowe, LW; Opie, WRJ. *Metals*, 1955, 1189 – 1193.

[26] Verhulst, D; Sabachy, B; Spitler, T; Prochazka, J. New developments in the Altair hydrochloride TiO_2 pigment process. Hydrometallurgy 2003, 5th International Conference in Honor of Professor Ritchie, Vancouver, 2003, 1, 565 – 575.

[27] Duyvesteyn, W; Sabachy, B; Edmund, V; Verhulst, D; West-sells, PG; Spitler, TM; Vince, A; Burkholder, JR; Paulus, BJ; Huls, M. Processing titaniferrous ore to titanium dioxide pigment. U. S. Patent, No. 2002/6375923 B1.

[28] Lakshmanan, VI; Sridhar, R; Rishea, MM; Joseph, DeLaat, R. Production of titanium metal from titanium-bearing ores involves leaching and selective removal of iron values by solvent extraction. U. S. Patent, 2001, No. 007646A1.

[29] Lakshmanan, VI; Sridhar R; Rishea, M; DeLaat, R. Methods of separation of titanium from ore. U. S. Patent, 2004, No. 6699446.

[30] Lakshmanan, VI; Sridhar, R; Harris, GB; Puvvada, G. Process for the recovery of titanium in mixed chloride media. U. S. Patent, 2005, No. 0142051A1.

[31] Lakshmanan, VI; Sridhar, R; Harris, GB; Puvvada, G. Process for the Recovery of Titanium in Mixed Chloride Media. U. S. Patent, 2010, No. 7803336B2.

[32] Lakshmanan, VI; Sridhar, R; Roy, R. Synthesis of TiO_2 by an innovative mixed chloride leach process. Monteiro, S. N., Verhulst, D. E., Anyalebechi, P. N., Pomykala, J. A. (Eds.). Chloride Symposium-EPD Congress, 2011, TMS, 425– 432.

[33] Lakshmanan, VI; Sridhar, R; Sheikhzeinoddin, T; Halim, MA; Roy, R. Extraction of titanium and vanadium by chloride leach processes. Shijie Wang; John E. Dutrizac; Michael L. Free; James Y. Hwang; Daniel Kim (Eds.)T. T. Chen, Honorary Symposium on Hydrometallurgy. Electrometallurgy and Materials Characterization TMS 2012, 295-301.

[34] Harris, B; White, C. Process for the recovery of titanium dioxide and value metals and system for same. *U. S. Patent Application Publication*. 2013, No. 0195738A1.

[35] Das, GK; Pranolo, Y; Zhu, Z; Cheng, CY. *Hydrometallurgy*, 2013, 133, 94–99.

[36] Mostafa, NY; Mahmoud, MHH; Heiba, ZK. *Hydrometallurgy*, 2013, 139, 88–94.

[37] Sole, KC. Hydrometallurgy, 1999, 51, 239–253.

[38] Kislik, V; Eyal, A. *Sol. Extra. Ion Ex.*, 1993, 11, 259–283.

[39] Cservenyak, I; Kelsall, GH; Wang, W. *Electrochem. Acta.*, 1996, 41, 563–572.

[40] Jansz, JJC. *Hydrometallurgy*, 1983, 11, 13-31.

[41] Zhu, Z; Zhang, W; Cheng, CY. *Hydrometallurgy*, 2011, 105, 304–313.

[42] Biswas, RK; Begum, DA. *Hydrometallurgy*, 1998, 49, 263–274.

[43] Remya, PN; Reddy, MLPJ. *Chem. Technol. Biotechnol*, 2004, 79, 734–741.

[44] Baur Von Werner H. *Acta Cryst.*, 1961, 14, 214—216.

[45] Cromer, DT; Herrington, KJ. *Am. Chem. Soc.*, 1955, 77, 4708—4709.

[46] Shang-Di Mo; Ching, WY. Phys. Rev. B., 1995, 51(19), 13023—13032.

[47] Lan Yucheng; Lu Yalin; Ren Zhifeng. *Nano Energy*, 2013, 2, 1031—1045.

[48] Marchand, Rene; Brohan Luc; Tournoux, Michel. *Mat. Res. Bull.*, 1980, 15, 1129—1133.

[49] Latroche, M; Brohan, L; Tournoux, MJ. *Solid State Chem.*, 1989, 81, 78—82.

[50] Akimoto, J; Gotoh, Y; Oosawa, Y; Nonose, N; Kumagai, T; Aoki, KJ. *Solid State Chem.*,1994, 113, 27—36.

[51] Simons, P Y ; Dachille, F. *Acta Cryst.*, 1967, 23, 334—336.

[52] Sato, H; Endo, S; Sugiyama, M; Kikegawa, T; Shimomura, O; Kusaba, K. *Science*,1991, 251, 786—788.

[53] Goresy Ahmed El; Chen Ming; Dubrovinsky Leonid; Gillet Philippe; Graup Günther. *Science*, 2001, 293, 1467—1469.

[54] Dubrovinskaia Natalia, A; Dubrovinsky Leonid, S; Ahuja, Rajeev; Prokopenko, Vitaly B; Dmitriev, V; Weber, HP; Osorio-Guillen, JM; Johansson Borje. *Phys. Rev. Lett.*, 2001, 87, 275501—275504.
[55] Dubrovinsky, LS; Dubrovinsky, NA; Swamy, V; Muscat, J; Harrison, NM; Ahuja, R; Holm, B; Johansson, B. *Nature*, 2001, 410, 653—654.
[56] Mattesini, M; de Almeida, JS; Dubrovinsky, L; Dubrovinsky, N; Johansson, B ; Ahuja, R. *Phys. Rev. B.*, 2004, 70, 212101—212104.
[57] Earle Marshall, D. *Phys. Rev.*, 1941, 61, 56—62.
[58] Calatayud, M; Mori-Sánchez, P; Beltrán, A; Pendás, AM; Francisco, E; Andrés, J; Recio JM. *Phys. Rev. B.*, 2001, 64, 184113 (1-9).
[59] Muscat J; Swamy, V; Harrison, NM. *Phys. Rev B.*, 2002, 65, 224112 (1-15).
[60] ZhangY; Lin, W; Li, Y; Ding, K; Li, JJ. *Phys. Chem. B.*, 2005, 109, 19270—19277.
[61] Hossain, FM; Sheppard, L; Nowotny, J; Murch, GEJ. *Phys. Chem. Solids.*, 2008, 69, 1820—1828.
[62] Wu, X; Holbig, E; Steinle-Neumann, GJ. *Phys. Condens. Matter.*, 2010, 22, 295501 (1—4).
[63] Shojaee, E; Mohammadizadeh, MRJ. *Phys. Condens. Matter.*, 2010, 22, 015401 (1—6).
[64] Landmann, M; Rauls, E; Schmidt, WGJ. *Phys. Condens. Matter.*, 2012, 24, 195503 (1—6).
[65] Amtout, A; Leonelli, R. *Phys. Rev. B.*, 1995, 51(11), 6842—6851.
[66] Tang, H; Levy, F; Berger, H; Schmid, PE. *Phys. Rev B.*, 1995, 52(11), 7771—7774.
[67] Winkler, J. Titanium Dioxide, European coatings literature, Vincentz, Hannover, Germany, 2003
[68] Fujishima, Akira; Honda Kenichi. *Nature*, 1972, 238, 37—38.
[69] Gayam UI; Abdullah, AHJ. *Photochem. Photobiol. C.*, 2008, 9, 1—12.
[70] Shan, AY; Ghazi, TIM; Rashid, SA. *Appl. Catal. A.*, 2010, 389, 1—8.
[71] Puma, GL; Bono, A; Krishnaiah, D; Collin, JGJ. Hazard. Mater, 2008, 157, 209—219.
[72] O'Regan, Brian; Grätzel Michael. *Nature*, 1991, 353, 737—740.
[73] Matsamura, M; Nomuraand, Y; Tsubomura, H. *Bull. Chem. Soc. Jap.*, 1977, 50, 2533—2537.
[74] Alonso, N; Beley, VM; Chartier, P; Ern, V. *Rev. Phys. Appl.*, 1981, 16, 5—10.
[75] Willig, F; Eichberger, R; Sundaresan, NS; Parkinson, BAJ. *Am. Chem. Soc.*, 1990, 112, 2702—2707.
[76] Han, F; Kambala, VSR; Srinivasan, M; Rajarathnam, D; Naidu, R. *Appl. Catal. A.*, 2009, 359, 25—40.
[77] Singh, Seema; Mahalingam Hari, Singh Promod Kumar. App. Cata. A: General, 2013, 462-463, 178—195
[78] Kohtani, S; Kudo, A; Sakata, T. *Chem. Phys. Lett.*, 1993, 206 (1—4), 166—170.
[79] Bang, JH; Kamat, PV. *Adv. Funct. Mater.*, 2010, 20(12), 1970—1976.
[80] Han, H; Ba, R. *Ind. Eng. Chem. Res.*, 2009, 48, 2891—2898.
[81] Hoffmann, MR; Martin, ST; Choi, W; et al. *Chem. Rev.*, 1995, 95, 69—96.
[82] Marcus, RAJ. *Phys. Chem.*, 1990, 94, 1050—1055.
[83] Mills, A; Hunte, AJJ. Photochem. Photobiol. A Chem, 1997, 108, 1—35.
[84] Gupta Shipra Mital; Tripathi Manoj. *Chinese Science Bulletin*, 2011, 56, 1639—1657.
[85] Hagfeld Anders; Grätzel Michael. *Chem. Rev.*, 1995, 95, 49—68.

[86] Bavykin, DV; Parmon, VN; Lapkin, AA; Walsh, FCJ. *Mater. Chem.*, 2004, 14, 3370—3377.
[87] Nakahira, A; Kato, W; Tamai, M; Isshiki, T; Nishio, K; Aritani, HJ. *Mater. Sci.*, 2004, 39, 4239—4245.
[88] Yu, J; Yu, H. *Mater. Chem. Phys.*, 2006, 100, 507—512.
[89] Yu, J; Yu, H; Cheng, B; Trapalis, CJ. *Mol. Catal. A: Chem.*, 2006, 249, 135—142.
[90] Chen, X; Cao, S; Weng, X; Wang, H; Wu, Z. *Catal. Commun.*, 2012, 26, 178—182.
[91] Tsai, CC ; Teng, H. *Chem. Mater.*, 2004, 16, 4352—4358.
[92] Mozia, S; Borowiak-Pale´n, E; Przepiórski, J; Grzmil, B; Tsumura, T; Toy-oda, M; Grzechulska-Damszel, J; Morawski, AWJ. *Phys. Chem. Solids.*, 2010, 71, 263—272.
[93] Gao, T; Fjellvåg, H; Norby, P. *Inorg. Chem.*, 2009, 48, 1423—1432.
[94] Bavykin, DV; Friedrich, JM; Lapkin, AA; Walsh, F C. *Chem. Mater.*, 2006, 18, 1124—1129.
[95] Song, F; Zhao, Y; Zhong, QJ. *Environ. Sci.*, 2013, 25, 554—560.
[96] Kim, M; Hwang, SH; Lim, SK; Kim, S. *Cryst. Res. Technol.*, 2012, 47, 1190—1194.
[97] Rajeshwar, K; Osugi, ME; Chanmanee, W; Chenthamarakshan, C R; Zanoni, MVB; Kajitvichyanukul, P; Krishnan-Ayer, RJ. *Photochem. Photobiol. C.*, 2008, 9, 171—192.
[98] Likodimos, V; Stergiopoulos, T; Falaras, P; Kunze, J; Schmuki, PJ. *Phys. Chem. C.*, 2008, 112, 12687—12696.
[99] Kasuga, T; Hiramatsu, M; Hoson, A; Sekino, T; Nithara, K. *Langmuir*, 1998, 14, 3160—3163.
[100] Hoyer, P. *Langmuir*, 1996, 12, 1411—1413.
[101] Armstrong Graham; Armstrong Robert, A.; Canales Jesus; Bruce Peter, G. *Chem. Commun.*, 2005, 2454—2456.
[102] Wu, D; Liu, J; Zhao, X; Li, A; Chen, Y; Ming, N. *Chem. Mater.*, 2005, 18, 547—553.
[103] Guo, Y; Lee, NH; Oh, HJ; Yoon, CR; Park KS; Lee, WH; Li, Y; Lee, HG; Lee, KS; Kim, SJ. *Thin Solid Films*, 2008, 516, 8363—8371.
[104] Liu, Nan; Chen, Xiaoyin; Zhang, Jinali; Schwank, Johannes, W. *Catalysis Today*, 2014, 225, 34—51.
[105] Bavykin, DV; Gordeev, SN; Moskalenko, AV; Lapkin, AA; Walsh, FCJ. *Phys. Chem. B.*, 2005, 109, 8565—8569.
[106] Sakai, N; Ebina, Y; Takada, K; Sasaki, TJ. *Am. Chem. Soc.*, 2004, 126, 5851—5858.
[107] Lee, CK; Lin, KS; Wu, CF; Lyu, MD; Lo, CCJ. *Hazard. Mater.*, 2008, 150, 494—503.
[108] Nah, YC; Paramasivam, I ; Schmuki, P. *Chem. Phys. Chem.*, 2010, 11, 2698—2713.
[109] Ou, HH; Lo, SL; Liao, CHJ. *Phys. Chem. C.*, 2011, 115, 4000—4007.
[110] Grandcolas, M; Cottineau, T; Louvet, A; Keller, N ; Keller, VR. *Appl. Catal. B.*, 2013, 138-139, 128—140.
[111] Zhang, YJ; Wang, YC; Yan, W; Li, T; Li, S; Hu, YR. *Appl. Surf. Sci.*, 2009, 255, 9508—9511.
[112] Bang, Jin Ho; Kamat Prashant, V. *Adv. Funct. Mater.*, 2010, 20, 1970—1976.
[113] Hu, Guoxin; Tang, Bo. *Mater. Chem. Phy.*, 2013, 138, 608—614.
[114] Gong, D; Grimes, CA; Varghese, OK; Hu, W; Singh, RS; Chen, Z; Dickey, ECJ. *Mater. Res.*, 2001, 16, 3331—3334.
[115] Zhong, Peng; Que, Wenxiu; Liao, Yulong; Zhang, Jin; Hu, XJ. *Alloys Compds*, 2012, 540, 159—164.

[116] Zhong, Peng; Que, Wenxiu; Zhang, Jin; Jia, Qiaoying; Wang, Wenjuan; Liao, Yulong; Hu, XJ. *Alloys Compds*, 2011, 509, 7808—7813.

[117] Wang, Jun; Lin, Zhiqun. *J. Phys. Chem. C.*, 2009, 113, 4026—4030.

[118] Wang, Jun; Lin, Zhiqun. *Chem. Mater.*, 2010, 22, 579—584.

[119] Paulose, M; Shankar, K; Yoriya, S; Prakasam, HE; Varghese, OK; Mor, GK; Latempa, TJ; Fitzgerald, A; Grimes, CAJ. *Phys. Chem. B.*, 2006, 110, 16179—16184.

[120] Prakasam, HE; Shankar, K; Paulose, M; Grimes, CAJ. *Phys. Chem. C.*, 2007, 111, 7235—7241.

[121] Varghese, OK; Gong, D; Paulose, M; Ong, KG; Grimes, CA; Dickey, ECJ. *Mater. Res.*, 2003, 18, 156—165.

[122] Cai, Q; Paulose, M; Varghese, OK; Grimes, CAJ. *Mater. Res.*, 2005, 20, 230—236.

[123] Mor, GK; Varghese, OK; Paulose, M; Grimes, CA. *Adv. Funct. Mater.*, 2005, 15, 1291—1296.

[124] De-Taconni, NR; Chenthamarakshan, CR; Yogeeswaran, G; Watcharenwong, A; de Zoysa, RS; Basit, NA; Rajeshwar, KJ. *Phys Chem B.*, 2006, 110, 25347—25355.

[125] Tauc, J. *Mater. Res. Bull.*, 1970, 5, 721—729.

[126] Crawford, GA; Chawla, N; Das, K; Bose, S ; Bandyopadhyay, A. *Acta, Biomater*, 2007, 3, 359—367.

[127] Kurtoglu, Murat Erdem; Longenbach, Travis; Gogotsi, Yury. *Int. J. App. Glass Sci.* Special Issue: Processing and Characterization, 2011, 2(2), 108-116.

[128] Fox, MA; Dulay, MT. *Chem. Rev.*, 1995, 93, 341—357.

[129] Mahmoodi, NM; Arami, MJ. Photochem. *Photobiol B.*, 2009, 94, 20–24.

[130] Konstantinou, IK; Sakellarides, TM; Sakkas, VA; et al. *Environ. Sci. Technol*, 2001, 35, 398–405.

[131] Seger, B; Kamat, PVJ. *Phys. Chem. C.*, 2009, 113, 18946–18952.

[132] Paulose, M; Varghese, OK; Mor, GK; Grimes, CA; Ong, KG. *Nanotechnology*, 17, 2006, 398–402.

[133] Ohkoshi, Shin-ichi; Tsunobuchi, Yoshihide; Matsuda, Tomoyuki; Hashimoto, Kazuhito; Namai, Asuka; Hakoe, Fumiyoshi; Tokoro, Hiroko. *Nature chemistry*, 2010, 2, 539–545.

[134] Mills, A. *Chem. Soc. Rev.*, 2005, 34, 1003–1011.

[135] Duncan Timothy, VJ. *Colloid and Interface Science*, 2011, 363, 1–24.

[136] Wang, R; Ruan, C; Kanayeva, D; Lassiter, K; Li, Y. *Nano Lett.*, 2008, 8, 2625–2631.

In: Titanium Dioxide
Editor: Jerri Brown
ISBN: 978-1-63321-391-3

Chapter 6

DYE-SENSITIZED TITANIUM DIOXIDE FOR GAS DETECTION

Xiaobo Zou**, Yanxiao Li, Xiaowei Huang, Jiyong Shi and Jiewen Zhao
School of Food and Biological Engineering, Jiangsu University,
Zhenjiang, Jiangsu, China

ABSTRACT

Porous Titanium Dioxide (TiO_2) electrodes having high dye adsorption characteristics is of interest biomolecule sensors and dye-sensitized solar cells. Dye adsorption characteristics directly affect the sensitivity and efficiency of devices. In recent years, the demand for gas sensors for safety control requirements and environmental monitoring has expanded enormously. There is an increasing need to detect low concentrations of air pollutants such as ammonia, amines, which are frequently encountered in everyday life. Films based on the semiconductor, titanium dioxide (TiO_2), have been investigated in this chapter. At low humidity, ammonia and amines can be ionized by the Cyanidin on the TiO_2 film and thereby decrease in the proton concentration at the surface. Consequently, more positively charged holes at the surface of the TiO_2 have to be extracted to neutralize the adsorbed cyanidin and water film. The resistance response to ammonia of the sensors was nearly independent on temperature from 10 to 50°C.

1. TITANIUM DIOXIDE AND GAS SENSOR

Titanium dioxide (TiO_2) is a semiconductor material that has received a lot of attention recently as a thin film sensor for gas detection[1]. It has been established that the sensing properties of the films are improved by decreasing the grain size of the TiO_2 so as to increase the surface to volume ratio of the films [2]. Films with nanometre-sized TiO_2 particles have

* Corresponding author Email address: zou_xiaobo@ujs.edu.cn (Z Xiaobo).

thus proved to be a leading candidate metal-oxide semiconductor for gas sensing. Films based on TiO_2 also respond to ammonia, amines and et al. [2, 3]. The TiO_2 sensing films described so far operate at high temperatures (250-450^{0}C) [4] [5] or use alternating currents. Maintenance of sensor at high temperature increases power consumption, reduces sensor life and complicates the design of the sensor due to need for integration of heater and temperature sensors with gas sensing film [6]. For operation at room temperature gas sensor based on polymer films have been reported [7]. However, these are difficult to fabricate reproducibly. Some rear metals films, such as tellurium [6] films, could be use as gas sensors operable at room temperature. The use of rare metal is expensive and difficult for industry produce. Thus, there is a need for gas sensors which are operable at room temperature, and easy producible.

Porous TiO_2 electrodes having high dye adsorption characteristics is of interest for bio-molecule sensors and dye-sensitized solar cells [1]. Nanocrystalline TiO_2 films sensitized with dyes were first used in solar cells [8].

Under a light beam, the dye absorbs photons and injects electrons into the semiconductor's conduction band. Similarly, under ammonia gas condition, the dye interaction with the ammonia and injects electrons into the semiconductor's conduction band. Therefore, the conductivity of the TiO2 films correlate with the concentration of the ammonia. In most of these conversion devices, the TiO_2 film is obtained by deposition of a colloidal suspension of commercial oxide particles (usually Degusa®), but the development of nanocrystalline films with large and porous surface areas has increased the efficiency of these devices [9-12]. Dye adsorption characteristics directly affect the sensitivity and efficiency of devices [1, 13].

The sol-gel process has proved to be a viable alternative to produce TiO_2 films, since it is a simple and versatile method for the production of oxide nanoparticles of controlled sizes [14, 15]. The conventional method uses the hydrolytic route, which involves the initial hydrolysis of the alkoxide precursor followed by continual condensations between the hydrolyzed particles forming the gel. This process is carried out at room temperature, and the desired morphological properties of the particles are obtained by controlling the conditions under which the synthesis is carried out [15]. Moreover, these materials are easily produced in film form.

Natural pigments of the flavonoid class, such as cyanidin, found in leaves and fruits and responsible for the colours of various vegetal tissues have been studied for application as sensitizers of solar cells [16-19]. Due to their cost efficiency, non-toxicity and complete biodegradation, natural pigments have been a popular subject of research.

In recent years, the demand for gas sensors for safety control requirements and environmental monitoring has expanded enormously. There is an increasing need to detect low concentrations of air pollutants such as ammonia, amines which are frequently encountered in everyday life, such as in agriculture where the ammonia are used in fertilizers, or in waste water where ammonia result from the surfactant, pharmaceutical and dye manufacturing industries [20, 21].

Ammonia has characteristic odors [7, 22]. Occupational safety and health administration (OSHA) has set a 15 min exposure limit for gaseous ammonia by a volume of 35 ppm in environmental air. In the past decade, many kinds of ammonia sensors have been developed based on different sensing mechanisms [2, 6, 20, 21, 23-25]. These include electrical [26], mass [27], or optical-based methods [28]. Though they comment on different measurement principles, nearly all commercial ammonia sensors are based on either electrochemical or chemoresistive metal oxide (doped SnO_2) principles. Sensors based on inorganic semiconductors such as SnO_2 or Fe_2O_3 are sensitive to low levels of a large number of gases but require high temperatures for operation, which increases complexity and needs high power of these devices [7].

On the other hand, sensors based on organic materials such as phthalocyanines, porphyrins, etc. can be operated at room temperature but lack specificity and sensitivity at lower concentration range [22]. For this, new types of the sensors related to room temperature are one of the main targets in developments and investigations now.

2. Pigment-Sensitized TiO_2

2.1. Film Preparation

The preparation of film consists of four steps as shown in Figure 1.

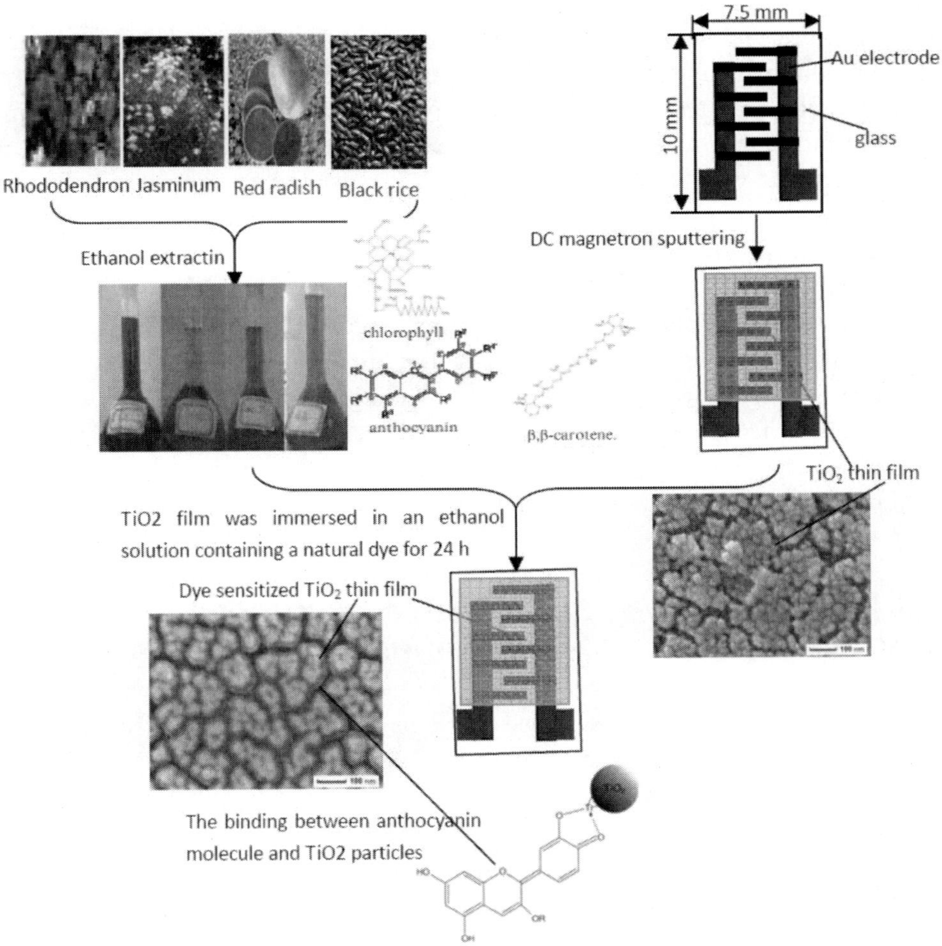

Figure 1. The diagram for assembling the dye-sensitized TiO_2 thin film gas sensor.

(1) Rinsing of the Microscope Glass Slides (Step 1)

The microscope glass slides were boiled in 200 ml H2SO4 and H2O2 solution with the volume ratio 1 : 3 for 30 min. Afterwards, the substrates were rinsed in acetone and ethanol in ultrasonic bath for 15 min, respectively. Two counter gold comb electrodes were prepared on the glass by sputtering as shown in Figure 1. Finally the microscope glass slides were cleaned by de-ionized water and dried by a N2 gas gun.

(2) Preparation and Characterization of TiO_2 Thin Films (Step 2)

Titanium oxide thin films were deposited onto the glass substrates with a sputtered-Au counter electrode using a home built DC magnetron system. pure titanium (99.999%) of 100 mm diameter and 6 mm thickness has been used as the sputtering target. Prior to the introduction of sputtering gas, the vacuum of the chamber was evacuated to lower than 8×10−4 Pa. Sputtering pressure was kept at 5.0 Pa. The flow rates of Ar (99.999%) and O2 (99.999%) were kept at constant values of 44.9 and 10 sccm (standard-state cubic centimeter per minute), respectively. The discharges were generated at constant power of 300 W. At first the Ti target was pre-sputtered in an argon atmosphere of 2.0 Pa in order to remove the surface oxide layer. When the discharge colour changes from pink to blue, it indicates that the oxide layer has been removed from the target surface. After that, oxygen and argon were introduced into the vacuum chamber and sputtering started. The temperature of the substrate was maintained at 200°C and the deposition time was 10 h. The thickness of the films was monitored by an α-step surface profiler. The adherence strength of the TiO2 thin film on the microscope glass slide was monitored by a coating adherence tester.

(3) Preparation of Natural Dye (Step 3)

Due to their cost efficiency, non-toxicity, and complete biodegradation, natural dyes have been a popular subject of research. The dyes from spinach (Spinacia oleracea), red radish (Raphanus sativus L), winter jiasmine (Jasminum nudiflorum), and black rice (Oryza sativa L. indica) were extracted with ethanol by the following steps: spinach, red radish, winter jiasmine, and black rice were washed with water and vacuum dried at 60 °C. After crushing into fine powder using a mortar, these materials were immersed in absolute ethanol at room temperature in the dark for one week. Then the solids were filtrated out, and the filtrates were concentrated at 40 °C for use.

(4) Preparation of Dye-Sensitized TiO2 Thin Films (Step 4)

The TiO2 porous film was solidified and sintered by firing the glass sheet at 450 °C in air for 30 min, and then cooled to around 80 °C. The glass solidified TiO2 was immerged in natural dye alcohol solution for 24 h and dried in moisture-free air. After that, a dye-sensitized TiO2 gas sensor was prepared.

2.2. Characterization Solution and Films

The adsorption spectrum and band gap energies of the samples were determined by using a UV-2450 UV–VIS spectrophotometer (Shimadzu Corporation, Japan) equipped with diffuse reflectance accessory (DRA). This technique allows the study of the reflectance spectra of the samples in the solid form. The crystalline structure of the films was measured using X-ray diffraction (XRD) with a Cu Ka source. The surface morphology of the fabricated films was examined using a scanning electron microscope (SEM: S-3000N, Hitachi, Japan) on gold-coated specimens.

2.2.1. Characterization of Natural Dye

Figure 2 shows the UV–VIS absorption spectra of spinach （Spinacia oleracea）, red radish （Raphanus sativus L）, winter jiasmine (Jasminum nudiflorum), and black rice (Oryza sativa L. indica) extracts in alcohol. From Figure 2, it can be seen that there is the same absorbent peak in about 525 nm for the extracts of black rice, and red radish, which is ascribed to the same dye cores (anthocyanin) contained in the two extracts. Because anthocyanin shows the color in the range of visible light from red to blue, it is prospected to become a high efficient sensitizing compound for wide band gap semiconductors. The molecular structure of anthocyanin is shown in Figure 3 (a). Due to carbonyl and hydroxyl groups presented on anthocyanin molecule, it can be bound with the surface of TiO2 film (Figure 3 (b)), which is favor of electric conversion effect.

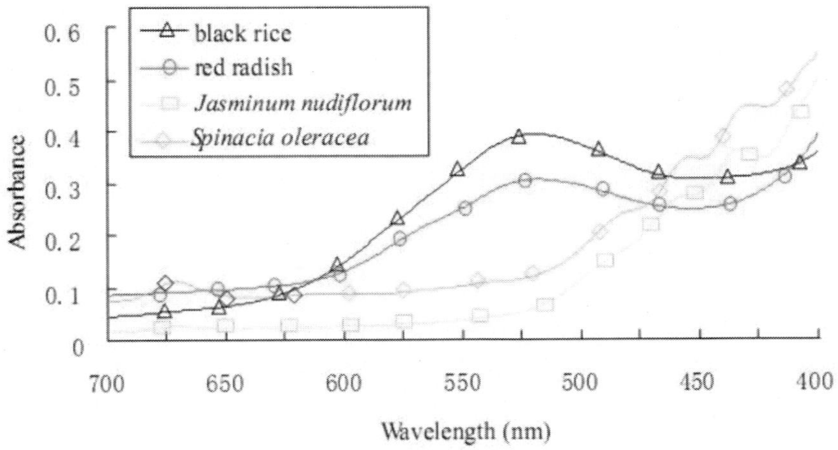

Figure 2. The absorption spectra of the extracts of spinach (*Spinacia oleracea*, red radish *(Raphanus sativus L),* winter jiasmine (*Jasminum nudiflorum*), and black rice (*Oryza sativa L. indica*) in alcohol.

Spinach and winter jiasmine show green and yellow color, the extracts also show similar colors. From Figure 2, it can be seen that character absorption peak of chlorophyll with wavelength about 660 nm and character absorption of carotenoid with wavelength about 455 nm appear in the extract of spinach and winter jiasmine. Figures 3 (c-d) present the basic molecular structure of chlorophyll and β-carotene. The alkyl group rather than carboxyl group or hydroxyl group on the dye molecule cannot form chemical bond with TiO2 porous film. In addition, strong steric hindrance of long chain alkane of chlorophyll and carotenoid also prevent the dye molecules from arraying on TiO2 film efficiently. Due to the two factors, the extracts of spinach and winter jiasmine are nearly absorbed on TiO2 film and sensitizing effect to TiO2 film is low, which is in accordance with our experimental results.

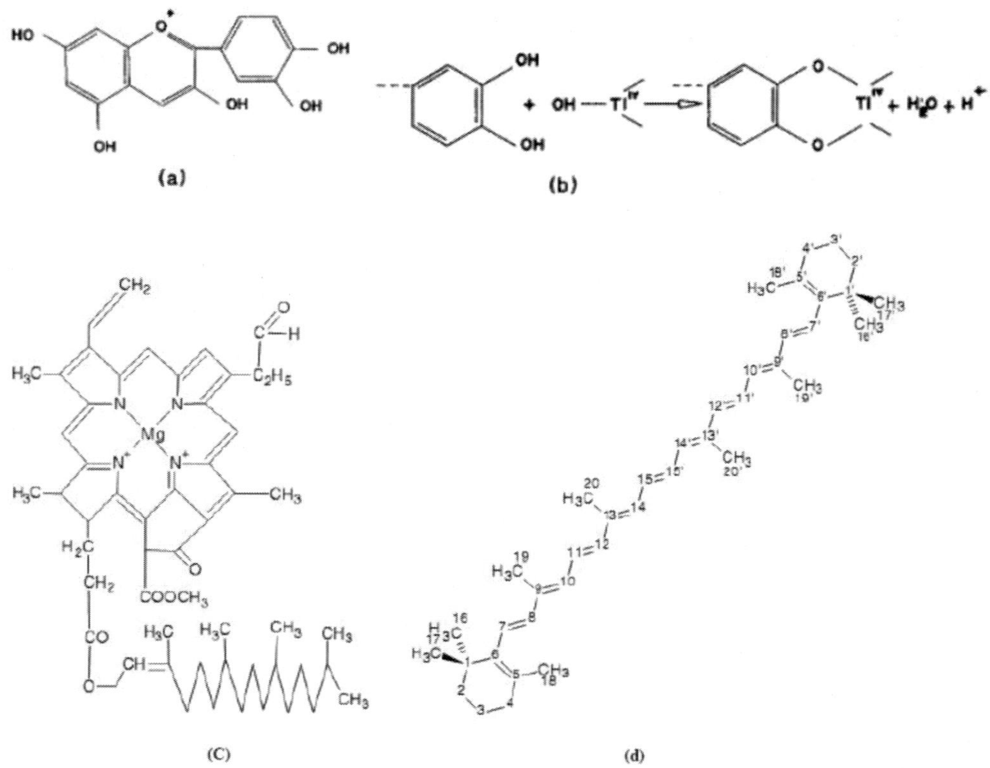

Figure 3. (a) Molecular structure of anthocyanin anion. (b) Mechanism of surface complexation of anthocyanin and TiO_2. (c) The chemical structure of chlorophyll. (d) The chemical structure of *β*-carotene.

2.2.2. Characterization of TiO2 Thin Film

The thickness of the films was ca. 0.5 μm monitored by a α-step surface profiler. The adherence strength of the TiO2 thin film on the microscope glass slide as deposited was 9.5 N monitored by a coating adherence tester. The crystalline microstructure of the thin films was characterized using XRD. Figure 4 shows the XRD spectra of TiO2 thin films annealed at different temperature (as deposited, 450°C) in air for 5 h. The TiO2 thin film still remained in a complete anatase phase. From Figure 4, we can see that the crystallinity of the TiO2 thin film annealed at 450°C became better than as deposeted. Because anatase films have the (1 0 1) plane as the most exposed face in nanocrystals [29].

Figure 5 illustrates the absorption spectra of black rice extract in ethanol solution and on TiO2 film. Absorption peak wavelength of black rice extract on TiO2 film (around 590nm) is greatly red-shift, compared to that in ethanol solution (around 525 nm). Anthocyanin (structural formula given in Figure 5(a)) is strongly adsorbed on TiO2 as a result of complexation with TiIV ions on the surface.

Surface complexation can readily occur via elimination of a proton (Figure 5(b)). The shift to the infrared region at the absorption maxima of the dye sensitized film confirmed the reaction between the TiO2 and the Anthocyanin, since the interaction caused a reduction of the electron density in the chromophore group, thus reducing its polarity (bathochromic effect). These observations are in agreement with values reported in natural dye sensitized solar cells [30-33].

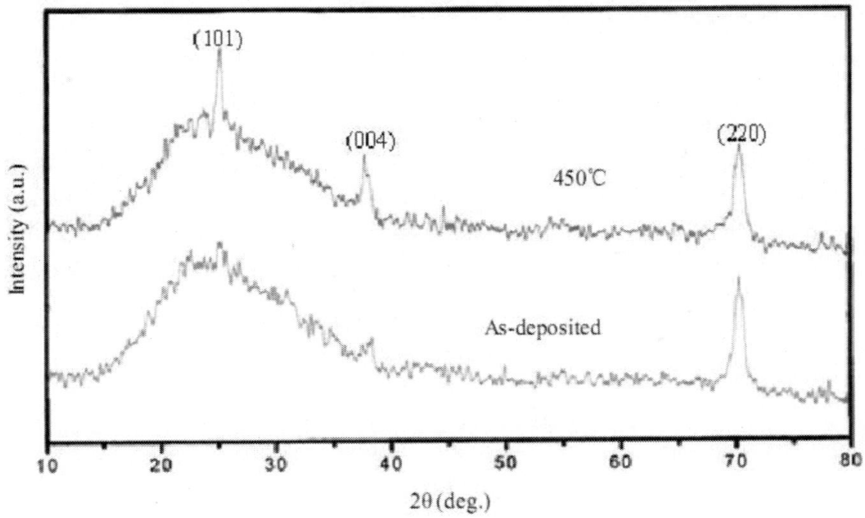

Figure 4. XRD patterns of the series of TiO_2 thin films.

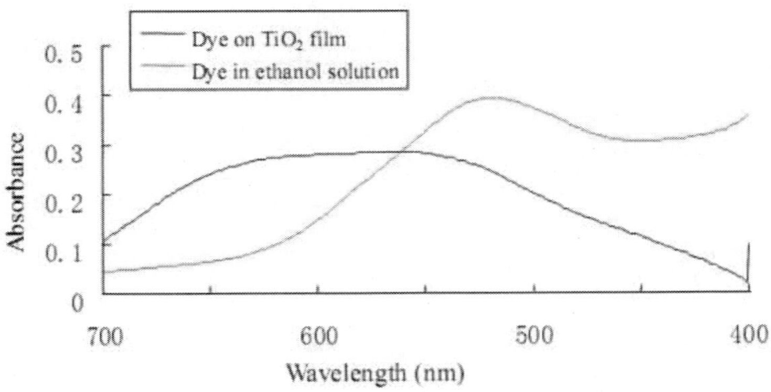

Figure 5. The absorption spectra of black rice dye in ethanol solution and on TiO_2 film.

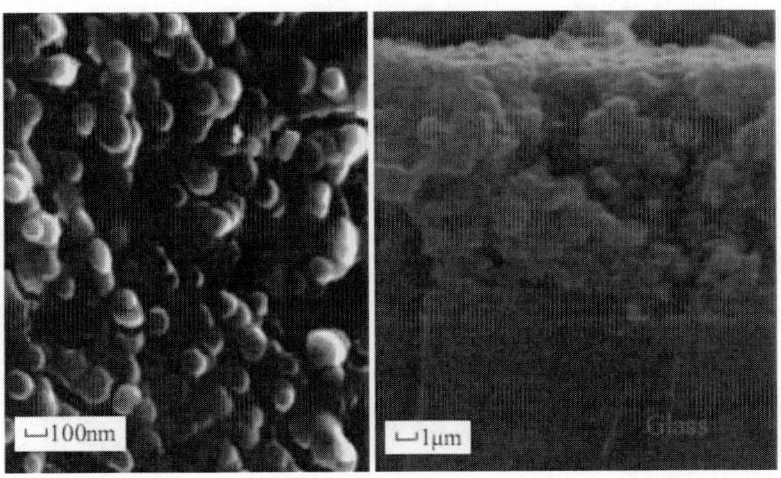

Figure 6. SEM micrographs of a sol-gel TiO_2 layer with a thickness of 10 μm on a glass substrate annealed for 1 h at 450 °C. (a)surface. (b) cross section.

2.2.3. Morphology of Films

The microstructure and the surface morphology of films were examined using a SEM as shown in Figure 6 (a) and (b). The sol-gel layers show a high porosity, a slightly more compact material close to the substrate, and a gradient of the pore distribution (Figure 6b).

This structure is reached within the whole layer. Such a morphology is advantageous for efficient electrochemical cells, because a fast and uninterrupted ionic charge transport can, in part, prevent recombination losses [9]. Furthermore, the sol-gel layers show noticeably larger particles with a size of 100 nm, few have a size of up to 200 nm (see Figure 6a). The thickness of the film is about 10μm.

3. MEASUREMENTS

3.1. Measurements System Based on The Sensor

The electrical resistance of the film was measured by an Agilent digital electrometer (34401A) with data acquisition capability using the IntuiLink software (Figure 1).

All the sensors were first tested in humid atmosphere, and then in NH3 one and their electrical characterizations were evaluated by comparing the gas response with the R.H. and the ammonia concentration. The response of the sensor is defined as the relative change of the resistance of the films in ambient air and in the analytes:

$$S = \frac{|R_g - R_0|}{R_0} \times 100\% \qquad (1)$$

where R_g is the resistance of the sensor in ammonia with air background and R_o is the resistance in air.

The devices were mounted in a specially constructed glass chamber (about 10 ml) where a carrier gas (air) containing known concentrations of the ammonia vapours was passed over the films. The concentration of the ammonia vapours could be changed rapidly for response time measurements. All the measurements were made at room temperature and under normal atmospheric pressure.

As shown in Figure 7, the carrier gases with known concentrations of the ammonia vapours were produced in two stages. In the first stage, air gas at a flow rate of 100 ml/min was passed over a permeation vial containing a known quantity of ammonia and held at a constant temperature of 25°C±0.5. The concentration of the ammonia vapour (C_1) in the stream of air with a flow rate of F_1 was calculated using a technique published in Ref. [34-36]:

$$C_1 = \frac{P\rho}{F_1} \qquad (2)$$

where P is the permeation rate and ρ the reciprocal vapour density of the ammonia. The permeation rate was calculated by determining how much of the liquid was lost from the vial over a known time period. The concentration of the vapours in the air stream in this stage was determined to be 50 ppm. In the second stage, the air flowing from the vial was mixed with a second air stream and a range of concentrations between 10 and 50 ppm obtained. The resulting concentration was calculated from [36]:

$$C = \frac{C_1 F_1}{F_1 + F_2} \tag{3}$$

where C is the concentration of the measured gas and F_2 the flow rate of the dilution air stream.

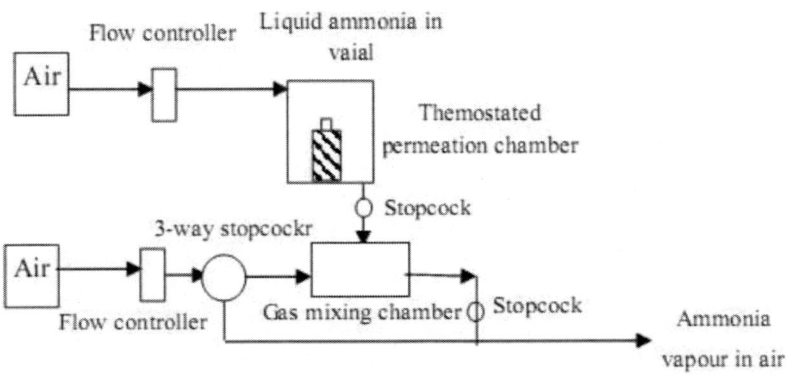

Figure 7. Schematic diagram of the flow rig for the generation of ammonia vapour standards in air.

3.2. Gas Measurements

Figure 8 (a) shows the response of the pigment sensitized TiO_2 film based sensor to different concentrations of ammonia in the background of ambient air with R.H. ~40% at room temperature. The resistance (R_0) of the sensor after manufacture in air is 4.78 M Ohm. The sensors indicated responses to ammonia within a concentration range from 10 to 50 ppm in steps of 10 ppm. After each increase in concentration of the vapours, the glass sample chamber was flushed with air. The resistance of the TiO_2 film decreased with exposure to ammonia (Figure 8 a).

This analysis was carried out three times: the sensor exhibited a good stability, due to the fact that the curves referred to the different cycles overlapped between them. The calibration curves for the films were obtained by plotting the relative variation, S and R, of the sensor resistance against the concentration of the vapours in the carrier gas. Figure 8 (b) shows the results of 5 sensors with different R_0 exposed to different concentrations of ammonia. Figure 8 (c) shows the changes in S for a film exposed to different concentrations of ammonia. The relative resistance of the sensor increased almost linearly with increasing concentrations of ammonia (r = 0.92) over the concentration range 10–50 ppm. This linearity changes of the films also indicates that the gold layers make good Ohmic contacts with the films. Even if metal–semiconductor Schottky-type junctions are formed then they are of no consequence in the sensors as they are well away from where the sensing of the vapours occurs on the surface of the films.

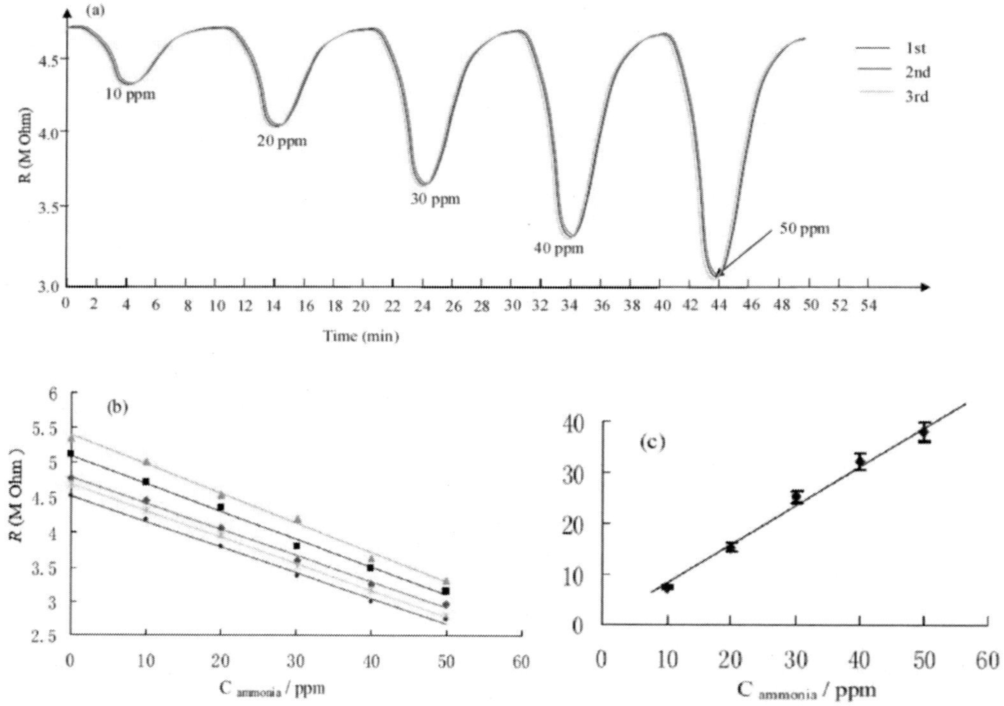

Figure 8. Typical response signals of ammonia in air for the fabricated pigment sensitized TiO_2 film based sensors at room temperature (22 ± 1 °C) with R.H.40%, n=3. (a) Variation of conductance with the flow of 10-50 ppm ammonia (response) and air (recovery), (b) the resistance of 5 sensors on relative ammonia concentration, (c) The response vs. concentration.

The response time, defined as the time taken for the sensor's resistance to reach the 90% of the steady-state resistance, is around 180-220s. Likewise, the recovery time represents the time required by the sensitivity factor to return to 10% below its equilibrium value in air following the zeroing of the test gas ammonia and it was found to be around 240-270s. The response and recovery times were not as high as those reported in the literature [37] and [38], indicating that NH_3 molecules are weakly attached to the sensing materials, or previously adsorbed ammonia molecules. These results indicate that cyanidin is responsible for the sensing action. However, the sensors appear to have some form of drift. This is mainly caused by the same response and recovery times set in the cycles. In the dynamic flow changes, there is little ammonia remain in the chamber or on the film when the air flux is restored after the gas test. This caused the incomplete recovery of the sensor. Therefore, the response and recovery times in Figure 8 (a) are 4 min and 6 min respectively.

The sensing mechanism for the metal oxides working in the humidified air at room temperatures (usually 0–30 °C) is often proposed to be related to the electrolytic dissociations of the gas species in the adsorbed water molecules covered on the surface of the metal oxides (39, 40). The ionized species in the water molecule would ensure a change in the pH value on the oxide surface and thus change the electrons/holes densities at the conduction or valence bands of the oxide semiconductor. This consequently would result in the sensor signals observed (*40*). However, as pointed out by Helwig et al.(*39*), this "dissociative gas sensing mechanism" requires that the analyte gases should have a good water solubility and be easily ionized in water. Moreover, the sensor signals induced by such type of sensing mechanism are usually small. Therefore, to explain, the following sequence of reactions can be considered:

$$NH_3 + H_2O \leftrightarrow NH_4^+ + OH^- \quad (4)$$

$$\text{Cyanidin} + OH^- \leftrightarrow \text{Cyanidin}^- + H_2O \quad (5)$$

where Cyanidin⁻ is anion of the Cyanidin. For form (4) and (5), the total reaction for the lower humidity condition is rearranged as shown in Figure 9.

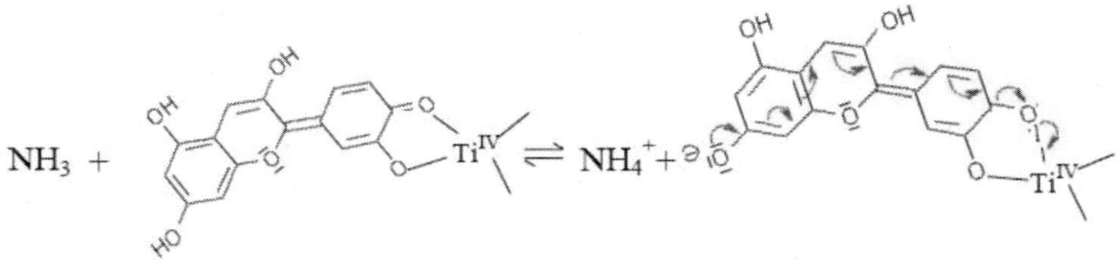

Figure 9. The reaction of ammonia (NH_3) and cyanidin sensitized TiO_2 film for the lower humidity condition.

An influence of the humidity on the clean air resistance of our samples is illustrated in Figure 10 (a). The resistance decreased significantly if the R.H. increased from 20 to 90%. The low humidity is necessary to ionized cyanidin molecules adsorved on the surface of TiO_2. Firstly, at room temperature, ammonia solves very well in water under formation of NH_4^+ and OH^- ions. A similar reaction of ammonia may occur on the sample surface with the adsorbed water. Secondly, at low humidity, ammonia could be ionized by the Cyanidin on the TiO_2 film and thereby decrease in the proton concentration at the surface. Consequently, more positively charged holes at the surface of the TiO_2 have to be extracted to neutralize the adsorbed cyanidin and water film. However the presence of water promotes desorption of the pigment from TiO_2 film which is undesirable. High humidity in the flow cell (above 50%) could lower the efficiency of the composite sensor.

Temperature tests of the ammonia gas sensors were performed with mini-heater. The temperature was fixed at some selected values within the interval from 10 to 50°C. The dependencies were obtained from the measurements of the resistance response to ammonia gas (R.H. 40%, 30 ppm of NH_3). The resistance response to ammonia was slightly decreased on temperature from 10 to 50°C, as shown in Figure 10 (b). The solubility of ammonia in water decreases as well as the amount of water itself on the surface. Therefore, the amount of ammonia which is present at the film and which is able to react with adsorption sites is reduced and the sensitivity decreases [25]. Long-term aging tests of the ammonia gas sensors have been performed. The aging tests were performed in the dark and in air. After the sensor element was stored in a container at room temperature over for 4 weeks, the response to 50 ppm ammonia was statistically indistinguishable from that prior to the storage period.

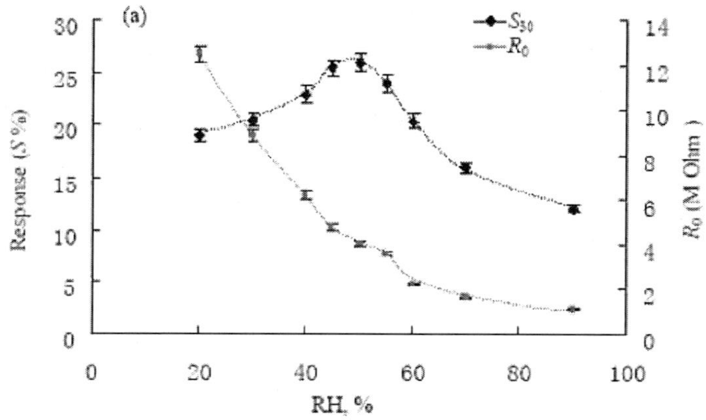

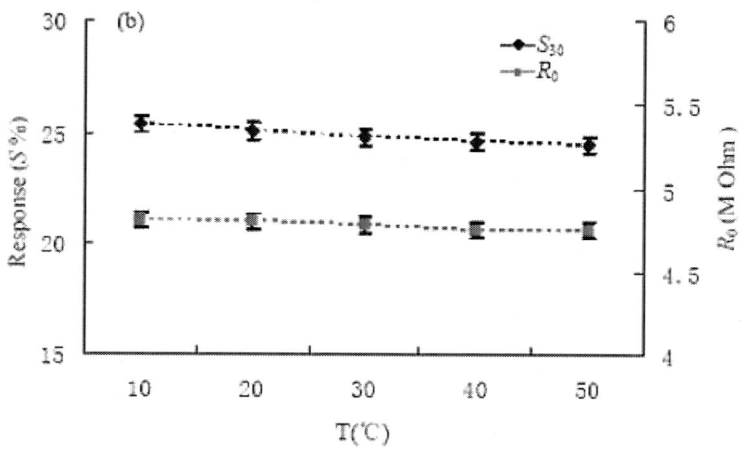

Figure 10. The dependence of the resistance response to NH_3 gas of 30 ppm (S_{30}) and the clean air resistance (R_0), (a) on relative humidity in air at the ambient temperature (22 ± 1 °C). (b) S_{30} and R_0 on relative temperature in the ambient humidity air (R.H. 40%).

The work described above demonstrates that a sensor for ammonia can be made using cyanidin sensitized TiO_2 film. The sensor can be operated at room temperature, which is an advantage over other TiO_2 sensors that operate at high temperatures. The resistance of the films decreases as the concentrations of the ammonia vapours in a carrier gas increases. The relative resistance of the films increases almost linearly as the concentration of the vapours increases over the concentration range 10–50 ppm. The response time for the sensor to increasing concentrations of ammonia is about 180-220s, and for decreasing concentrations of the vapours is 240-270s. The low humidity could improve the sensitivity of the sensors. The humidity dependence on the R_0 and response makes this sensor needs calibration in practical application. Furthermore, the sensor must be used in dark, TiO_2 in air and under illumination act as photocatalyst. Therefore illumination would cause de degradation of the adsorbed pigment and consequently a reduction of sensitivity and degradation of the sensor is expected. These results are not actually as good as those reported in the literature, but this preliminary work proposes simpler and cheaper processes to realize NH_3 sensor for room temperature applications.

REFERENCES

[1] Masuda, Y., Bekki, M., Sonezaki, S., Ohji, T., and Kato, K. (2009) Dye Adsorption Characteristics of Anatase TiO2 Film Prepared in an Aqueous Solution, *Thin Solid Films 518*, 845-849.

[2] Zhu, X., Nanny, M. A., and Butler, E. C. (2007) Effect of inorganic anions on the titanium dioxide-based photocatalytic oxidation of aqueous ammonia and nitrite, *Journal of Photochemistry and Photobiology A: Chemistry 185*, 289-294.

[3] Nho, Y.-C., Park, J.-S., Youn, J.-K., Lim, Y.-M., and Shin, J.-H. (2009) Photocatalytic activity of titanium dioxide prepared by a gamma-ray irradiation, *Radiation Physics and Chemistry 78*, 509-512.

[4] Karunagaran, B., Uthirakumar, P., Chung, S. J., Velumani, S., and Suh, E. K. (2007) TiO2 thin film gas sensor for monitoring ammonia, *Materials Characterization 58*, 680-684.

[5] Bernacka-Wojcik, I., Senadeera, R., Wojcik, P. J., Silva, L. B., Doria, G., Baptista, P., Aguas, H., Fortunato, E., and Martins, R. (2010) Inkjet printed and "doctor blade" TiO2 photodetectors for DNA biosensors, *Biosensors and Bioelectronics 25*, 1229-1234.

[6] Sen, S., Muthe, K. P., Joshi, N., Gadkari, S. C., Gupta, S. K., Jagannath, Roy, M., Deshpande, S. K., and Yakhmi, J. V. (2004) Room temperature operating ammonia sensor based on tellurium thin films, *Sensors and Actuators B: Chemical 98*, 154-159.

[7] Saxena, V., Choudhury, S., Gadkari, S. C., Gupta, S. K., and Yakhmi, J. V. (2005) Room temperature operated ammonia gas sensor using polycarbazole Langmuir–Blodgett film, *Sensors and Actuators B: Chemical 107*, 277-282.

[8] Regan, B., and Grätzel, M. (1991) A low-cost, high-efficiency solar cell based on dye-sensitized colloidal TiO2 films, *Nature 353*, 737-740.

[9] Baia, M., Danciu, V., Cosoveanu, V., and Baia, L. (2008) Porous nanoarchitectures based on TiO2 aerogels and Au particles as potential SERS sensor for monitoring of water quality, *Vibrational Spectroscopy 48*, 206-209.

[10] Moradian, A., Mohr, G. J., Linnhoff, M., Fehlmann, M., and Spichiger-Keller, U. E. (2000) Continuous optical monitoring of aqueous amines in transflectance mode, *Sensors and Actuators B: Chemical 62*, 154-161.

[11] Baia, L., Diamandescu, L., Barbu-Tudoran, L., Peter, A., Melinte, G., Danciu, V., and Baia, M. (2011) Efficient dual functionality of highly porous nanocomposites based on TiO2 and noble metal particles, *Journal of Alloys and Compounds 509*, 2672-2678.

[12] Dong, Q., Su, H., Cao, W., Zhang, D., Guo, Q., and Lai, Y. (2007) Synthesis and characterizations of hierarchical biomorphic titania oxide by a bio-inspired bottom-up assembly solution technique, *Journal of Solid State Chemistry 180*, 949-955.

[13] Mor, G. K., Varghese, O. K., Paulose, M., Shankar, K., and Grimes, C. A. (2006) A review on highly ordered, vertically oriented TiO2 nanotube arrays: Fabrication, material properties, and solar energy applications, *Solar Energy Materials and Solar Cells 90*, 2011-2075.

[14] König, B., Kohls, O., Holst, G., Glud, R. N., and Kühl, M. (2005) Fabrication and test of sol-gel based planar oxygen optodes for use in aquatic sediments, *Marine Chemistry 97*, 262-276.

[15] Beltrán-Pérez, G., López-Huerta, F., Muñoz-Aguirre, S., Castillo-Mixcóatl, J., Palomino-Merino, R., Lozada-Morales, R., and Portillo-Moreno, O. (2006) Fabrication and characterization of an optical fiber pH sensor using sol-gel deposited TiO2 film doped with organic dyes, *Sensors and Actuators B: Chemical 120*, 74-78.

[16] Reijnders, L. (2010) Design issues for improved environmental performance of dye-sensitized and organic nanoparticulate solar cells, *Journal of Cleaner Production 18*, 307-312.

[17] Bruder, I., Karlsson, M., Eickemeyer, F., Hwang, J., Erk, P., Hagfeldt, A., Weis, J., and Pschirer, N. (2009) Efficient organic tandem cell combining a solid state dye-sensitized and a vacuum deposited bulk heterojunction solar cell, *Solar Energy Materials and Solar Cells 93*, 1896-1899.

[18] Lee, D. H., Lee, M. J., Song, H. M., Song, B. J., Seo, K. D., Pastore, M., Anselmi, C., Fantacci, S., De Angelis, F., Nazeeruddin, M. K., Gräetzel, M., and Kim, H. K. (2011) Organic dyes incorporating low-band-gap chromophores based on π-extended benzothiadiazole for dye-sensitized solar cells, *Dyes and Pigments 91*, 192-198.

[19] Cheng, X., Sun, S., Liang, M., Shi, Y., Sun, Z., and Xue, S. (2012) Organic dyes incorporating the cyclopentadithiophene moiety for efficient dye-sensitized solar cells, *Dyes and Pigments 92*, 1292-1299.

[20] López de Mishima, B. A., and Mishima, H. T. (2008) Ammonia sensor based on propylene carbonate, *Sensors and Actuators B: Chemical 131*, 236-240.

[21] Yimit, A., Itoh, K., and Murabayashi, M. (2003) Detection of ammonia in the ppt range based on a composite optical waveguide pH sensor, *Sensors and Actuators B: Chemical 88*, 239-245.

[22] Timmer, B., Olthuis, W., and Berg, A. v. d. (2005) Ammonia sensors and their applications—a review, *Sensors and Actuators B: Chemical 107*, 666-677.

[23] Murugappan, K., Lee, J., and Silvester, D. S. (2011) Comparative study of screen printed electrodes for ammonia gas sensing in ionic liquids, *Electrochemistry Communications 13*, 1435-1438.

[24] Wang, S. G., Zhang, Q., Yang, D. J., Sellin, P. J., and Zhong, G. F. (2004) Multi-walled carbon nanotube-based gas sensors for NH3 detection, *Diamond and Related Materials 13*, 1327-1332.

[25] Ostrick, B., Pohle, R., Fleischer, M., and Meixner, H. (2000) TiN in work function type sensors: a stable ammonia sensitive material for room temperature operation with low humidity cross sensitivity, *Sensors and Actuators B: Chemical 68*, 234-239.

[26] Tulliani, J.-M., Cavalieri, A., Musso, S., Sardella, E., and Geobaldo, F. (2011) Room temperature ammonia sensors based on zinc oxide and functionalized graphite and multi-walled carbon nanotubes, *Sensors and Actuators B: Chemical 152*, 144-154.

[27] Varghese, O. K., Gong, D., Dreschel, W. R., Ong, K. G., and Grimes, C. A. (2003) Ammonia detection using nanoporous alumina resistive and surface acoustic wave sensors, *Sensors and Actuators B: Chemical 94*, 27-35.

[28] Shang, Y., Wang, X., Xu, E., Tong, C., and Wu, J. (2011) Optical ammonia gas sensor based on a porous silicon rugate filter coated with polymer-supported dye, *Analytica Chimica Acta 685*, 58-64.

[29] Zheng, S. K., Wang, T. M., Xiang, G., and Wang, C. (2001) Photocatalytic activity of nanostructured TiO2 thin films prepared by dc magnetron sputtering method, *Vacuum 62*, 361-366.

[30] Hao, S., Wu, J., Huang, Y., and Lin, J. (2006) Natural dyes as photosensitizers for dye-sensitized solar cell, *Solar Energy 80*, 209-214.

[31] Calogero, G., and Marco, G. D. (2008) Red Sicilian orange and purple eggplant fruits as natural sensitizers for dye-sensitized solar cells, *Solar Energy Materials and Solar Cells 92*, 1341-1346.

[32] Gómez-Ortíz, N. M., Vázquez-Maldonado, I. A., Pérez-Espadas, A. R., Mena-Rejón, G. J., Azamar-Barrios, J. A., and Oskam, G. (2010) Dye-sensitized solar cells with natural dyes extracted from achiote seeds, *Solar Energy Materials and Solar Cells 94*, 40-44.

[33] Zhou, H., Wu, L., Gao, Y., and Ma, T. (2011) Dye-sensitized solar cells using 20 natural dyes as sensitizers, *Journal of Photochemistry and Photobiology A: Chemistry 219*, 188-194.

[34] Mabrook, M., and Hawkins, P. (2001) A rapidly-responding sensor for benzene, methanol and ethanol vapours based on films of titanium dioxide dispersed in a polymer operating at room temperature, *Sensors and Actuators B: Chemical 75*, 197-202.

[35] Choi, M. F., and Hawkins, P. (1997) The development of optical chemical sensors for the detection of volatile compounds from spoiled hams, *Sens. Actuators B 38/39*, 390–394.

[36] Namiesik, J. (1984) Generation of standard gaseous mixtures, *J. Chromatogr. 300*, 79–108.

[37] Joshi, A., Gangal, S. A., and Gupta, S. K. (2011) Ammonia sensing properties of polypyrrole thin films at room temperature, *Sensors and Actuators B: Chemical 156*, 938-942.

[38] Capone, S., Siciliano, P., Quaranta, F., Rella, R., Epifani, M., and Vasanelli, L. (2000) Analysis of vapours and foods by means of an electronic nose based on a sol–gel metal oxide sensors array, *Sensors and Actuators B: Chemical 69*, 230-235.

[39] Helwig, A., Müller, G., Sberveglieri, G., and Eickhoff, M. (2009) On the low-temperature response of semiconductor gas sensors, *J. Sens 2009*, 1-7.

[40] Ostrick, B., Muhlsteff, J., Fleischer, M., Meixner, H., Doll, T., and Kohl, C.-D. (1999) Adsorbed water as key to room temperature gas-sensitive reactions in work function type sensors: the carbonate–carbon dioxide system, *Sens. Actuators B 57*, 115–119.

In: Titanium Dioxide
Editor: Jerri Brown

ISBN: 978-1-63321-391-3

Chapter 7

NATURAL ZEOLITE SUPPORTED IONS DOPED NANO TiO_2 PHOTOCATALYST: A PHOTOCATALYST WITH HIGHER RECYCLING EFFICIENCY, ADSORPTION AND PHOTOCATALYTIC ACTIVITIES

Cheng Wang[1,2,*] and Huisheng Shi[3]
[1]School of Materials Science and Engineering,
Shaanxi University of Science and Technology, Xi'an, China
[2]Department of Chemical and Materials Engineering,
University of Alberta, Edmonton, Canada
[3]Key Laboratory of Advanced Civil Engineering Materials
(Tongji University), Ministry of Education, Shanghai, China

ABSTRACT

Titanium dioxide, as a chemically stable, nontoxic, highly efficient, and relatively inexpensive photocatalyst, has been attached much attention in recent years. However, practical applications of TiO_2 are limited because of the problems of low recycling efficiency and visible-light photocatalytic activity of TiO_2 fine particles. Natural zeolite supported ions doped nano TiO_2 photocatalyst, a photocatalyst loading nano TiO_2 on the surface of zeolite and doping different ions into the crystallite of TiO_2, which shows higher recycling efficiency, adsorption and photocatalytic activities than that of pure TiO_2, seems to be a promising photocatalyst for practical application. In this chapter, we introduced and concluded the preparation, characterization, photocatalytic activities, and the related mechanisms of natural zeolite supported ions doped TiO_2 photocatalysts based on our previous works. It shows that i) the photocatalysts can be successfully prepared by $TiCl_4$ hydrolysis method; ii) the zeolite substrate facilitates the formation of anatase,

* Corresponding author: Cheng Wang, School of Materials Science and Engineering, Shaanxi University of Science and Technoligy, Weiyang District, Xi'an 710021, China; Email: wangchengtj@163.com(Cheng Wang). Tel: +86-29-86168688

restrains the transformation of anatase to rutile and the growth of TiO_2 crystallite, and also improves the adsorption activity of the as-prepared photocatalysts; iii) the doped ions can diffuse into TiO_2 lattice and give TiO_2 the capacity to absorb light at lower energy levels, and certain ions with proper doping concentration can evidently improve photocatalytic activities of the photocatalysts; iv) Cr doped sample shows the highest photocatalytic activity than that of the other ions doped samples due to its lower energy gap and recombination probability of photo-generated electron-hole pair; v) the photocatalysts show photocatalysis-adsorption co-effect on the organic molecules, and this co-effect is helpful for photocatalytic degradation of the organic contaminant. The current problems and future developments of this photocatalyst are also prospected.

Keywords: Natural zeolite, ions doping, TiO_2, photocatalyst

INTRODUCTION

Semiconductor photocatalysts and their photocatalytic oxidation technical have been rapidly attached much attention since the discovery of photocatalytic splitting of water on a TiO_2 electrode by Fujishima and Honda in 1972[1]. Titanium dioxide, with the properties of chemically stable, nontoxic, highly efficient, and relatively inexpensive etc., has been believed to be one of the most promising photocatalyst for practical applications. However, practical applications of TiO_2 are limited because of the problems of low recycling efficiency and visible-light photocatalytic activity of TiO_2 fine particles. In order to solve the above problems, researchers have been devoted in searching for suitable substrates to support TiO_2 and suitable strategies to sensitize TiO_2.

Zeolites, aluminosilicate micro-porous solids known as molecular sieves, have been most favorable to be used in supporting TiO_2, due to their uniform pores and channels, large surface areas, excellent cation exchange and absorption abilities [2-11]. In recent years, researchers[2-4, 9-11] used natural zeolites instead of synthetic zeolites to support nano TiO_2 due to their low cost and abundant reserves, and it has been confirmed that natural zeolites supported TiO_2 photocatalysts display higher photocatalytic efficiency at low initial concentration in comparison with pure TiO_2.

Previous researches have proved that the photocatalytic activity of TiO_2 can be improved by using kinds of strategies, such as semiconductor compound[12,13], ions doping[9-11, 14], novel metal deposition[3,15], surface dye-sensitization[16,17] etc., to sensitize TiO_2. It has also been confirmed that photocatalytic activity and recovery efficiency can be simultaneously improved by using certain strategies to sensitize the supported TiO_2[3,9-11,18].

Based on the above background, we used natural zeolite as the substrate and different ions such as Fe[9], Cr[10], S[11] etc., as the dopants to prepare natural zeolite supported ions doped nano TiO_2 photocatalysts. The preparation, characterization, photocatalytic activities, and the related mechanisms of the photocatalysts were investigated and concluded in this chapter.

1. PREPARATION OF THE PHOTOCATALYSTS

Up to now, different kinds of methods such as solid state dispersion(SSD)[19], papermaking[8,20], sol-gel[3-6], $TiCl_4$ hydrolysis[2,9-11] etc. have been adopted to prepare (ions doped) TiO_2/zeolite photocatalysts. Among these methods, $TiCl_4$ hydrolysis method seems to be more suitable due to its relatively environment-friendly procedure and strong combination between TiO_2 and zeolite. The detailed procedure for preparation and formation of the photocatalysts are as follows:

i. Raw zeolite, as a aluminosilicate mineral, the H^+ can be attracted to aluminium by static force due to the replacement of Si with Al [5,6].
ii. $TiCl_4$ can hydrolyze and form titanium polycation in acid aqueous solution, the detailed procedure is shown as follows:

$$TiCl_4 + H_2O \longrightarrow [Ti\text{-}O\text{-}Cl\ldots TiO\text{-}]^{n+} \longrightarrow [(TiO)_8(OH)_{12}]^{4+}$$

iii. Titanium polycations can exchange with H^+ and then load onto the surface of zeolite[2,6].
iv. The titanium polycations can be oxidized and then crystallize to anatase or rutile after calcined at certain temperature, and simultaneously, TiO_2 can combine with zeolite by forming Ti-O-Si(Al) bond which has been proved by using FT-IR analysis[2].

2. CHARACTERIZATION OF IONS DOPED TiO_2/ZEOLITE PHOTOCATALYSTS

2.1. Crystal Structure of the Photocatalysts

2.1.1. Effect of Zeolite on the Crystal Structure of TiO_2

Zeolite, as the substrate of photocatalyst, can definitely affect the crystal structure of TiO_2. In order to investigate this affect, we prepared pure TiO_2 and TiO_2/zeolite calcined at 200°C-600°C and compared their cystallite sizes and crystal forms. As shown in Figure 1 and Table 1, it can be seen that crystallite sizes of TiO_2(TiO_2/zeolite) are obviously smaller than that of pure TiO_2.

The crystal forms of TiO_2(TiO_2/zeolite) are mainly anatase, and there is just a little amount of rutile contained for 600°C sample. However, the crystal forms of pure TiO_2 are totally different with that of TiO_2(TiO_2/zeolite), which are brookite and anatase for 200°C-300°C samples, anatase and rutile for 400°C-600°C samples. It can be concluded that zeolite facilitates the formation of anatase, restrains the transformation of anatase to rutile, and also restrains the growth of TiO_2 crystallite.

Table 1. Cystallite sizes and crystal forms of pure TiO_2 and TiO_2/zeolite calcined at 200℃-600℃

Calcination temperature/℃	Pure TiO_2 Crystallite size/nm	Crystal form	TiO_2/zeolite Crystallite size/nm	Crystal form
200	4.5	brookite, anatase	5.9	anatase
300	5.2	brookite, anatase	6.1	anatase
400	8.2	anatase, rutile	6.5	anatase
500	11.7	anatase, rutile	8.4	anatase
600	16.5	anatase, rutile	9.8	anatase, rutile

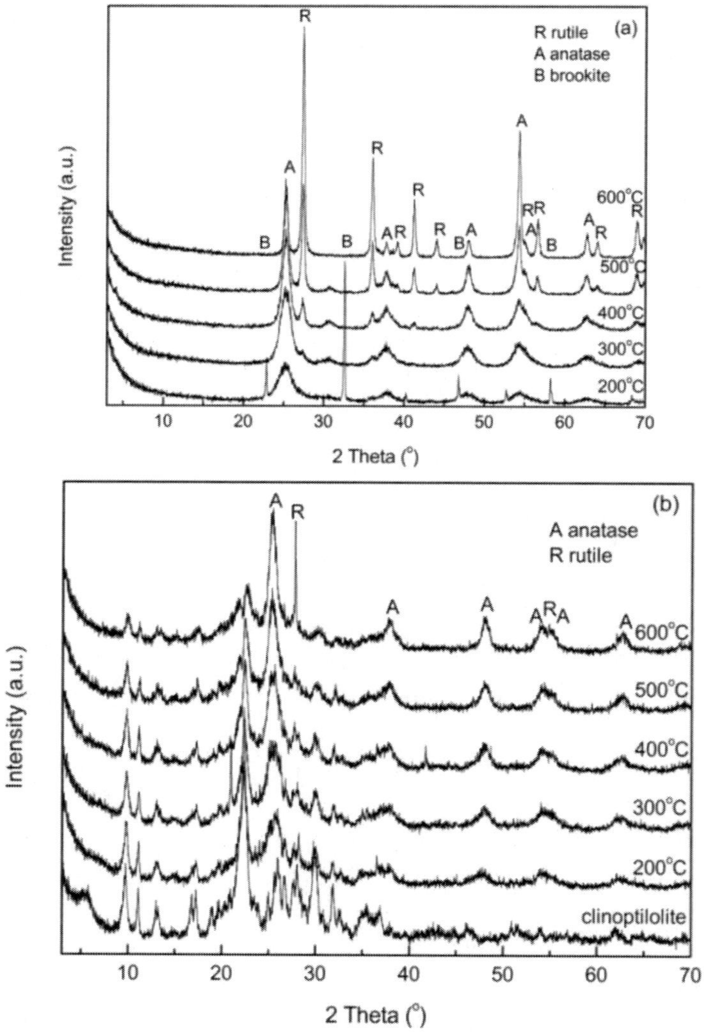

Figure 1. XRD patterns of pure TiO_2(a) and TiO_2/zeolite(b) calcined at 200℃-600℃.

2.1.2. *The Effect of Doped Ions on Crystal Structure of TiO_2*

Lots of researches have proved that the doped ions with proper ionic radius can be incorporated into the TiO_2 lattice via displacing Ti^{4+} sites, and thus give TiO_2 the capacity to absorb light at lower energy levels. We have investigated the effect of Fe, Cr, S ions on crystal structure of TiO_2 by using XPS analysis. As shown in the XPS spectra in Figure 2, it can be seen that all the Ti2p binding energies of doped samples shift to higher energies in comparison with that of undoped sample, which infers that the doped ions in the forms of Fe^{3+}, Cr^{6+}/Cr^{3+}, S^{6+} diffuse into TiO_2 lattices to form the Fe(Cr, S)-O-Ti bond. This also can be confirmed by XRD analysis. As an example, we have investigated the structure of Cr doped TiO_2/zeolite at 200℃ to 600℃ (Figure 3), and it shows that the (101) peak at 25.4° of anatase gradually shifts to lower angle as the calcination temperatures ascend, which suggests that Cr ions enter into the TiO_2 crystal and consequently induce distortion to crystal lattice.

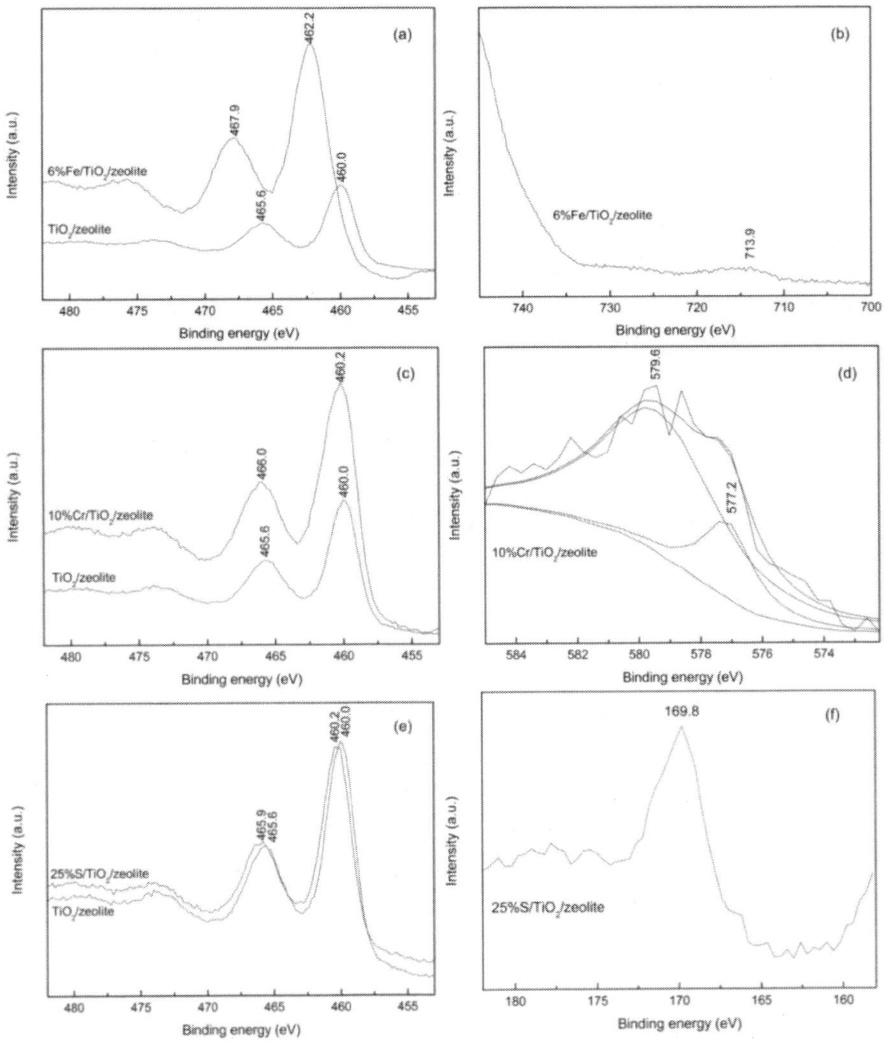

Figure 2. XPS spectra of Ti2p(a) and $Fe2p_{3/2}$(b), Ti2p(c) and $Fe2p_{3/2}$(d) and Ti2p(e) and $Fe2p_{3/2}$(f) of undoped, Fe, Cr, S doped TiO_2/zeolite.

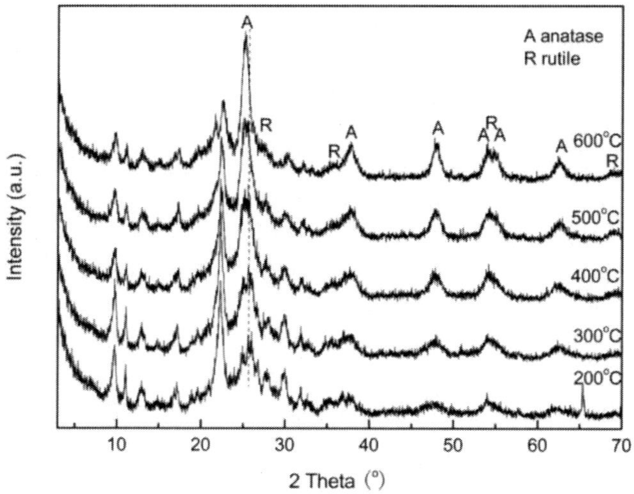

Figure 3. XRD patterns of Cr doped TiO_2/zeolite calcined at 200°C-600°C.

2.2. The Pore Structure and Specific Surface Area of Ions Doped TiO_2/Zeolite

The pore structure and specific surface area of zeolite can be changed after loading with TiO_2 nano-particles, and it can definitely affect properties of the photocatalysts. As an example, we have investigated the average pore diameter and specific surface area of the raw zeolite and Cr ions doped TiO_2/zeolite calcined at 200°C-600°C[10]. From Table 2, it can be concluded that 1) average pore diameters of the supported photocatalysts decrease while the specific surface areas increase as comparison with that of raw zeolite, which is attributed to the loading of TiO_2 nano-particles with high specific surface area; 2) specific surface areas of the supported photocatalysts decrease with the ascent of calcination temperature, which is attributed to growth of TiO_2 particles; 3) average pore diameters of the supported photocatalysts increase with the ascent of calcination temperature, which may be attributed the decomposition of organic or impurity components contained in the pores of zeolite.

Table 2. Average pore diameter (APD) and specific surface area (SSA) of the raw zeolite, 10%Cr/TiO_2/zeolite calcined at 200℃-600℃

Samples	APD (nm)	SSA (m^2/g)
raw zeolite	8.7	56
Cr/TiO_2/zeolite 200℃	3.1	205
Cr/TiO_2/zeolite 300℃	3.5	181
Cr/TiO_2/zeolite 400℃	4.1	187
Cr/TiO_2/zeolite 500℃	4.9	152
Cr/TiO_2/zeolite 600℃	6.7	92

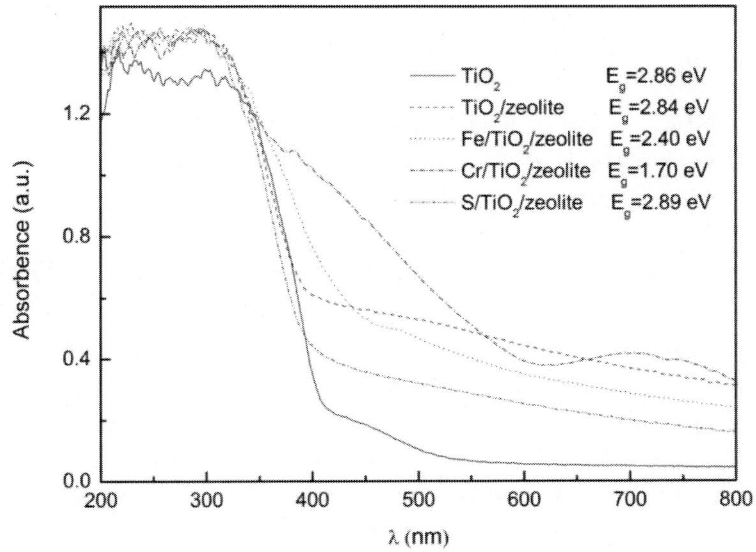

Figure 4. UV-vis DRS spectra of TiO_2, TiO_2/zeolite, Fe, Cr, S, doped TiO_2/zeolite and their correspondent energy band gaps (Eg).

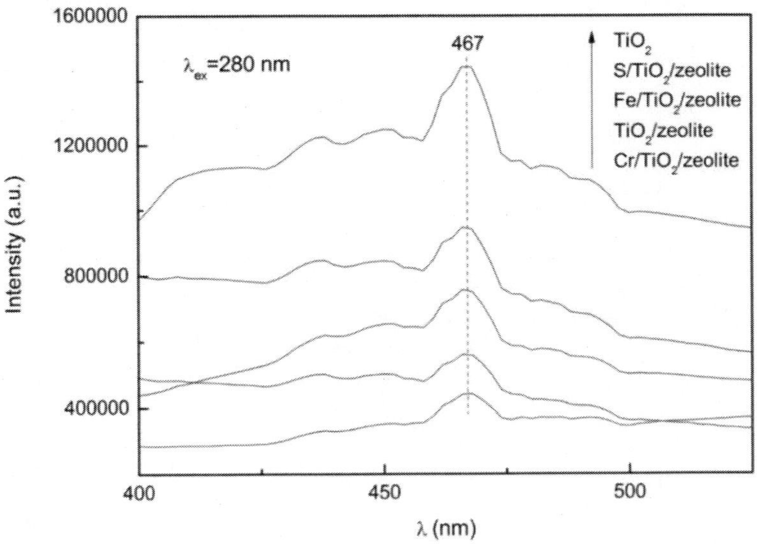

Figure 5. Photoluminescence of TiO_2, TiO_2/zeolite, Fe, Cr, S, doped TiO_2/zeolite.

2.3. The Optical Absorption and Photoluminescence Properties of Ions Doped TiO_2/Zeolite

The purpose of ions doping is to reduce the energy gap, extend the visible light response region of TiO_2, restrain the recombination probability of photo-generated electron-hole pair, and thus improve its visible light photocatalytic activity. The energy gap and photoluminescence properties of the photocatalysts can be determined by using UV-vis DRS and PL analysis, respectively. Based on the above techniques, we have investigated the energy gap and photoluminescence properties of pure TiO_2, TiO_2/zeolite, Fe, Cr and S ions doped TiO_2/zeolite, respectively. The results show that the energy gap of TiO_2 obviously reduces after doping with Fe and Cr ions(as shown in Figure 4) and the recombination probability of photo-generated electron-hole pair of Cr doped sample is the lowest(as shown in Figure 5), which suggest that Cr doped sample may has the highest photocatalytic activity. The detailed mechanisms can be seen in our previous papers [9-11].

3. The Photocatalytic Activity of Ions Doped TiO_2/Zeolite

The photocatalytic activity of photocatalysts can be evaluated by the degradation of organic molecular especially azo dye in an aqueous solution. As shown in Figure 6, it can be observed that all the decolorization rate of MO solution are rather low which is due to the weak UV light(a 40W UV lamp) adopted in our photodegradation experiment[9-11]. However, it still can be seen that the decolorization rates of MO solution catalyzed by the zeolite supported TiO_2 are obviously higher than that of pure TiO_2, which can be attributed to the excellent adsorption property of zeolite. This adsorption property leads to more organic molecules attach onto the surface of photocatalysts and is helpful for photocatalytic degradation of organic contaminant.

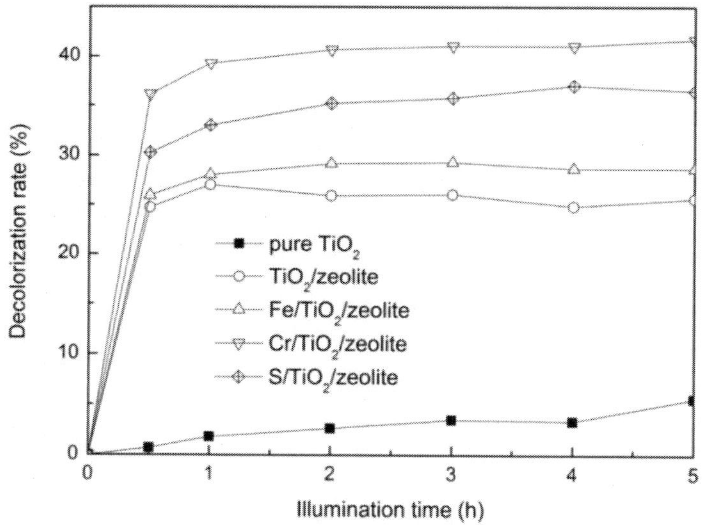

Figure 6. Decolorization rate of MO catalyzed by pure TiO_2, TiO_2/zeolite, Fe, Cr, S doped TiO_2/zeolite.

n addition, all the decolorization rate catalyzed by doped samples are higher than that of the undoped sample, which show that photocatalytic activities of the samples can be improved after doping with proper ions. The Cr doped TiO_2/zeolite shows the highest photocatalytic activity than that of other doped samples which further verifies the UV-vis DRS and PL results. It has not been accurately investigated but obviously can be seen in the photodegradation experiment that TiO_2/zeolite powders can be easily recovered from the solution while pure TiO_2 powder disperse in the solution and form a milky dispersion which is hard to recover.

Conclusion and Prospect

Natural zeolite supported ions doped TiO_2 photoctalysts have been investigated and proved to have higher recycling efficiency, adsorption and photocatalytic acitivies, and seems to be a promising photocatalysts for practical application. However, further studies on how to i) improve visible-light photocatalytic activity of the photocatalysts, for example, choosing proper doped or co-doped ions, and other strategies to sensitize TiO_2, ii) improve the recycling efficiency of the photocatalysts, for example, combining magnetic material with zeolite and then using this magnetic zeolite to support TiO_2, iii) improve the stability of the photocatalysts, for example, finding methods to enhance the binding force of zeolite substrate with TiO_2, etc. should be investigated.

Acknowledgments

The project supported by the Hebei Natural Science Foundation, China (Grant No. E2008000537), State Key Laboratory of Advanced Technology for Materials Synthesis and Processing (Wuhan University of Technology) (No. 2013-KF-13), Foundation of Shaanxi University of Science and Technology (No. BJ12-18) and Shaanxi Natural Science Foundation, China (Grant No. 2013JQ6007).

References

[1] A. Fujishima and K. Honda, Photolysis-decomposition of water at the surface of an irradiated semiconductor, *Nature* 5385 (1972) 37-38.
[2] Y. S. Jiang, Q. Jin, J. Zhang, S. S. Fang, Composite of TiO_2/zeolite structure and photocatalytic reactivity, *Journal of Inorganic Materials* 17 (2002) 1301-1305.
[3] M.L. Huang, C.F. Xu, Z.B. Wu, Y.F. Huang, J.M. Lin, J.H. Wu, Photocatalytic discolorization of methyl orange solution by Pt modified TiO_2 loaded on natural zeolite, *Dyes Pigmments* 77 (2008) 327-334.
[4] F. F. Li, S. M. Sun, Y. S. Jiang, M. S. Xia, M. M. Sun, B. Xue, Photodegradation of an azo dye using immobilized nanoparticles of TiO_2 supported by natural porous mineral, *Journal of Hazardous Materials* 152 (2008) 1037-1044.
[5] F. F. Li, Y. S. Jiang, L. X. Yu, Z. W. Yang, T. Y. Hou, S. W. Sun, Surface effect of natural zeolite (clinoptilolite) on the photocatalytic activity of TiO_2, *Applied Surface Science* 252 (2005) 1410-1416.
[6] H. Chen, A. Matsumoto, N. Nishimiya, K. Tsutsumi, Preparation and characterization of TiO_2 incorporated Y-zeolite, *Colloids and Surfaces A: Physicochemical and Engineering Aspects* 157 (1999) 295-305.
[7] M. Alvaro, E. Carbonell, V. Fornés, and H. García, Enhanced photocatalytic activity of zeolite-encapsulated TiO_2 clusters by complexation with organic additives and N-doping, *Journal of Chemical Physics and Physical Chemistry* 7 (2006) 200-205.
[8] Seonghyuk Koa, Jan Pekarovica, Paul D. Fleminga, Pnina Ari-Gur, High performance nano-titania photocatalytic paper composite. Part I: Experimental design study for TiO_2

composite sheet using a natural zeolite microparticle system and its photocatalytic property, *Materials Science and Engineering B* 166 (2010) 127-131.

[9] C. Wang, H. S. Shi, Y. Li Synthesis and characteristics of natural zeolite supported Fe^{3+}-TiO_2 photocatalysts, *Applied Surface Science*, 257 (2011) 6873-6877.

[10] C. Wang, H. S. Shi, Y. Li, Synthesis and characterization of natural zeolite supported Cr-doped TiO_2 photocatalysts, *Applied Surface Science*, 258 (2012) 4328-4333.

[11] C. Wang, Y. Li, Preparation and characterisation of S doped TiO2/natural zeolite with photocatalytic and adsorption activities, *Materials Technology: Advanced Performance Materials*, 2014, DIO.1753555714Y.0000000127.

[12] J. Y. Wang, Z. H. Liu, Q. Zheng, Z. K. He, R. X. Cai, Preparation of photosensitized nanocrystalline TiO2 hydrosol by nanosized CdS at low temperature, *Nanotechnology* 17 (2006) 4561-4566.

[13] S. H. Elder, F. M. Cot,Y. Su, S. M. Heald, A. M. Tyryshkin, M. K. Bowman, Y. Gao, A. G. Joly, M. L. Balmer, Ana C. Kolwaite, K. A. Magrini, D. M. Blake, The discovery and study of nanocrystalline TiO_2-(MoO_3) core-shell materials, *Journal of the American Chemical Society* 122 (2000) 5138-5146.

[14] J. Zhang, Y. Wu, M. Xing, S. A. K. Leghari, S. Sajjad, Development of modified N doped TiO_2 photocatalyst with metals, nonmetals and metal oxides, *Energy & Environmental Science* 3 (2010) 715-726.

[15] V. Subramanian, E. E. Wolf, P. V. Kamat, Catalysis with TiO_2/gold nanocomposites. effect of metal particle size on the fermi level equilibration, *Journal of the American Chemical Society* 126 (2004) 4943-4950.

[16] Y. Cho, W. Choi, C. H. Lee, T. Hyeon, H. I. Lee, Visible light-induced degradation of carbon tetrachloride on dye-sensitized TiO_2, *Environmental science & technology* 35 (2001) 966-970.

[17] L. Pan, J. J. Zou, X. W. Zhang, L. Wang, Water-mediated promotion of dye sensitization of TiO_2 under visible light, *Journal of the American Chemical Society* 133 (2011) 10000-10002.

[18] Y. Li, X. Zhou, W. Chen, L. Li, M. Zen, S. Qin, S. Sun, Photodecolorization of Rhodamine B on tungsten-doped TiO_2/activated carbon under visible-light irradiation, *Journal of hazardous materials* 227 (2012) 25-33.

[19] V. Durgakumari, M. Subrahmanyama, K.V. Subba Rao, A. Ratnamala, M. Noorjahan, Keiichi Tanaka, An easy and efficient use of TiO_2 supported HZSM-5 and TiO_2 + HZSM-5 zeolite combinate in the photodegradation of aqueous phenol and p-chlorophenol, *Applied Catalysis A: General* 234 (2002) 155-165.

[20] S. Fukahori, H. Ichiura, T. Kitaoka, H. Tanaka, Capturing of bisphenol A photodecomposition intermediates by composite TiO_2-zeolite sheets, *Applied Catalysis B: Environmental* 46 (2003) 453-462.

In: Titanium Dioxide
Editor: Jerri Brown
ISBN: 978-1-63321-391-3

Chapter 8

TITANIUM DIOXIDE-INDUCED PHOTOCTALYTIC REDUCTION FOR ORGANIC SYNTHESIS

Shigeru Kohtani and Hideto Miyabe*
School of Pharmacy, Hyogo University of Health Sciences,
Chuo-ku, Kobe, Japan

ABSTRACT

This chapter summarizes the synthetically useful photocatalytic reduction on a semiconductor particle titanium dioxide (TiO_2). TiO_2 has the potential to induce the reductive chemical transformation in the absence of oxygen (O_2), especially by choosing the suitable sacrificial hole scavenger such as alcohols or amines, since electrons photo-generated in the conduction band (CB) or those trapped at surface defect sites can transfer into an organic substrate adsorbed on a photocatalyst. The TiO_2-induced reduction of aldehydes or ketones gives the corresponding alcohols in the presence of alcohols as hole scavengers. The photocatalytic reduction of nitro compounds on TiO_2 gives the corresponding amines *via* the formation of hydroxylamine intermediates. Additionally, the reduction of nitro compounds was successfully applied into the synthesis of benzimidazoles or tetrahydroquinolines. Pt/TiO_2 is an effective catalyst for the reduction of imines. The reduction of imines was applied into several unique transformations involving both reductive and oxidative processes. The reduction of aromatic cyanides in ethanol gives toluene derivative and triethylamine *via* redox processes.

1. INTRODUCTION

Heterogeneous photocatalysis on semiconductor particulate systems has become an active area of research in photochemistry such as water-purification [1], storage of solar energy [2, 3], environmental purification [3, 4], and so on. However, the use of semiconductors such as TiO_2 in organic chemistry has generally received much less attention [5-9]. Particularly, the reported applications into organic chemistry have mainly concentrated

* Author to whom correspondence should be addressed; Fax: +81-78-304-2794. E-Mail: miyabe@huhs.ac.jp.

on oxidation reactions; thus, less known about reductive organic reactions. TiO_2 has the potential to induce the reductive chemical transformation, especially by choosing the suitable sacrificial hole scavenger such as alcohols or amines. Since the photocatalytic hydrogenation of ethene and ethyne on TiO_2 was reported in 1975 [10], the reductive photocatalysis of alkenes and alkynes was well studied [11-21]. In this review, the recent progress in synthetically useful TiO_2-catalyzed reductive transformation of organic compounds is summarized. Particularly, we summarized the reduction of carbonyl compounds such as aldehydes or ketones, the reduction of nitro compounds, the reduction of imines and same chemical transformations involving redox processes.

2. PHOTOCATALYSIS ON *TITANIUM DIOXIDE*

The photocatalysis on TiO_2 is defined as a light-driven redox reaction at a solid/liquid or a solid/gas interface. Electrons (e^-) generated in conduction band (CB) and holes (h^+) simultaneously generated in valence band (VB) induce redox reactions (Figure 1). In general, the photocatalytic reduction of an electron acceptor (**A**) can be carried out in the presence of a large excess amount of electron donor (**D**) such as alcohols or amines. The aim using the electron donor (**D**) is to scavenge holes (h^+) generated in VB. Additionally, oxygen (O_2) acts as a competitive electron acceptor; thus, the reductive chemical transformations are generally performed in the absence of O_2. Under these conditions, electrons (e^-) in CB or those trapped at surface defect sites effectively transfer into an organic substrate adsorbed on TiO_2.

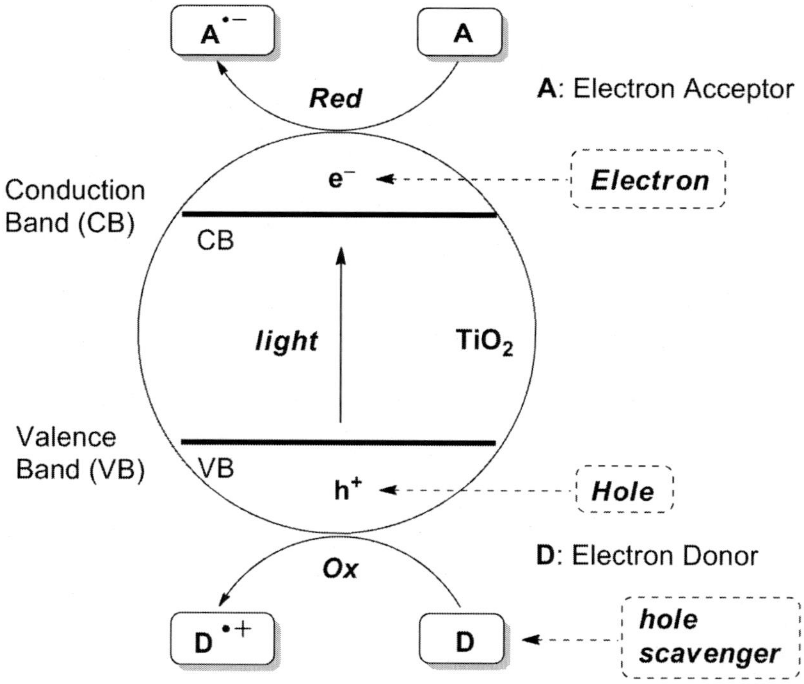

Figure 1. Mechanistic principle of photocatalysis on TiO_2.

3. Reduction of Carbonyl Compounds

The first report on photocatalytic reduction of carbonyl compound is, to our knowledge, the hydrogenation of pyruvate reported by Cuendet and Grätzel [22]. In the absence of oxygen, pyruvate was converted to lactate under irradiation of aqueous suspension of TiO_2. Other keto carboxylic acids were also reduced under the similar conditions. They reported that carbonyl groups adjacent to carboxyl functionality can be catalytically photo-reduced at the TiO_2 surface.

The photocatalytic reduction of aldehydes was studied by Li and co-workers [23]. In the absence of oxygen, aromatic aldehydes were effectively reduced to the corresponding alcohols using TiO_2 as a photocatalyst. The photo-induced reduction was conducted in alcohol media such as ethanol, methanol, and 2-propanol. The reduction was most efficient when ethanol was employed as the solvent. In these transformations, alcohol solvent acts as hole scavenger and was oxidized to aldehydes or ketones on TiO_2. In the case of benzaldehyde **1**, benzyl alcohol **2** was obtained in ca. 80% yield (Scheme 1). The reaction of benzaldehyde **1** in *O*-deuterated ethanol (CH_3CH_2OD) gave a product labeled with two deuterium atoms; thus, the reaction proceeds through two stages of electron-transfer from CB of TiO_2 and protonation (Scheme 2). This reduction of aldehydes was recently applied to a micro-reaction system by Matsushita and co-workers [24].

TiO_2-induced reduction of aldehydes **3** and **5** was studied by Kohtani, Miyabe and co-workers (Scheme 3) [25]. The reaction carried out in deaerated ethanol under UV irradiation. Although aldehydes **3** and **5** have shown the good reactivity and the reaction proceeds within 1 hour, the yields of products **4** and **6** were 76% and 33% yields, respectively. In the case of **5**, the competitive formation of acetal was observed in ethanol, leading to erosion of chemical efficiency.

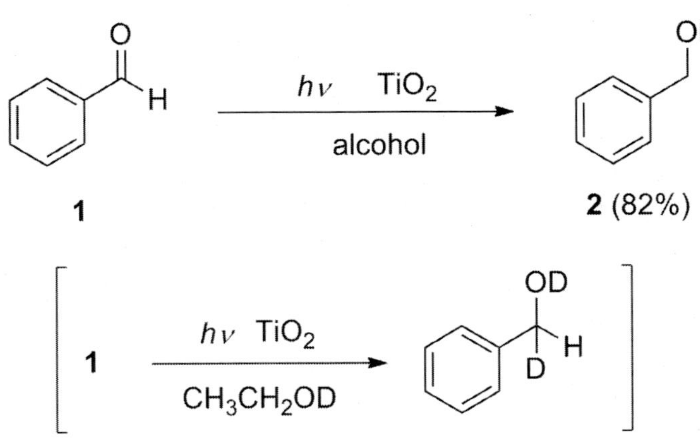

Scheme 1. TiO_2-catalyzed reduction of benzaldehyde 1.

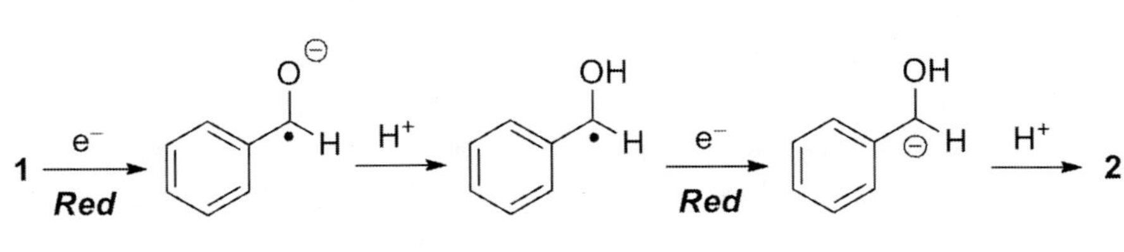

Scheme 2. Reaction pathway.

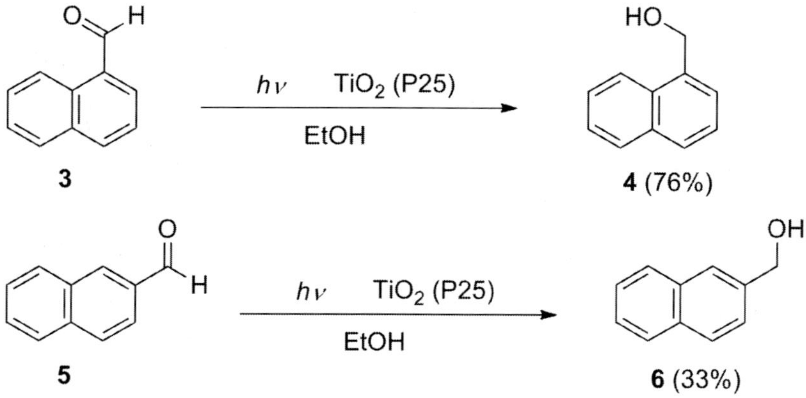

Scheme 3. TiO_2-catalyzed reduction of aldehydes 3 and 5.

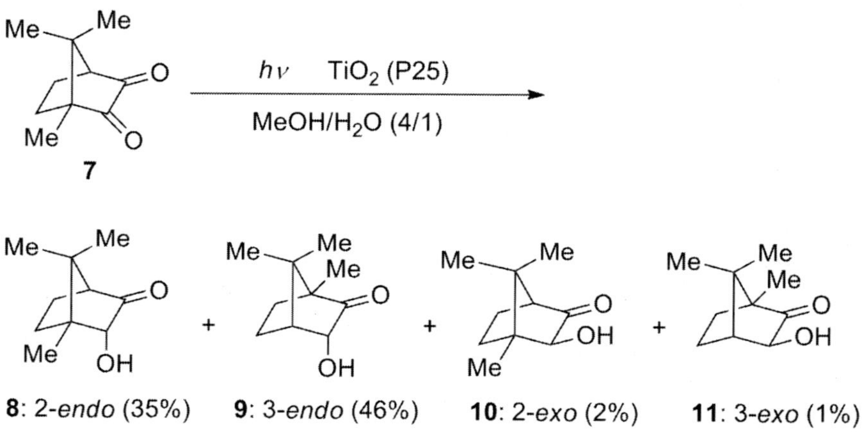

Scheme 4. Reduction of camphorquinone 7.

More recently, a molecular-level investigation in the reduction of furfuraldehyde on TiO_2 supported platinum nanoparticles was reported by Baker, Somorjai and co-workers [26]. In this study, sum frequency generation (SFG) vibrational spectroscopy shows that a furfuryl-oxy intermediate forms on TiO_2 and is the selective precursor to furfuryl alcohol.

The photocatalytic reduction of 1,2-diketones **7** and **12** was reported by Park and co-workers [27]. TiO_2-catalyzed reaction of camphorquinone **7** gave the 2-*endo*-, 3-*endo*-, 2-*exo*- and 3-*exo*-hydroxycamphors **8-11** (Scheme 4). *Endo*-products **8** and **9** were formed much more favorably than *exo*-products **10** and **11**, but there is little selectivity between 2 and 3 positions. When methanol/H_2O (4:1, v/v) was employed as the solvent, *endo*-products **8** and **9** were obtained in 35% and 49% yields, respectively.

Next, the reaction of 1-phenyl-1,2-propanedione **12** was investigated (Scheme 5). In the absence of TiO_2, the photo-irradiation of **12** gave the complex mixture including benzoic acid **15** as a major product, due to photocleavage. In contrast, the formation of desired α-hydroxyketones **13** and **14** was observed as the main pathway when TiO_2 was employed as a photocatalyst. The best yield of **13** and **14** was obtained by using methanol solvent containing 0.5 M triethylamine (TEA) as a sacrificial electron donor. The high preference for 1-hydroxy-1-phenyl-2-propanone **13** over 2-hydroxy-1-phenyl-1-propanone **14** is also noteworthy.

The photochemical reduction of benzil **16** to benzoin **17** was studied by Park and co-workers (Scheme 6) [28]. Although the photolysis of **16** was less effective in the absence of TEA and TiO_2, the addition of TEA and TiO_2 promoted the reduction. In the reaction medium of MeCN/MeOH/H_2O/TEA (88/7/2/3), the yield of **17** was as high as 85% at 50% conversion of **16**. However, the competitive over-reduction of benzoin **17** into hydrobenzoin **18** also proceeded. Indeed, benzoin **17** was reduced to **18** in 45% yield under similar photochemical conditions.

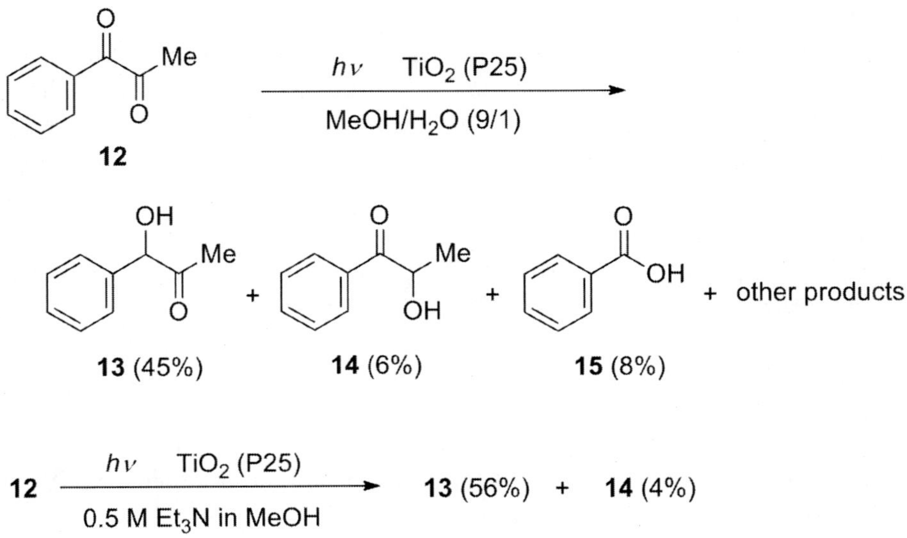

Scheme 5. Reduction of 1-phenyl-1,2-propanedione 12.

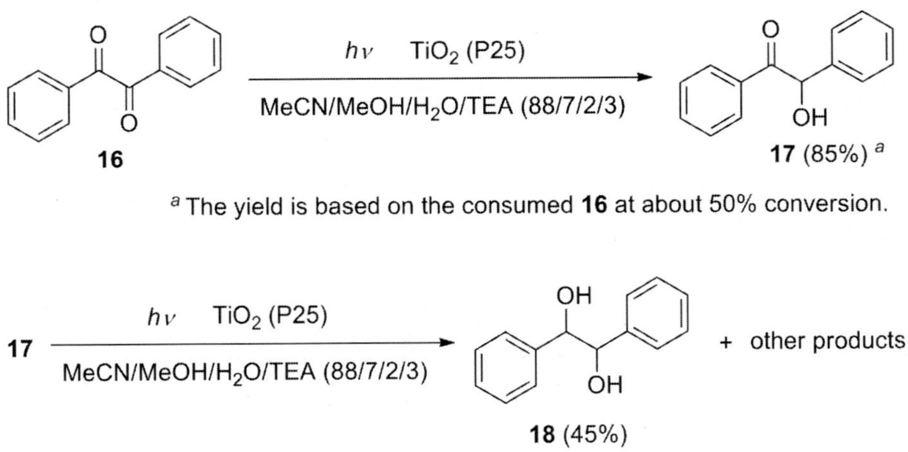

Scheme 6. Reduction of benzil 16.

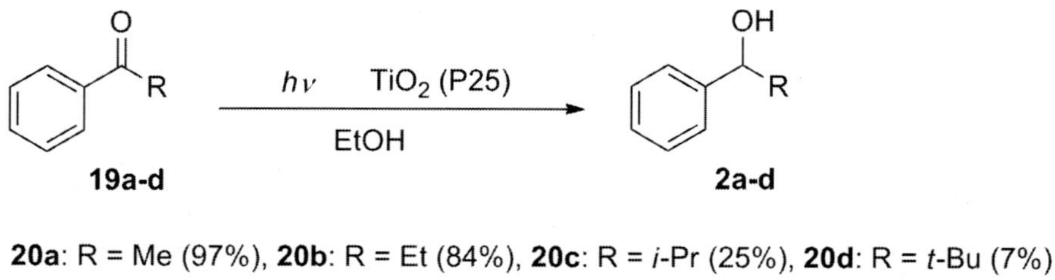

Scheme 7. TiO_2-catalyzed reducrion of acetophenone derivatives 19a-d.

The efficient photocatalytic reduction of aromatic ketones under UV light irradiation was reported by Kohtani, Miyabe and co-workers [25]. The TiO_2-induced reduction of acetophenone **19a** in deaerated ethanol was completed within 4 h to give alcohol **20a** in 97% yield (Scheme 7). Although the reduction of **19b** also proceeded effectively, the reaction of sterically hindered ketones **19c** and **19d** resulted in a decrease of chemical yields. The decrease of yields is due to negative reduction potentials of **19c** and **19d** compared to those of **19a** and **19b** [25]. Additionally, the reduction of aliphatic ketones did not occur under these conditions, because the reduction potential of aliphatic ketones is too negative.

The reaction of acetophenone derivatives **19e-k** having the substituted aromatic ring was investigated (Scheme 8). Highly efficient conversions of **19f-k** into alcohols **20f-k** were observed, expect for bulky substrate **19e** having 2-methyl group on aromatic ring. The adsorption of **19e** on TiO_2 would be suppressed by steric effect. The sterically hindered benzophenone **19l** has shown the good reactivity comparable to acetophenone derivatives. Hydrogenation of cyclic ketones **19m**, **19n** and **19o** was also facile.

Adsorptive and kinetic properties for photocatalytic reduction of acetophenone **19a** have been investigated by Kohtani, Miyabe and co-workers [29]. The schematic models to explain the experimental data are proposed (Figure 2). Interestingly, ca. 75% of electrons accumulated at CB and/or surface defect Ti sites (Ti_{st}^{4+} sites) took part in the reduction of **19a**. The residual 25% electrons could not react with **19a** and remained at deeper Ti_{st}^{4+} sites.

Recently, Kohtani, Miyabe and co-workers have shown that the photocatalytic reduction of acetophenone derivatives proceeds *via* the surface defect Ti (Ti_{st}) sites on the TiO_2 surface where is not only the adsorption sites but also electron trap sites [30]. A reasonable reaction mechanism is proposed (Figure 3). The first electron transfer to the adsorbed acetophenone derivatives is strongly affected by the relative position of E_{red}. In this step, the trapped electron seems to be transferred from Ti3d orbital to π* orbital on the carbonyl moiety. The second electron transfer should be faster than the first one because a predicted reduction potential of acetophenone ketyl radical, –1.59 V vs. SHE in CH_3CN calculated by a quantum chemical calculation [31], is ca. +0.3 V more positive than that of acetophenone **19a** (–1.89 V).

The reaction rate for 2,2,2-trifluoroacetophenone **19p** having an electro-withdrawing trifluoromethyl group was much slower than that for acetophenone **19a**, though E_{red} for **19p** is sufficiently positive compared to that for **19a** (Scheme 9) [29]. The amount of adsorption of **19p** on TiO_2 was too small compared to **19a**. This is due to the formation of hemiketal and ketal from **19p** in ethanol. The hemiketal and ketal species cannot interact with the surface Ti^{4+} sites on TiO_2 because of the lack of the lone pair on the carbonyl oxygen atom.

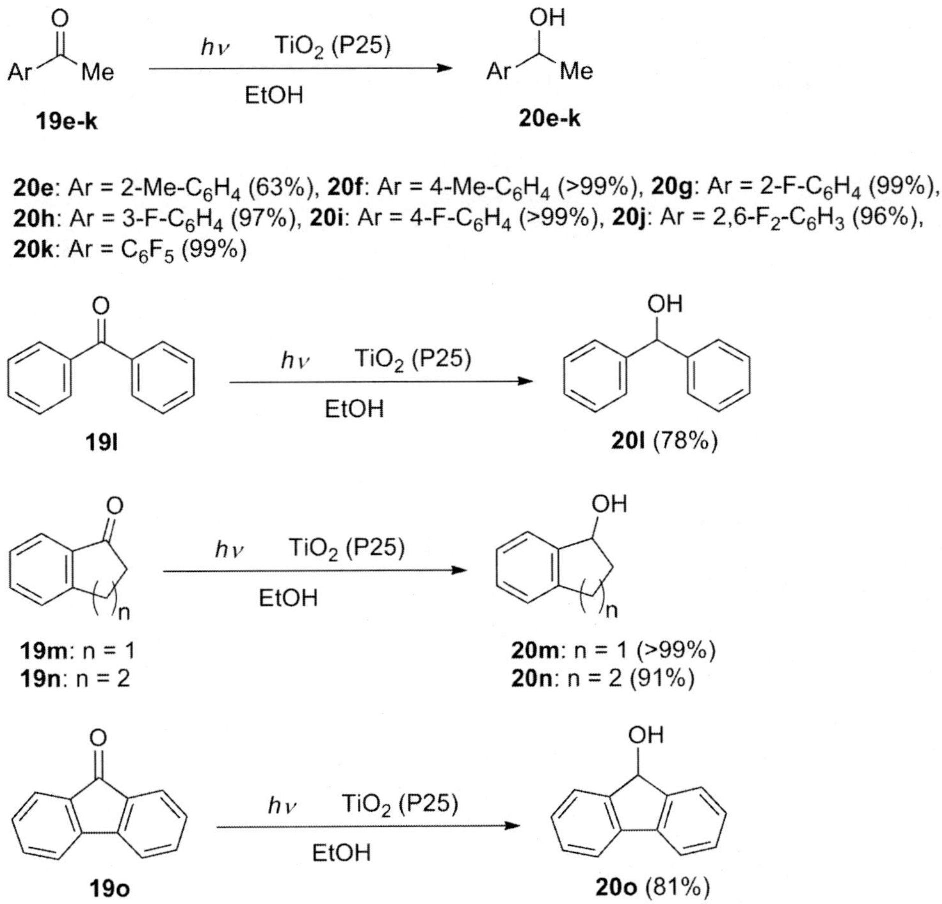

Scheme 8. Reduction of ketones 19e-o.

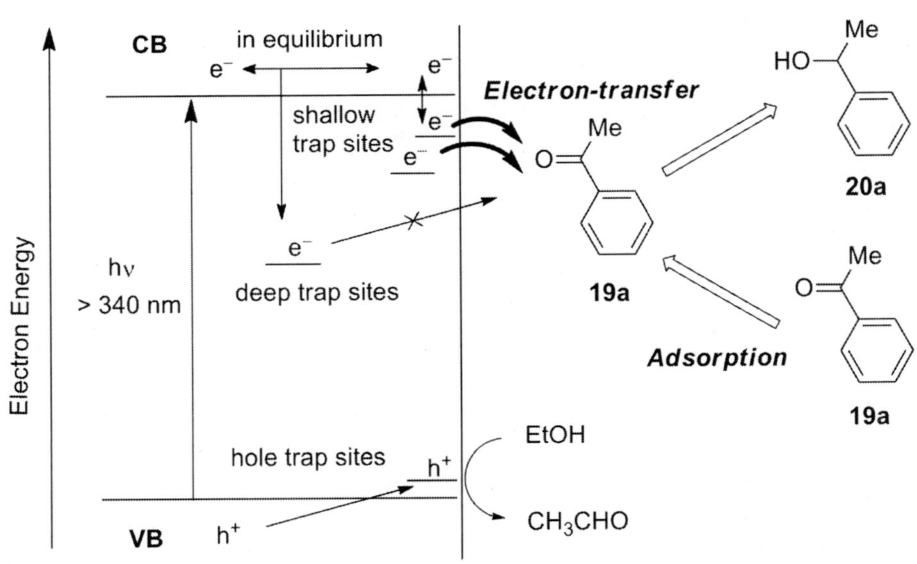

Figure 2. The earlier stage of reduction.

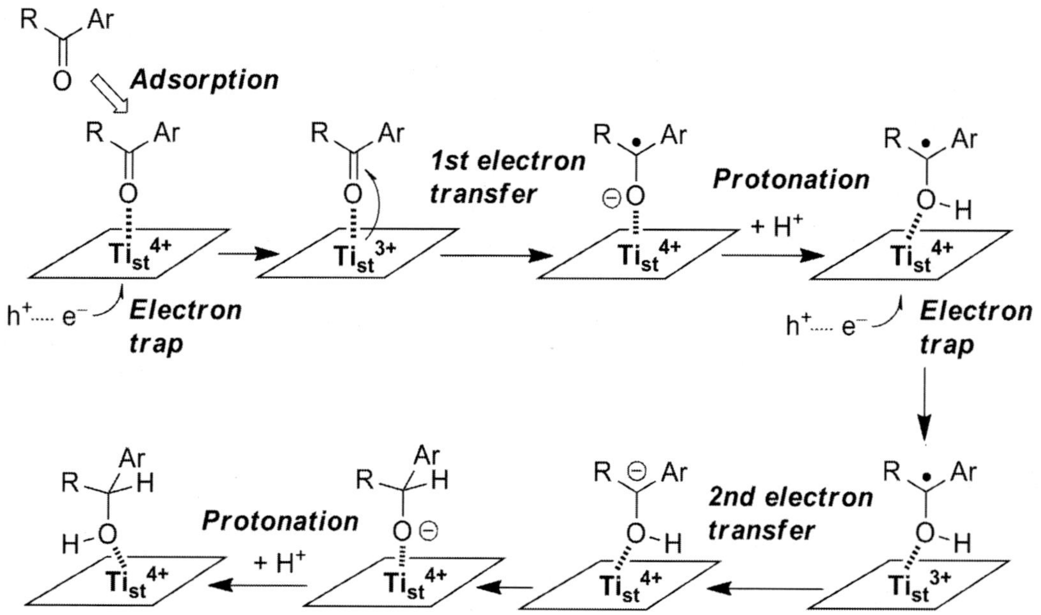

Figure 3. Proposed reaction mechanism of reduction on TiO_2 surface.

Dye-sensitization of TiO_2 is receiving increased attention to extend UV light response of TiO_2 toward visible light region. Kohtani, Miyabe and co-workers reported that acetophenone derivatives were reduced on P25 TiO_2 powder modified with metal-free organic dyes under visible light irradiation [32]. In their study, fluorescein (Fl) was employed as a dye (Table 1). In the presence of TEA as a sacrificial electron donor, the use of Fl-TiO_2 successfully extended the photocatalytic UV response of TiO_2 toward visible light region. The reaction of **19a** proceeded almost quantitatively to give **20a** in 59% conversion and 56% yield after 24 h irradiation (entry 1).

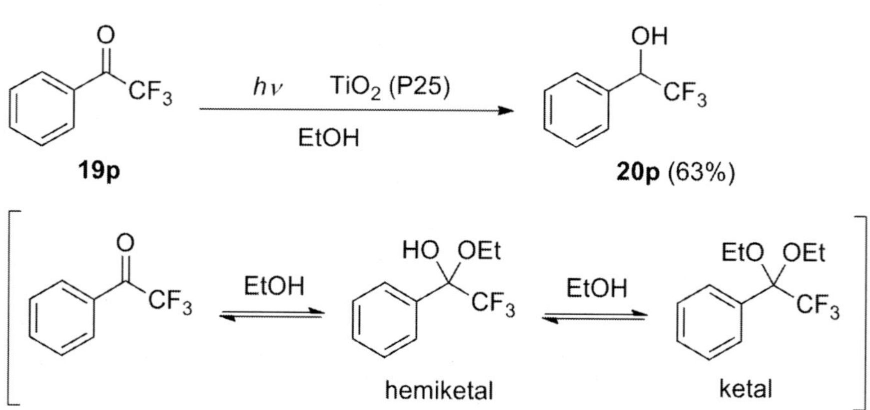

Scheme 9. TiO_2-catalyzed reduction of 2,2,2-trifluoroacetophenone 19p.

The reaction of **19a** completed after 96 h to give **20a** in 99% yield (entry 2). Conversions and yields of three substrates **19g**, **19h** and **19i** after 24 h irradiation are shown in entries 3-5. Schematic illustration of fluorescein-sensitized photo-reduction on TiO_2 is shown as Figure 4.

Table 1. Hydrogenation using fluorescein-TiO_2

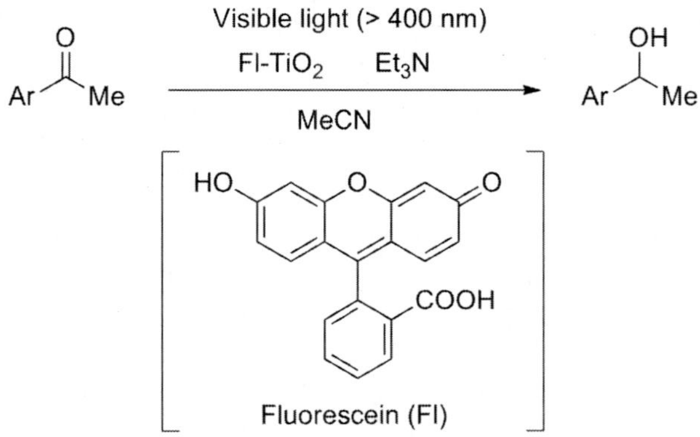

Entry	Ketone	Time (h)	Conversion (%)	Yield (%)[a]
1	**19a**: Ar = Ph	24	59	56
2	**19a**: Ar = Ph	96	100	99
3	**19g**: Ar = 2-F-C_6H_4	24	78	74
4	**19h**: Ar = 3-F-C_6H_4	24	69	69
5	**19i**: Ar = 4-F-C_6H_4	24	52	48

[a] Determined by GC analysis.

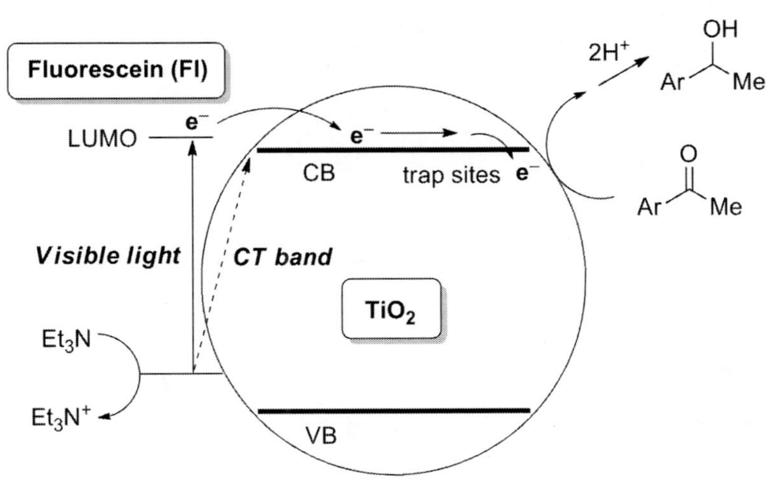

Figure 4. Fluorescein-sensitized reaction on TiO_2.

4. Reduction Of Nitro Compounds

The photocatalytic reduction of nitro compounds on TiO_2 was widely studied [33-47]. TiO_2-induced reduction of nitro compounds was first studied by Li and co-workers (Scheme 10) [33]. The reduction of 6-nitrocoumarin **21** gave 6-aminocoumarin **22** in 79% yield under the irradiation of a suspension of TiO_2 in ethanol containing **21**. This simple method was applicable to the reduction of a wide variety of nitro compounds.

The photocatalytic reduction of *p*-nitroacetophenone **23** was studied at a shorter time interval (Scheme 11). After 2 min irradiation, *p*-nitroacetophenone **23**, *p*-acetophenylhydroxylamine **24** and *p*-acetoaniline **25** were obtained in 40%, 28% and 31% yields, respectively. In contrast, *p*-acetoaniline **25** was obtained in 99% yield after 15 min irradiation. These results indicate that the reduction of nitro compound **23** proceeded *via* hydroxylamine **24**. The reaction mechanism is shown in Scheme 11. The effect of aliphatic alcohols as solvent was also investigated [34]. The rate of reduction increased with increasing polarity of solvent. In the TiO_2-catalyzed reduction of 4-nitrophenol into 4-aminophenol, the best yield of 92% was obtained in methanol suspensions. The role of TiO_2 and mechanism of reduction were well studied by Ferry and Glaze [35, 36]. The formation of by-product in reduction of nitrobenzene was also studied [37].

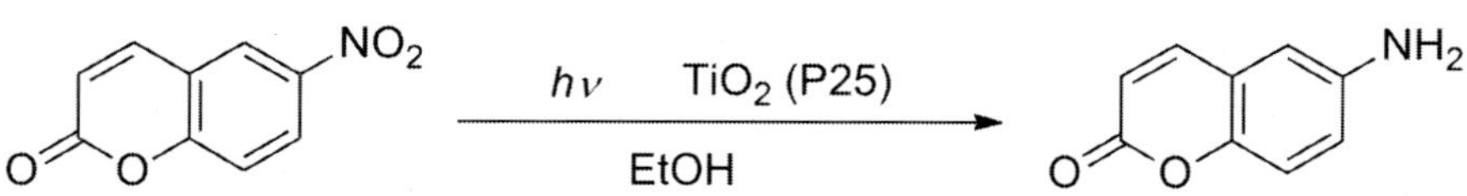

21 **22** (79%)

Scheme 10. TiO_2-catalyzed reduction of 6-nitrocoumarin 21.

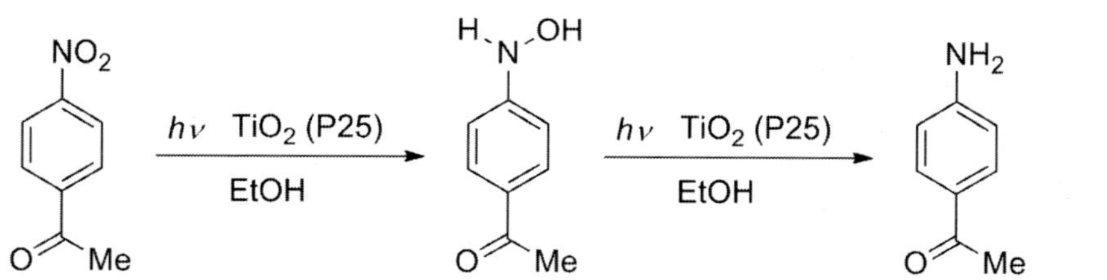

23 **24** **25**

after 2 min: **23** (40%) + **24** (31%) + **25** (16%)

after 15 min: **23** (0%) + **24** (0%) + **25** (99%)

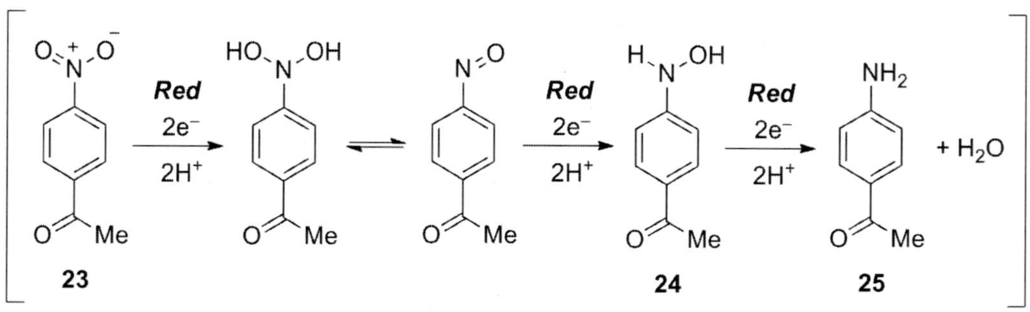

Scheme 11. TiO_2-catalyzed reduction of *p*-nitroacetophenone 23 and reaction mechanism.

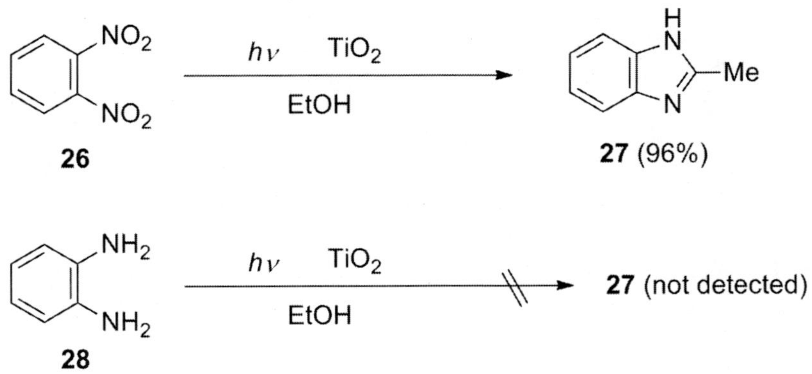

Scheme 12. Synthesis of benzimidazole 27 from o-dinitrobenzene 26.

Li and co-workers reported that benzimidazoles were prepared from an *o*-dinitrobenzene and alcoholic solvents under TiO_2-catalytic conditions [40]. For an example, 2-methylbenzimidazole **27** was obtained by TiO_2-mediated photocatalysis of dinitrobenzene **26** in ethanol (Scheme 12). To verify the reaction mechanism, the reaction of *o*-phenylenediamine **28** was investigated under similar reaction conditions. Irradiation of **28** in ethanol failed to produce 2-methylbenzimidazole **27**. From these results, the second nitro group would not be reduced in the pathway to 2-methylbenzimidazole **27** as illustrated in Scheme 13. At first, a nitro group of dinitrobenzene **26** is reduced to give 2-nitroaniline. At the same time, ethanol solvent is oxidized to aldehyde. The 2-methylbenzimidazole **27** would be formed *via* the formation of imine, the reduction of second nitro group into hydroxylamine and the ring formation.

Park and co-workers reported that tetrahydroquinoline **30** was prepared from nitroarene **29** and ethanol under TiO_2-catalyzed conditions (Scheme 14) [41]. 4-Ethoxy-1,2,3,4-tetrahydroquinoline **30** was obtained in 71% accompany with a small amount of *m*-methylaniline. Recently, the synthesis of quinolones from nitroarenes using iridium clusters supported on TiO_2 was reported by Cao and co-workers [42].

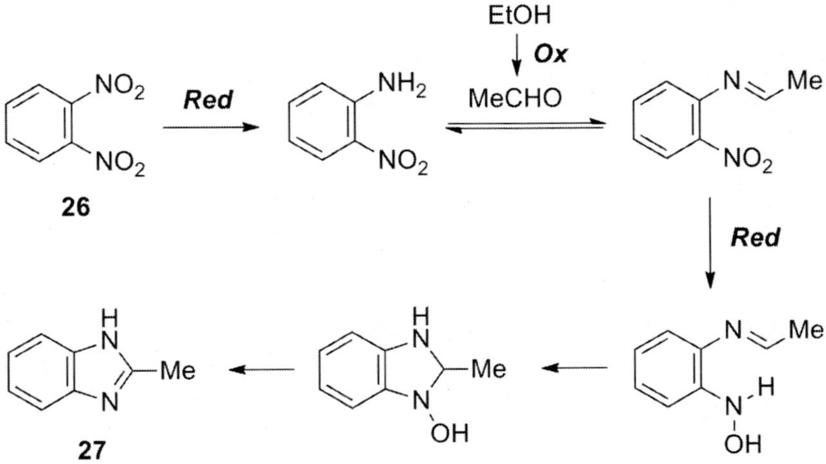

Scheme 13. Reaction pathway.

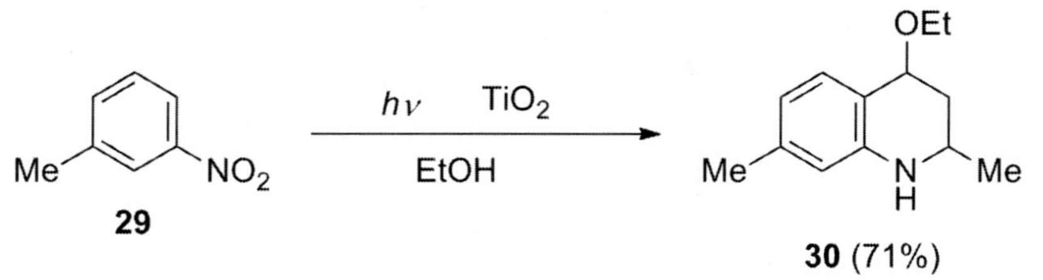

Scheme 14. Synthesis of tetrahydroquinoline 30.

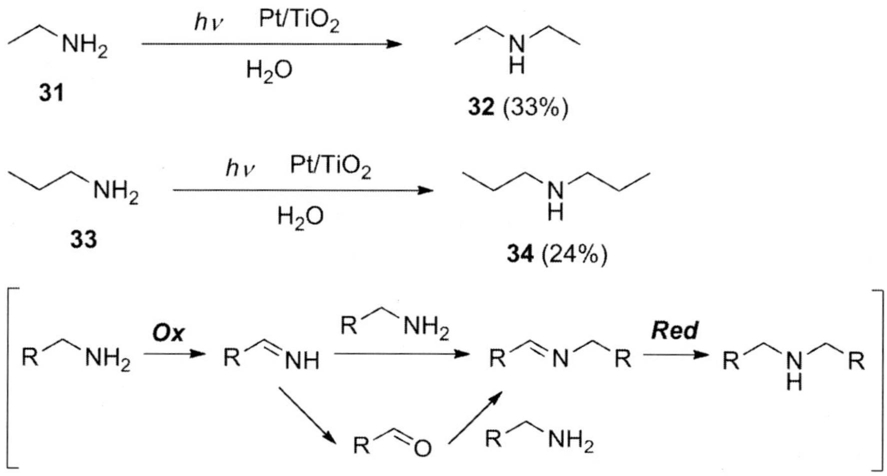

Scheme 15. Conversion of primary amines to symmetrical secondary amines.

The reduction of nitrobenzene is used as a model reaction for developing the modified TiO_2 catalysts [43-47]. Tada and co-workers reported that the loading of Ag nanoparticles on TiO_2 enhanced the activity of a photocatalyst [43, 44]. Zhang and co-workers studied the reactive N-doped TiO_2 catalyst prepared by a modified sol-gel method using urea as nitrogen source [45]. Aromatic nitro compounds were effectively reduced by using this N-doped TiO_2 and potassium iodide as photocatalysts in the presence of methanol. König and co-workers studied the TiO_2 catalyst, which was surface-functionalized by a Ru-complex, to absorb green light [46]. Nitrobenzene was cleanly reduced to aniline using Ru-sensitized TiO_2 under green LED light or sun light irradiation.

5. Reduction of Imines

Pt/TiO_2 catalyst is effective for the reduction of the C=N bond of imines. The conversion of primary amines to secondary amines investigated in aqueous media by Ohtani and co-workers [48]. The use of a mixture of TiO_2 and platinum black as a catalyst promoted the conversion of primary amines **31** and **33** to symmetrical secondary amines **32** and **34** (Scheme 15). In these transformations, a primary amine was firstly oxidized by holes to form an imine. This imine was hydrolyzed to give NH_3 and a corresponding aldehyde, which was condensed with primary amine to yield the imine intermediate. Finally, the imine intermediate was photocatalytically reduced to produce the symmetrical secondary amine.

This one-pot deamino-condensation was applied into the cyclization of diamines **35**, **37** and **39** (Scheme 16). Under similar aqueous conditions, the cyclic secondary amines **36**, **38** and **40** were obtained in reasonable yields. The cyclization of L-lysine to pipecolinic acid was examined using TiO_2 or CdS [49, 50].

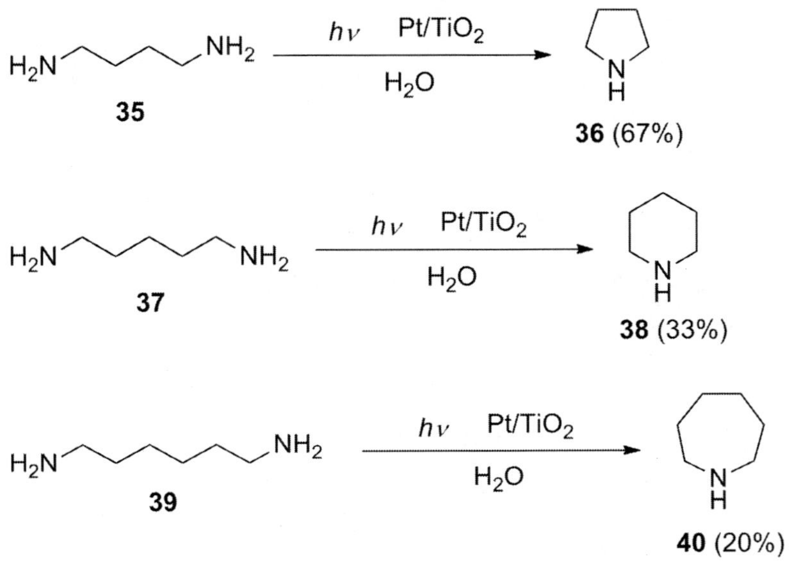

Scheme 16. Cyclization of diamines.

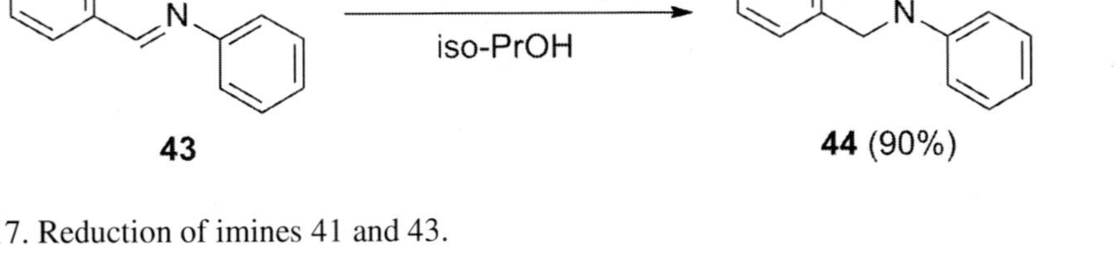

Scheme 17. Reduction of imines 41 and 43.

The direct reduction of imines such as *N*-benzylidenebenzylamine **41** and *N*-benzylideneaniline **43** to the corresponding secondary amines was also investigated (Scheme 17) [51].

Recently, Shiraishi and co-workers reported the photocatalytic condensation of amines with alcohols *via* the reduction of imines [52]. TiO_2-catalyst loaded with Pd nanoparticles promoted the condensation of aniline **46** with benzylalcohol **45** (Scheme 18). In this reaction, the irradiation of Pd/TiO_2 produced H_2 and benzaldehyde which reacted with aniline **46** to give imine **43**. The rate-determining step of this transformation is the reduction of imine **43**.

Under these photocatalytic conditions, *N*-monoalkylation of amines with alcohols took place effectively (Scheme 19).

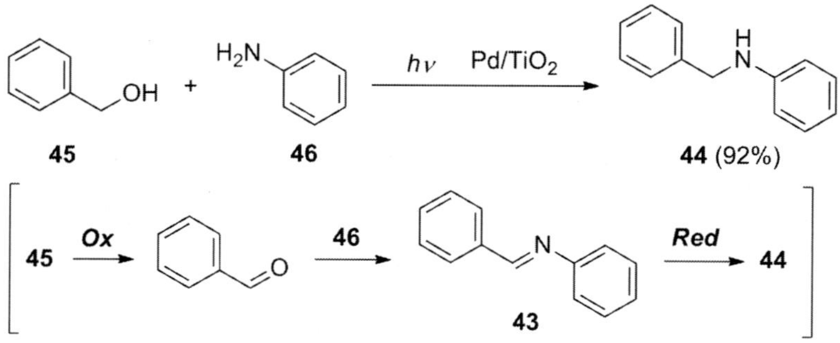

Scheme 18. Condensation of aniline 46 with benzylalcohol 45 *via* imine 43.

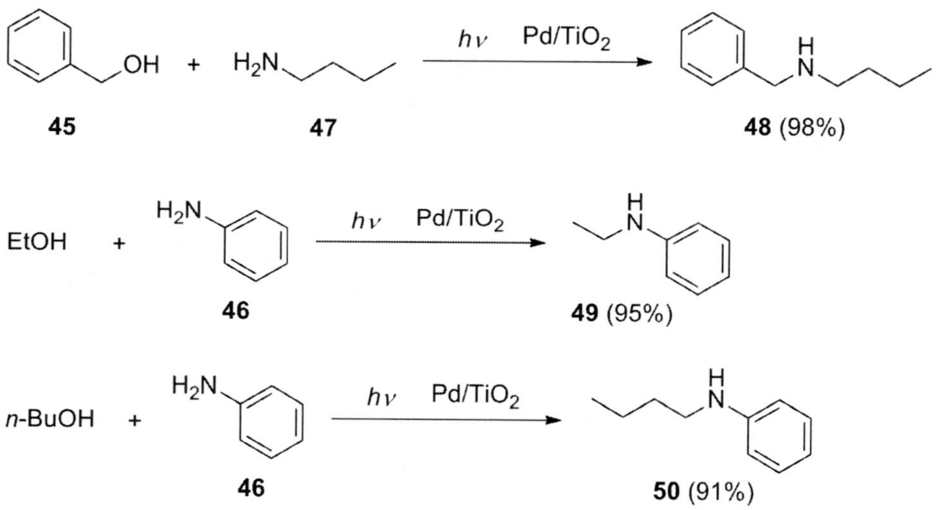

Scheme 19. *N*-Monoalkylation of amines with alcohols.

6. Reduction of Cyanides

Shiraishi and co-workers reported the photocatalytic reduction of aromatic cyanides on TiO_2 loaded with Pd nanoparticles in the presence of ethanol [53]. Pd/TiO_2-catalyzed reduction of benzonitrile **51** gave the toluene **52** and triethylamine in high chemical yields (Scheme 20).

To understand the reaction pathway, the reduction of various substrates was investigated under similar photocatalytic conditions (Scheme 21). Interestingly, the photocatalytic reactions of three substrates **53**, **54** and **55** produced toluene **52** and triethylamine with almost quantitative yields. Additionally, trace amounts of **54** and **55** were detected by GC analysis during the photoreduction of benzonitrile **51**. Therefore, they proposed the reaction pathway involving the condensations of amines with aldehyde, generated by the oxidation of ethanol, and the reductions of imine intermediates (Scheme 22).

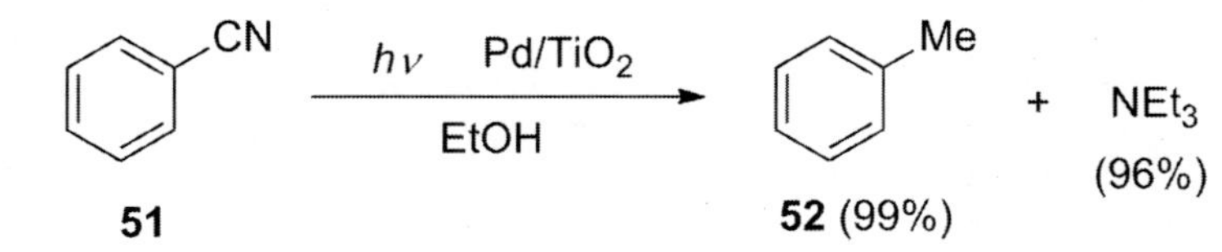

Scheme 20. Reduction of benzonitrile 51.

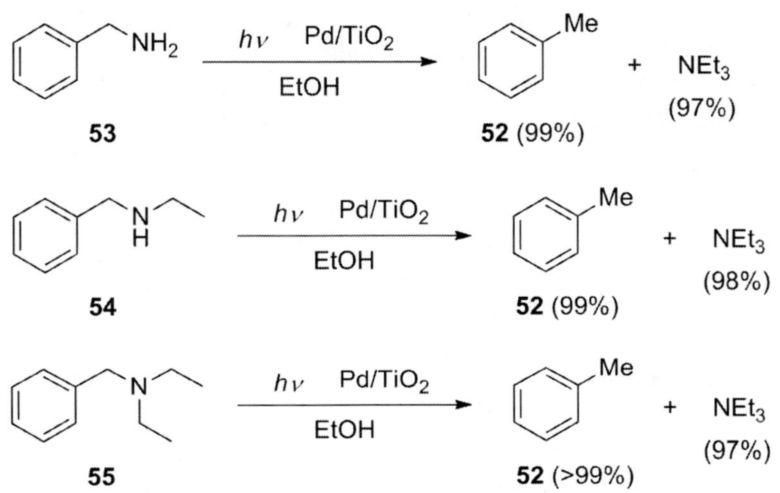

Scheme 21. Reduction of various substrates.

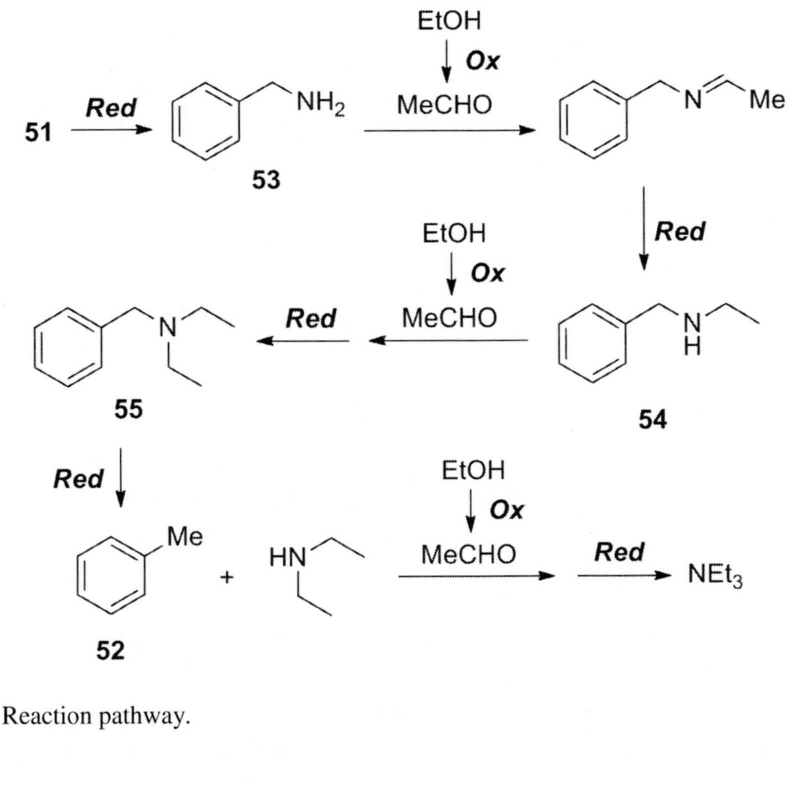

Scheme 22. Reaction pathway.

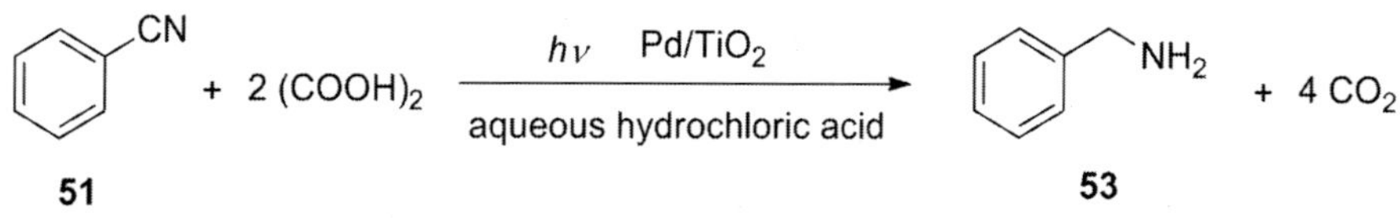

Scheme 23. Reduction of benzonitrile 51 to benzylamine 53.

Recently, the reduction of benzonitrile **51** to benzylamine **53** was achieved by Kominami and co-workers [54]. In the presence of oxalic acid as a hole scavenger, the reduction of benzonitrile **51** in acidic aqueous suspensions of Pd-loaded TiO_2 gave **53** with reasonable chemical efficiency (Scheme 23). The combination of oxalic acid and acidic water is important for this reductive transformation.

CONCLUSION

The photocatalysis on semiconductor particles has become an active area of research in organic chemistry as well as photochemistry. Although the TiO_2-mediated photocatalysis has mainly developed in oxidative chemical transformations, the use of TiO_2 can induce the reductive chemical transformation by choosing the suitable sacrificial hole scavenger as mentioned in this chapter. One of the most significant features of TiO_2-mediated photocatalysis is that we can utilize both oxidation and reduction processes; thus, the combination of redox reactions affords the unique chemical transformations.

ACKNOWLEDGMENT

The works performed by our group was partially supported by JSPS KAKENHI Grant-in-Aid for Scientific Research (C) Grant Number 24590067 (to S. Kohtani) and 25460028 (to H. Miyabe).

REFERENCES

[1] Fujishima, A., & Honda, K. (1972). Electrochemical Photolysis of Water at a Semiconductor Electrode. *Nature*, 238, 37–38.

[2] Kudo, A., & Miseki, Y. (2009). Heterogeneous Photocatalyst Materials for Water Splitting. *Chem. Soc. Rev.*, 38, 253–278.

[3] Fujishima, A., Zhang, X., & Tryk, D. A. (2007). Heterogeneous Photocatalysis: From Water Photolysis to Applications in Environmental Cleanup. *Int. J. Hydrogen. Energy*, 32, 2664–2672.

[4] Hoffmann, M. R., Martin, S. T., Choi, W., & Bahnemann D. W. (1995). Environmental Applications of Semiconductor Photocatalysis. *Chem. Rev.*, 95, 69–96.

[5] Fox, M. A. (1987). Selective Formation of Organic Compounds by Photoelectrosynthesis at Semiconductor Particles. *Top. Curr. Chem.*, 142, 72–99.

[6] Kisch, H. (2001). Semiconductor Photocatalysis for Organic Synthesis. *Adv. Photochem.*, 26, 93–143.

[7] Palmisano, G., Augugliaro, V., Pagliaro, M., & Palmisano, L. (2007). Photocatalysis: A Promising Route for 21st Centyry Organic Chemistry. *Chem. Commun.*, 3425–3437.

[8] Shiraishi, Y., & Hirai, T. (2008). Selective Organic Transformation on Titanium Oxide-based Photocatalysts. *J. Photochem. Photobiol. C: Photochem. Rev.*, 9, 157–170.

[9] Palmisano, G., García-López, E., Marcí, G., Toddo, V., Yurdakal, S., Augugliaro, V., & Palmisano, L. (2010). Advances in Selective Conversion by Heterogeneous Photocatalysis. *Chem. Commun.*, 46, 7074–7089.

[10] Boonstra, A. H., & Mutsaers, C. A. H. A. (1975). Photohydrogenation of Ethyne and Ethene on the Surface of Titanium Dioxide. *J. Phys. Chem.*, 79, 2025–2027.

[11] Yun, C., Anpo, M., Kodama, S., & Kubokawa, Y. (1980). U.V. Irradiation-induced Fission of a C=C or C≡C Bond Adsorbed on TiO_2. *J. Chem. Soc., Chem. Commun.*, 609.

[12] Anpo, M., Aikawa, N., Kodama, S., & Kubokawa, Y. (1984). Photocatalytic Hydrogenation of Alkynes and Alkenes with Water over TiO_2: Hydrogenation Accompanied by Bond Fission. *J. Phys. Chem.*, 88, 2569–2572.

[13] Anpo, M., Aikawa, N., & Kubokawa, Y. (1984). Photocatalytic Hydrogenation of Alkynes and Alkenes with Water over TiO_2: Pt-Loading Effect on the Primary Processes. *J. Phys. Chem.*, 88, 3998–4000.

[14] Yamataka, H., Seto, N., Ichihara, J., Hanafusa, T., & Teratani S. (1985). Reduction of C-C Multiple Bonds Using an Illuminated Semiconductor Catalyst. *J. Chem. Soc., Chem. Commun.*, 788–789.

[15] Kodama, S., Nakaya, H., Anpo, M., & Kubokawa, Y. (1985). A Common Factor Determining the Features of the Photocatalytic Hydrogenation and Isomerization of Alkenes over Ti-Si Oxides. *Bull. Chem. Soc. Jpn.*, 58, 3645–3646.

[16] Anpo, M., Nakaya, H., Kodama, S., Kubokawa, Y., Domen, K., & Onishi, T. (1986). Photocatalysis over Binary Metal Oxides: Enhancement of the Photocatalytic Activity of TiO_2 in Titanium-silicon Oxides. *J. Phys. Chem.*, 90, 1633–1636.

[17] Baba, R., Nakabayashi, S., Fujishima, A., & Honda, K. (1987). Photocatalytic Hydrogenation of Ethylene on the Bimetal-deposited Semiconductor Powders. *J. Am. Chem. Soc.*, 109, 2273–2277.

[18] Anpo, M., Shima, T., Kodama, S., & Kubokawa, Y. (1987). Photocatalytic Hydrogenation of CH_3CCH with H_2O on Small-particle TiO_2: Size Quantization Effects and Reaction Intermediates. *J. Phys. Chem.*, 91, 4305–4310.

[19] Lin, L., & Kuntz, R. R. (1992). Photocatalytic Hydrogenation of Acetylene by Molybdenium-sulfur Complex Supported on TiO_2. *Langmuir*, 8, 870–875.

[20] Kuntz, R. R. (1997). Comparative Study of $Mo_2O_xS_y(cys)_2^{2-}$ Complexes as Catalysts for Electron Transfer from Irradiated Colloidal TiO_2 to Acetylene. *Langmuir*, 13, 1571–1576.

[21] Yamashita, H., Kawasaki, S., Ichihashi, Y., Takeuchi, M., Harada, M., Anpo, M., Louis, C., & Che, M. (1998). Characterization of Ti/Si Binary Oxides Prepared by the Sol-gel Method and their Photocatalytic Properties: The Hydrogenation and Hydrogenolysis of CH_3CCH with H_2O. *Korean J. Chem. Eng.*, 15, 491–495.

[22] Cuendet, P., & Grätzel, M. (1987). Direct Photoconversion of Pyruvate to Lactate in Aqueous TiO_2 Dispersions. *J. Phys. Chem.*, 91, 654–657.

[23] Joyce-Pruden, C., Pross, J. K., & Li, Y. (1992). Photoinduced Reduction of Aldehydes on Titanium Dioxide. *J. Org. Chem.*, 57, 5087–5091.

[24] Matsushita, Y., Kumada, S., Wakabayashi, K., Sakeda, K., & Ichimura, T. (2006). Photocatalytic Reduction in Microreactors. *Chem. Lett.*, 35, 410–411.

[25] Kohtani, S., Yoshioka, E., Saito, K., Kudo, A., & Miyabe, H. (2010). Photocatalytic Hydrogenation of Acetophenone Derivatives and Diaryl Ketnoes on Polycrystalline Titanium Dioxide. *Catal. Commun.*, 11, 1049–1053.

[26] Baker, L. R., Kennedy, G., Van Spronsen, M., Hervier, A., Cai, X., Chen, S., Wang, L.-W., & Somorjai, G. A. (2012). Furfuraldehyde Hydrogenation on Titanium Oxide-supported Platinum Nanoparticles Studied by Sum Frequency Generation Vibrational Spectroscopy: Acid-base Catalysis Explains the Molecular Origin of Strong Metal-support Interactions. *J. Am. Chem. Soc.*, 134, 14208–14216.

[27] Park, J. W., Hong, M. J., & Park, K. K. (2001). Photochemical Reduction of 1,2-Diketones in the Presence of TiO_2. *Bull. Korean Chem. Soc.*, 22, 1213–1216.

[28] Park, J. W., Kim, E. K., & Park, K. K. (2002). Photochemical Reduction of Benzil and Benzoin in the Presence of Triethylamine and TiO_2 Photocatalyst. *Bull. Korean Chem. Soc.*, 23, 1229–1234.

[29] Kohtani, S., Yoshioka, E., Saito, K., Kudo, A., & Miyabe, H. (2012). Adsorptive and Kinetic Properties on Photocatalytic Hydrogenation of Aromatic Ketones upon UV Irradiated Polycrystalline Titanium Dioxide: Differences between Acetophenone and its Trifluoromethylated Derivative. *J. Phy. Chem. C*, 116, 17705–17713.

[30] Kohtani, S., Kamoi, Y., Yoshioka, E., & Miyabe, H. (2014). Kinetic Study on Photocatalytic Hydrogenation of Acetophenone Derivatives on Titanium Dioxide. *Catal. Sci. Technol.*, 4, 1084–1091.

[31] Fu, Y., Liu, L., Yu, H.-Z., Wang, Y.-M., & Guo Q.-X. (2005). Quantum-chemical Predictions of Absolute Standard Redox Potentials of Diverse Organic Molecules and Free Radicals in Acetonitrile *J. Am. Chem. Soc.*, 127, 7227–7234.

[32] Kohtani, S., Nishioka, S. Yoshioka, E, & Miyabe, H. (2014). Dye-sensitized Photo-hydrogenation of Aromatic Ketones on Titanium Dioxide under Visible Light Irradiation. *Catal. Commun.*, 43, 61–65.

[33] Mahadavi, F., Brudon, T. C., & Li, Y. (1993). Photoinduced Reduction of Nitro Compounds on Semiconductor Particulates. *J. Org. Chem.*, 58, 744–746.

[34] Brezová, V., Blažková, A., Šurina, I., & Havlínová, B. (1997). Solvent Effect on the Photocatalytic Reduction of 4-Nitrophenol in Titanium Dioxide Suspensions. *J. Photochem. Photobiol. A: Chem.*, 107, 233–237.

[35] Ferry, J. L., & Glaze, W. H. (1998). Photocatalytic Reduction of Nitroorganics over Illuminated Titanium Dioxide: Electron Transfer between Excited-state TiO_2 and Nitroaromatics. *J. Phys. Chem. B*, 102, 2239–2244.

[36] Ferry, J. L., & Glaze, W. H. (1998). Photocatalytic Reduction of Nitro Organics over Illuminated Titanium Dioxide: Role of the TiO_2 Surface. *Langmuir*, 14, 3551–3555.

[37] Flores, S. O., Rios-Bernij, O., Valenzuela, M. A., Córdova, I., Gómez, R., & Gutiérrez, R. (2007). Photocatalytic Reduction of Nitrobenzene over Titanium Dioxide: By-product Identification and Possible Pathways. *Top. Catal.*, 44, 507–511.

[38] Kominami, H, Iwasaki, S., Maeda, T., Imamura, K., Hashimoto, K., Kera, Y., & Ohtani, B. (2009). Photocatalytic Reduction of Nitrobenzene to Aniline in an Aqueous

Suspension of Titanium(IV) Oxide Particles in the Presence of Oxalic Acid as a Hole Scavenger and Promotive Effect of Dioxygen in the System. *Chem. Lett.*, 38, 410–411.

[39] Shiraishi, Y., Togawa, Y., Tsukamoto, D., Tanaka, S., & Hirai, T. (2012). Highly Efficient and Selective Hydrogenation of Nitroaromatics on Photoactivated Rutile Titanium Dioxide. *ACS Catal.*, 2, 2475–2481.

[40] Wang, H., Partch, R. E., & Li, Y. (1997). Synthesis of 2-Alkylbenzimidazoles via TiO_2-mediated Photocatalysis. *J. Org. Chem.*, 62, 5222–5225.

[41] Park, K. H., Joo, H. S., Ahn, K. I, & Jun, K. (1995). One Step Synthesis of 4-Ethoxy-1,2,3,4-tetrahydroquinoline from Nitroarene and Ethanol: A TiO_2 Mediated Photocatalytic Reaction. *Tetrahedron Lett.*, 36, 5943–5946.

[42] He, L., Wang, J.-Q., Gong, Y., Liu, Y.-M., Cao, Y., He, H.-Y., & Fan, K.-N. (2011). Titania-supported Iridium Subnanoclusters as an Efficient Heterogeneous Catalyst for Direct Synthesis of Quinolines from Nitroarenes and Aliphatic Alcohols. *Angew. Chem. Int. Ed.*, 50, 10216–10220.

[43] Tada, H., Ishida, T., Takao, A., & Ito, S. (2004). Drastic Enhancement of TiO_2-photocatalyzed Reduction of Nitrobenzene by Loading Ag Clusters. *Langmuir*, 20, 7898–7900.

[44] Tada, H., Ishida, T., Takao, A., Ito, S., Mukhopadhyay, S., Akita, T., Tanaka, K., & Kobayashi, H. (2005). Kinetic and DFT Studies on the Ag/TiO_2-photocatalyzed Selective Reduction of Nitrobenzene to Aniline. *ChemPhysChem*, 6, 1537–1543.

[45] Wang, H., Yan, J., Chang, W., & Zhang, Z. (2009). Practical Synthesis of Aromatic Amines by Photocatalytic Reduction of Aromatic Nitro Compounds on Nanoparticles N-doped TiO_2. *Catal. Commun.*, 10, 989–994.

[46] Füldner, S., Mild, R., Siegmund, H. I., Schroeder, J. A., Gruber, M., & König, B. (2010). Green-light Photocatalytic Reduction Using Dye-sensitized TiO_2 and Transition Metal Nanoparticles. *Green Chem.*, 12, 400–406.

[47] Tanaka, A., Nishino, Y., Sakaguchi, S., Yoshikawa, T., Imamura, K., Hashimoto, K., & Kominami, H. (2013). Functionalization of a Plasmonic Au/TiO_2 Photocatalyst with an Ag Co-catalyst for Quantitative Reduction of Nitrobenzene to Aniline in 2-Propanol Suspensions under Irradiation of Visible Light. *Chem. Commun.*, 49, 2551–2553.

[48] Nishimoto, S., Ohtani, B., Yoshikawa, T., & Kagiya, T. (1983). Photocatalytic Conversion of Primary Amines to Secondary Amines and Cyclization of Polymethylene-α,ω-diamines by an Aqueous Suspension of TiO_2/Pt. *J. Am. Chem. Soc.*, 105, 7180–7182.

[49] Pal, B., Ikeda, S., Kominami, H., Kera, Y., & Ohtani, B. (2003). Photocatalytic Redox-Combined Synthesis of L-Pipecolinic Acid from L-Lysine by Suspended Titania Particle: Effect of Noble Metal Loading on the Selectivity and Optical Purity of the Product. *J. Catal.*, 217, 152–159.

[50] Ohtani, B., Pal, B., & Ikeda, S. (2003). Photocatalytic Organic Synthesis: Selective Cyclization of Amino Acids in Aqueous Suspensions. *Catal. Surv. Asia*, 7, 165–176.

[51] Ohtani, B., Goto, Y., Nishimoto, S., & Inui, T. (1996). Photocatalytic Transfer Hydrogenation of Schiff Bases with Propan-2-ol by Suspended Semiconductor Particles Loaded with Platinum Deposits. *J. Chem. Soc. Faraday Trans.*, 92, 4291–4295.

[52] Shiraishi, Y., Fujiwara, K., Sugano, Y., Ichikawa, S., & Hirai, T. (2013). *N*-Monoalkylation of Amines with Alcohols by Tandem Photocatalytic and Catalytic Reactions on TiO_2 Loaded with Pd Nanoparticles. *ACS Catal.*, 3, 312–320.

[53] Sugano, Y., Fujiwara, K., Shiraishi, Y., Ichikawab, S., & Hiraia, T. (2013). Photocatalytic Hydrodenitrogenation of Aromatic Cyanides on TiO_2 Loaded with Pd Nanoparticles. *Catal. Sci. Technol.*, 3, 1718–1724.
[54] Imamura, K., Yoshikawa, T., Nakanishi, K., Hashimoto, K., & Kominami, H. (2013). Photocatalytic Reduction of Benzonitrile to Benzylamine in Aqueous Suspensions of Palladium-loaded Titanium(IV) Oxide. *Chem. Commun.*, 49, 10911–10913.

In: Titanium Dioxide
Editor: Jerri Brown
ISBN: 978-1-63321-391-3

Chapter 9

NANOSIZED TITANIA WITH ANATASE STRUCTURE: SYNTHESIS, CHARACTERIZATION, APPLICATIONS AND ENVIRONMENTAL EFFECTS

G. M. Kuz'micheva, E. N. Domoroshchina, E. V. Savinkina and L. N. Obolenskaya
Lomonosov State University of Fine Chemical Technology,
Moscow, Russia

ABSTRACT

Nanosized titanium dioxide modifications are used in catalysis (for extraction of uranium compounds from sea water and producing hydrogen), sorption (for purification of aqueous media from a wide range of toxic ions), medicine (excellent cleaner of water and air environments from organic pollutants and bacteria), microelectronics (promising for solar cells) and so on. In comparison with other catalysts the semiconductor TiO_2 is chemically and biologically inert and stable to photodegradation, and unlike other disinfection technologies do not form dangerous (mutagenic and carcinogenic) compounds. Among other modifications of titania the most promising for application is anatase.

The known methods to obtain samples with nano-anatase were reviewed; their advantages and disadvantages were noted. The sulfate technique was selected as the easiest to performance and perspective. The samples obtained by this technique were characterized by a complex of methods: X-ray diffraction using X-ray (wide and small-angle scattering) and synchrotron radiation (distributing character of nanoparticles on sizes and the sizes of nano-objects, the composition of the samples, the sizes of coherent scattering regions and the quantity of nano-anatase microstressing), electron diffraction, scanning and transmission electron microscopy (the size of crystallite, nanoparticles, agglomerates and aggregates), quantitative electron probe microanalysis and X-ray fluorescence electron spectroscopy (elemental composition and charge state of components), low-temperature nitrogen adsorption (specific surface area, size of nano-objects, the volume of nano- and ultrananoporos), infrared spectroscopy and Raman spectroscopy.

It was established that the conditions of samples synthesis effect their composition, microstructure and substructure, electronic and surface properties. The possibility of directed synthesis of nano-anatase with necessary physical-chemical properties has been shown.

It was determined that the photocatalytic activity of the obtained samples is comparable to that of the commercial photocatalysts based on titanium dioxide, however, they can be multiple reused in a photocatalytic decomposition reaction of organic dyes without regenerating.

The systematic work on modifying of nano-anatase by various ways was carried out for the realization of photocatalytic activity in the visible region of the spectrum, the study of the characteristics of the modified samples and their photocatalytic and sorption properties.

We have found the correlations of the synthesis conditions, composition of the samples and photocatalytic properties which exceed those properties of known commercial products. Adsorption of Nb(V), Ta(V), Mo(VI) and W(VI) ions from water by pure and modified samples with various compositions and characteristics has been studied. The relationship between the sorption degree and the conditions of synthesis, the sorption process and the nature of the analytes has been established.

1. INTRODUCTION

The development of materials with fundamentally new characteristics is intimately associated with the preparation of nanoscale systems. Nanoscale modifications of titanium dioxide are of great interest due to their unique properties. These materials are widely used in photocatalysis, sorption, microbiology, nanobiotechnology, medicine, microelectronics, pigment production and so on. In addition, titanium dioxide is used as a component of the nanocomposites to improve substantially their properties (increase in the specific surface, mechanical strength, selectivity of the catalysts formed on their basis, etc.) or to demonstrate the synergy effect.

Any physical quantity (property) Ω (e.g., electronic, optical, mechanical, etc.) can be expressed for a nanoparticle as a function of its size Ω(P). Experimentally observed size effects of semiconductors, to which titanium dioxide pertain, include an increase in the band gap and the shift of energy of deep levels with decrease in the size of nanocrystallites (relationship between the value of band gap and particle size can be described by $\Delta E_g \sim P^{-n}$ with n=1.16÷1.37) [1]. Kormann et al. [2] investigated the relationship between the band gap of TiO_2 and sizes of particles. Lin et al. [3] found that the band gap energy decreases with decreasing of particle size from 29 to 17 nm and it increases by 0.05, 0.012 and 0.036 eV for ultra-fine TiO_2 particles of 3.8, 4.9 and 5.7 nm in size, respectively. Mechanical properties of nanomaterials (strength, ductility, etc.) have also changed significantly with change of crystallite size due to particular distribution of crystal defects, primarily surface ones (twins, stacking faults, grain boundaries). The energy-band structure becomes anisotropic with decrease in the size of nanocrystallites resulting in the differences in the power spectra and properties of nanocrystallites of various geometrical shapes [1].

Magnetic properties of nanosystems are determined not only by the size of nanoparticles (it should be noted that the magnetic properties of diamagnetics and paramagnetics do not depend on the size factor) and their morphology but also by the interaction between nanoparticles as well as between nanoparticles and matrix. Nonstoichiometry of titanium dioxide causes appearance of ferromagnetic interactions at room temperature and significantly increases the photocatalytic activity [1].

Optical properties of semiconductor nanoparticles depend greatly on adsorbed foreign atoms and crystal defects. The outer surface of the nanoparticles is characterized by high energy of formation and, besides, it contains atoms with incomplete coordination sphere. It facilitates both adsorption reducing the surface energy and aggregation reducing the total surface area. Surface defects may act as potential wells or barriers for holes and electrons, which usually lead to degradation of the optical properties [1].

Unique adsorptive properties of nano titanium dioxide are due to formation of large amount of surface hydroxyl groups with high reactivity with decreasing size of nanocrystals (down to 100 Å and below). For adsorption, is important that the quantity of these crystallites is close to 100%. The surface hydroxyl groups are self-generated by contact with water resulting in high adsorption capacity of the titanium dioxide surface [4].

Nano titanium dioxide is known as an effective photocatalyst [5]. Anatase titanium dioxide exhibits photocatalytic activity in the ultraviolet region of the solar spectrum and exerts full activity in the photo-stimulated catalytic and photoelectric reactions. It is characterized by high chemical stability, low cost, nontoxicity and, therefore, it is a promising material for the creation of solar cells and photocatalysts that are active under the visible light [6,7]. Photoactivity increases due to higher Fermi level in anatase with band gap ΔE_g=3.3 – 3.4 eV as compared to rutile with ΔE_g=3.1 – 3.2 eV [8]. The TiO_2 band gap decrease results in increase of photoabsorption edge shift to the visible region (to the longer wavelengths) and vice versa.

Therefore, structural and morphological characteristics of the hierarchical level of nano-objects (size, shape, composition, crystal defects) play a crucial role in the manifestation of the properties inherent to such kind of materials. The definition of these characteristics and the establishment of relationship between properties and specific characteristics of samples with anatase was the motivation of this work.

2. THE MAIN CHARACTERISTICS OF NANO-OBJECTS

Nanocrystals and paracrystalline substances can be placed between the crystalline solids with long-range order, which is extended to arbitrarily large distances (infinite continuum approximation) due to the presence of translational symmetry of space lattice, and amorphous state with short-range order, which is extended to distances comparable with the dimensions of the particles. For paracrystalline state, the strict crystalline order is violated gradually, periodicity exists approximately and depends on the nature of order and symmetry. For nanocrystalline state, the long-range order is violated abruptly and is lost at the boundary of crystallite or crystalline block.

Nanocrystals possess several important features that distinguish them from other solid states:

- they have sizes in the nanometer range (at least one size) (10^{-7}-10^{-9} m);
- they are characterized by the presence of crystal structure (the long-range order) which is described by periodic lattice like crystals, however nanocrystals cannot be considered within the framework of infinite continuum approximation due to their small sizes (1-100 nm) (i.e., they have long-range order as crystal but it is limited by much smaller boundaries);
- the structure of grain boundaries (the surface structure - for isolated nanoparticles), which differ from the structure of crystalline blocks to a greater or lesser extent, is characterized by its own short-range order. Properties of grain boundaries (width is 2-3 monolayers) differ greatly from those of corresponding compact materials;
- the number of bonds coming to the surface is greater than the number of bonds isolated in volume. The thinner is the crystallite, the more unclosed bonds it has. In this case nanoparticles become more reactive.

Figure 1 shows the facets of anatase, rutile and brookite with Ti-O bonds (the bonds on the surface are green colored; the bonds isolated in volume are blue colored; the bonds coming out of the surface are red colored.

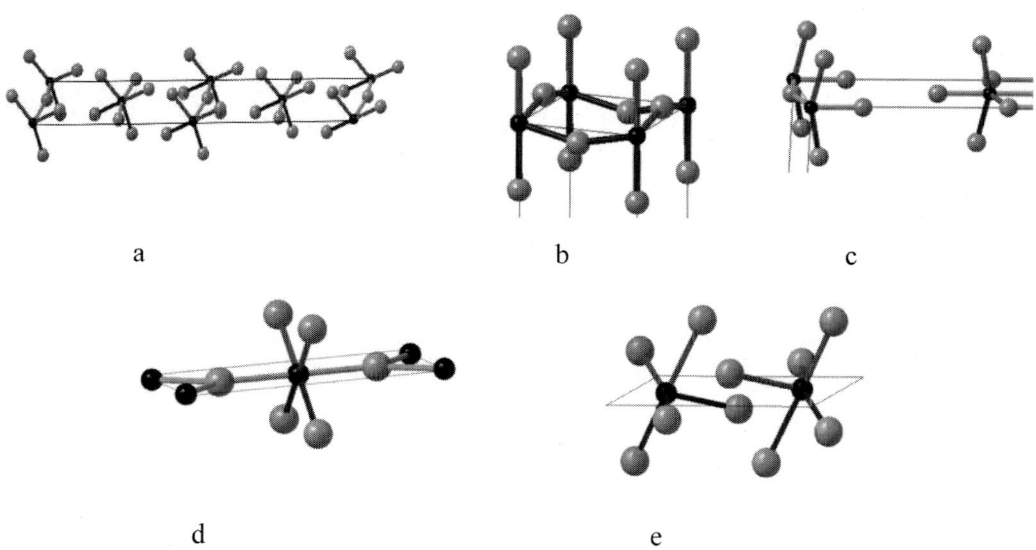

Figure 1. External surfaces of anatase - (101) (a), (001) (b) and (100) (c), rutile – (110) (d), and brookite – (110) (e).

On the (101) surface of anatase structure the ratio of free/closed Ti-O bonds is the greatest (7.5/6.5) as compared to other its surfaces. Moreover, this ratio is the greatest for anatase as compared to rutile (3/3) and brookite (5/5). This is another reason why anatase is more preferred, from a practical standpoint, than other titanium dioxide polymorphs [2,9,10].

Figure 2 shows the schematic model of a nano-object in the spherical approximation.

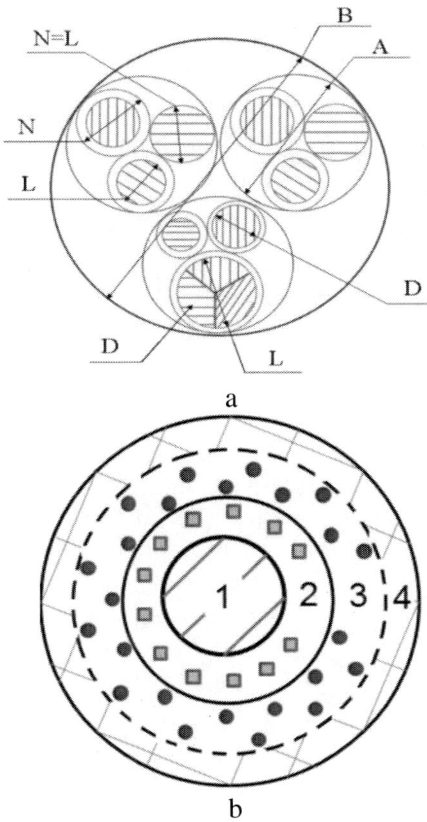

Figure 2. Schematic model of nano-object (a) and nanoparticle (b): *A*- agglomerate (nanoparticles held together by strong bonds), *B*- aggregate (agglomerates and nanoparticles held together by weak bonds), *N*- nanoparticle, *L*- crystallite, *D* – coherent scattering region.

There are several possible structures of nanoparticles which can consist of:

- core only (Figure 2a; Figure 2b – *region 1*): $N=L=D$.
- core and outer layer (*core – shell* structure): $TiO_2@X$: $(D=L)\neq N$ or $D\neq L\neq N$ (it is related to the defect structure of the coherent scattering region: twins, incoherent boundaries, small-angle regions). In these cases, X is the layer or layers (Figure 2b, *regions 2-4*) which may include amorphous TiO_2 and (or) $TiO_{2-x}(OH)_{2x}\times yH_2O$ (HD) [4], unreacted initial material, intermediate reaction products.

The composition and structure of nanoparticles, which can be changed by varying the conditions of synthesis as well as by modifying or further processing, rule the properties of nanomaterials. It should be noted that recently the definition of the nanoparticles is associated with their unique properties different from those of bulk objects rather than their sizes, i.e., the comparison between the size of the nanoparticles and correlation radius of any physical phenomenon (e.g., the mean free path of electrons or photons, the coherence length in superconductors, the magnetic domain size or solid phase nuclei, etc.) is performed. In other words, the "quantum size effects" definition is used. The occurrence of such effects depends both on the nature of the substances and on properties.

3. SYNTHESIS

Titania nanopowders can be prepared by a variety of techniques: hydrothermal synthesis, sol-gel processing, microemulsion synthesis, thermal decomposition of alkoxides, and others [9,11-18]. The samples for this investigation were synthesized using the well-known sulfate method [19], the main advantages of which are low process temperatures and the possibility of synthesizing compounds of a specified composition and morphology. $(TiO)SO_4 \times 2H_2O$ (I) and $(TiO)SO_4 \times xH_2SO_4 \times yH_2O$ (II) were used as starting reagents. The general scheme of the synthesis of samples is shown in Figure 3.

We varied titanyl sulfate concentration, reaction temperature and duration, and other process parameters (with or without stirring, with or without hydrochloric acid, dispersion of the precipitate in 2.5 M NaOH followed by filtration or without these steps) [20,21].

Commercial samples with nano-TiO_2 were used as reference samples: *Aeroxide Degussa (Evonik) P25* prepared by pyrolysis of titanium tetrachloride; *Hombifine N* and *Hombikat UV100* (Sachleben Chemie GmbH) synthesized by pyrolysis of titanyl sulfate.

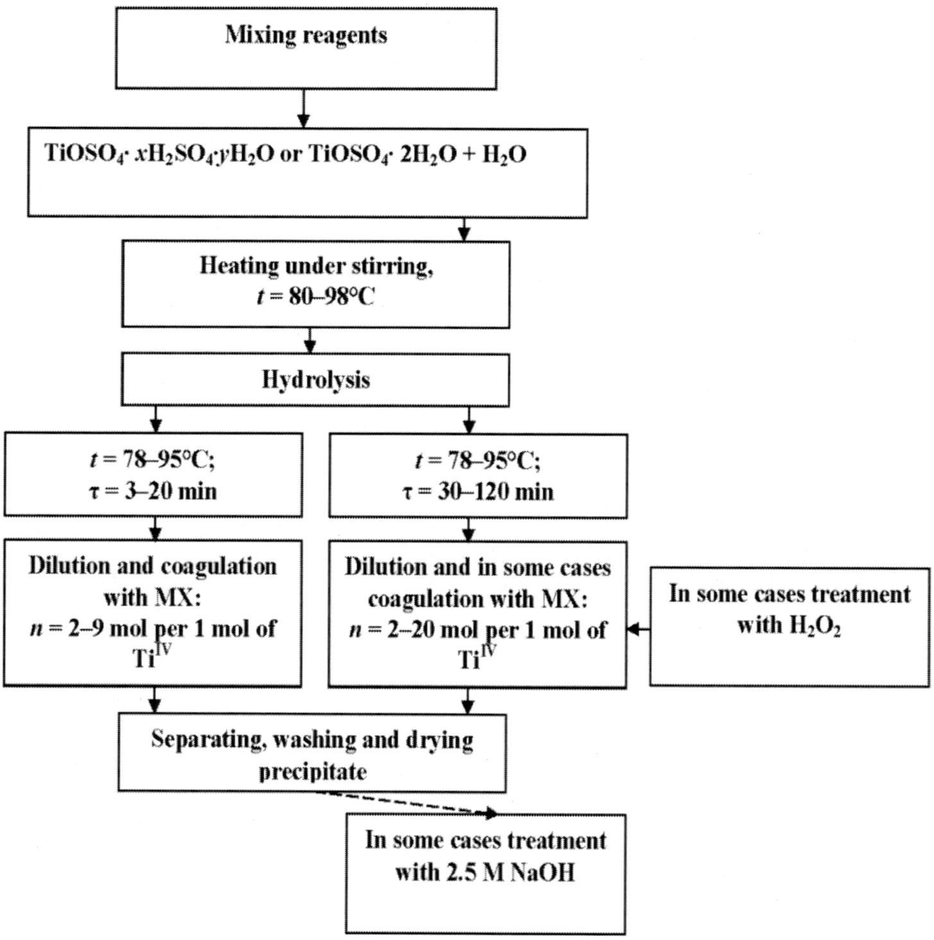

Figure 3. Scheme of the synthesis of samples with anatase.

4. CHARACTERIZATION

4.1. X-ray Diffraction Investigation

Technique

X-ray diffraction of the samples with rotation was performed on EMPYREAN, XRD-6000, HZG-4 diffractometers: Cu*K*α radiation; diffracted-beam monochromator; continuous or step (counting time per data point, 10 s; scan step, 0.02°; angular range, 2θ = 2°–50°) scan modes. The intensity data set was reduced using the PROFILE FITTING V 4.0 software package [22]. Qualitative XRD phase analyses were carried out using ICDD, ICSD, and our original data.

The *D* value (the average size of the coherent scattering region) was calculated using the Scherrer formula:

$D = 0.9\lambda/\beta_D\cos\theta$ (Equation 1), where $\beta_D = \sqrt{B^2 - b^2}$ (reflections are approximated by the Gaussian function) is the physical width of the diffraction peak, *B* is the integrated peak width, $b \sim 0.14°$ is the instrumental correction obtained with an α-Al_2O_3 standard (National Institute of Standards, United States);

$D = 0.9\lambda/\beta_F\cos\theta$ (Equation 2), where $\beta_F = F_{WHM}$ - full width at half maximum.

Microstresses - stresses of type II (the ε value) which are caused by excess surface energy and incomplete coordination of the surface atoms were estimated by the Stokes and Wilson formula: $\beta\varepsilon = \varepsilon tg\theta$ (Equation 3), where βε - broadening of reflection caused by stresses. In addition, simultaneous consideration of the influence of coherent scattering region size and microstresses on the broadening of reflection was carried out using the Williamson-Hall formula: $\beta = 0.9\lambda/D\cos\theta + 4\varepsilon tg\theta$ (Equation 4), where λ is the wavelength, 2θ ~ 25°.

The synchrotron X-ray diffraction techniques were used for the investigation of samples. All the measurements were carried out at the Structural Materials Science beamline of Kurchatov synchrotron radiation source [23,24]. X-ray diffraction patterns were measured in transmission geometry using photosensitive plates with FujiFilm Imaging Plates as the coordinate detectors (X-ray beam size, 200×200 μm; wavelength, 0.68886 Å; the sample-to-detector distance, 150 mm; exposure time, 15 min). Digital imaging was performed using the Fuji BAS- 5000 scanner with step of 100 microns. The integration of two-dimensional diffraction patterns to the standard I(2θ) form was carried out using Fit2D program [25]. On the integration stage, the corrections were made: for the not strictly perpendicularity of photosensitive plate with respect to the X-ray beam; for the exact shape of the integration region; for the linear polarization of synchrotron radiation. The polycrystalline standard Si powder (NIST SRM640c) was used to calibrate the angular scale of the diffraction patterns.

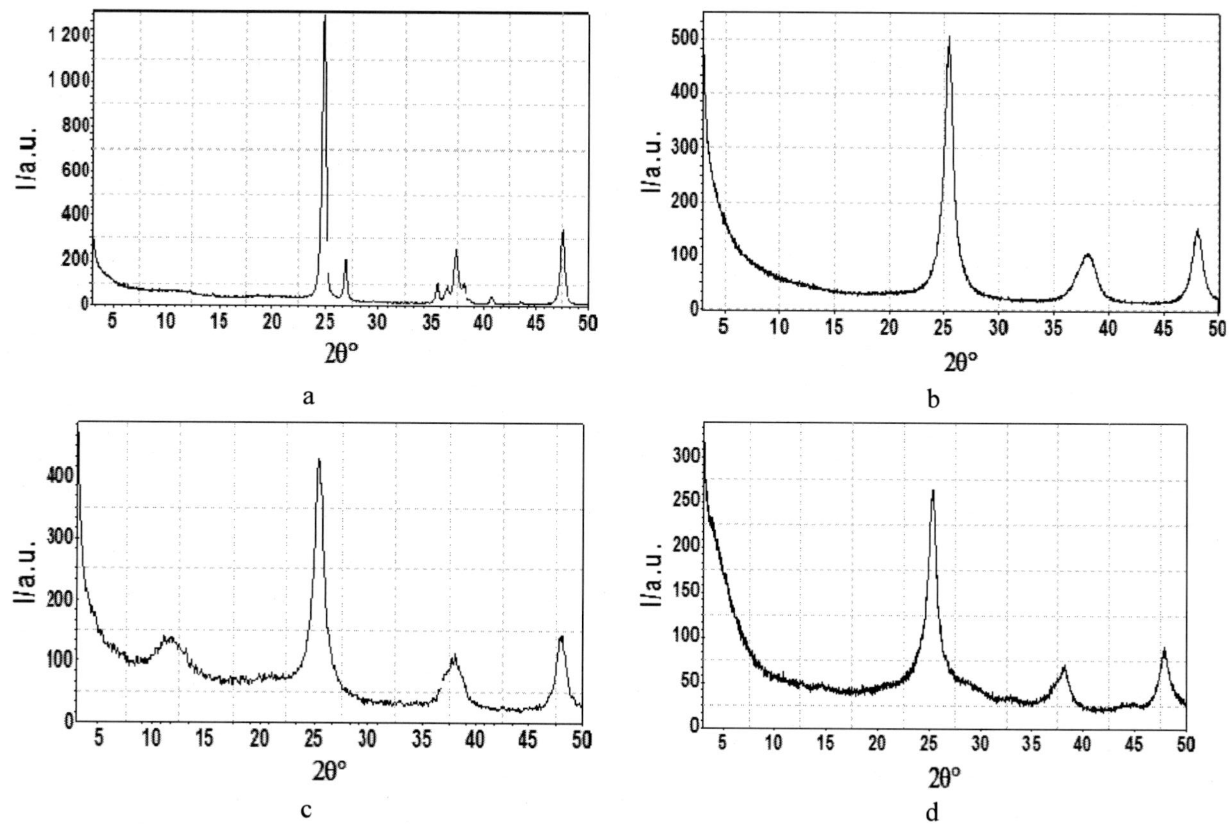
I/a.u.
2θ°
a
b
c
d

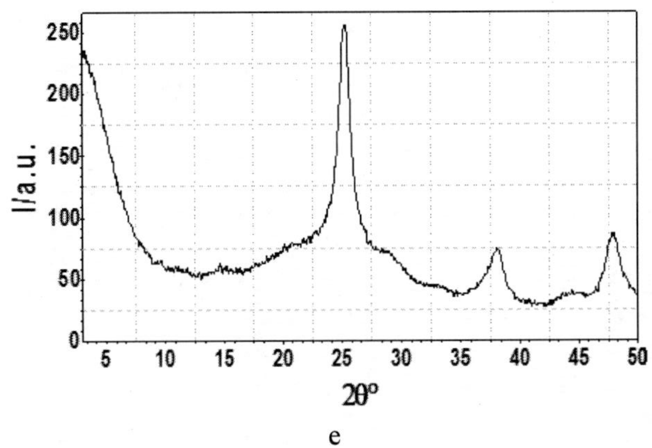

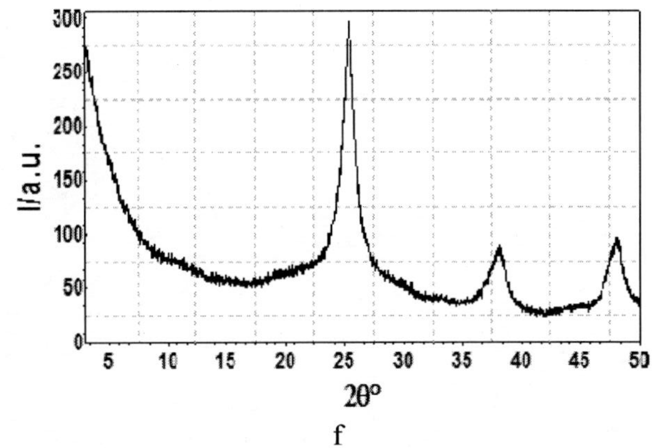

Figure 4. X-ray diffraction patterns of *Degussa P25* (a), *Hombifine N* (b), *Hombikat UV100* (c), *sample 1*(d), *sample 2* (e), *sample 3* (f).

Table 1. Results of calculation of the coherent scattering regions and microstresses

Calculation technique	D, Å	ε,%
The Scherrer formula (Equation 1):		
Degussa P25 (85% of anatase +15% rutile)	191(8)/600(3)	
Hombifine N (anatase)	83(4)	
Hombikat UV100 (anatase+HD*)	81(4)	
Sample 2 (Figure 4e)	67(3)	
The Scherrer formula (Equation 2):		
Degussa P25 (85% of anatase +15% rutile)	250(1)/640(3)	
Hombifine N (anatase)	83(4)	
Hombikat UV100 (anatase+HD*)	82(4)	
Sample 2 (Figure 4e)	58(3)	
The Williamson-Hall formula (Equation 4):		
Degussa P25 (85% of anatase +15% rutile)	185(9)/550(3)	0.075
Hombifine N (anatase)	74(4)	0.003
Hombikat UV100 (anatase+HD*)	73(4)	0.003
Sample 2 (Figure 4e)	65(3)	0.030

* HD- $TiO_{2-x}(OH)_{2x}{\times}yH_2O$.

For the samples under investigation, the small angle X-ray scattering (SAXS) experiments were carried out. X-ray diffraction patterns were measured in vacuum on specialized SAXSess diffractometer (Anton Paar, Austria; CuKα radiation; linear collimation; measurement time, 3 min; low-angle region, 2θ=0÷40°). The size distribution function of particles and particle aggregates was calculated based on the curves of small angle scattering in the angular region 2θ=0÷7° by GNOM program using the spherical particle approximation [26].

Results

According to the X-ray diffraction patterns of commercial samples (Figure 4), the *Hombifine N* (Figure 4b) contains anatase (space group $I4_1/amd$; JCPDS № 89-4921), *Hombikat UV100* (Figure 4c) contains anatase and amorphous phase with general formula $TiO_{2-x}(OH)_{2x}{\times}yH_2O$ (HD) with the reflection at 2θ~11-12°, and *Degussa (Evonic) P25* is a mixture of anatase (~ 85%) and rutile (~ 15%) (space group $P4_2/mnm$; JCPDS № 89-4920) (Figure 4a).

Weight percent of anatase (x_A,%) and rutile (x_R,%) phases in the samples was calculated from the XRD patterns using Equations 5 and 6 as follows: $x_R=1/[1+0.8(I_A/I_R)$,% (Equation 5), $x_A=100-x_R$ (Equation 6).

Samples with anatase synthesized using sulfate method were single-phase (*sample 3*, Figure 4f) or contained various quantityies of β-TiO_2, according to the reflections found at 2θ ~15°, ~ 28° and ~ 45° (JCPDS № 46-1238) (*sample 2* and *sample 3*, Figure 4 d,e). Therefore, commercial samples and our samples differ in their compositions, average values of coherent scattering regions (the *D* value), and microstresses (the ε value) (Table 1).

The *D* values calculated with different reflections *hkl* (101, 004, 200) of anatase were approximately the same. It does not exclude the spherical shape of crystallites (in the case of equality of *L* and *D* values) for *Degussa P25, Hombifine N* and our samples.

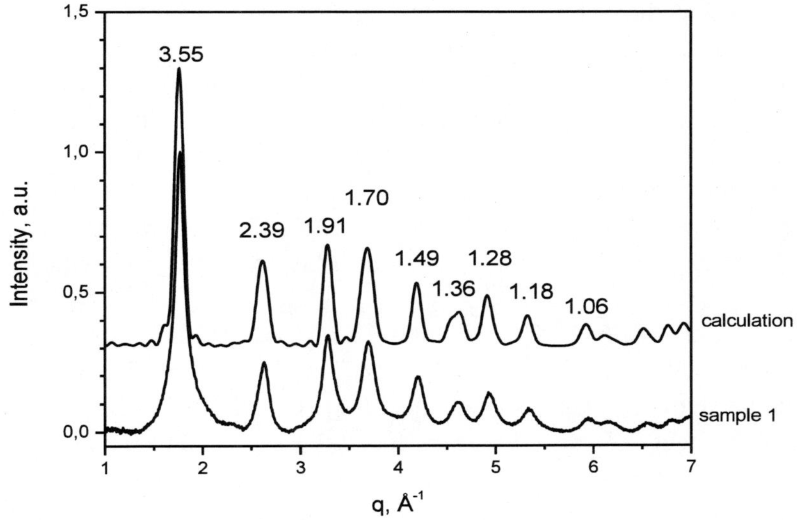

Figure 5. Calculation of diffraction pattern according to Debye formula, and experimental curve for *sample 1* (Figure 4d). Interplanar distances (Å) are given for the major peaks.

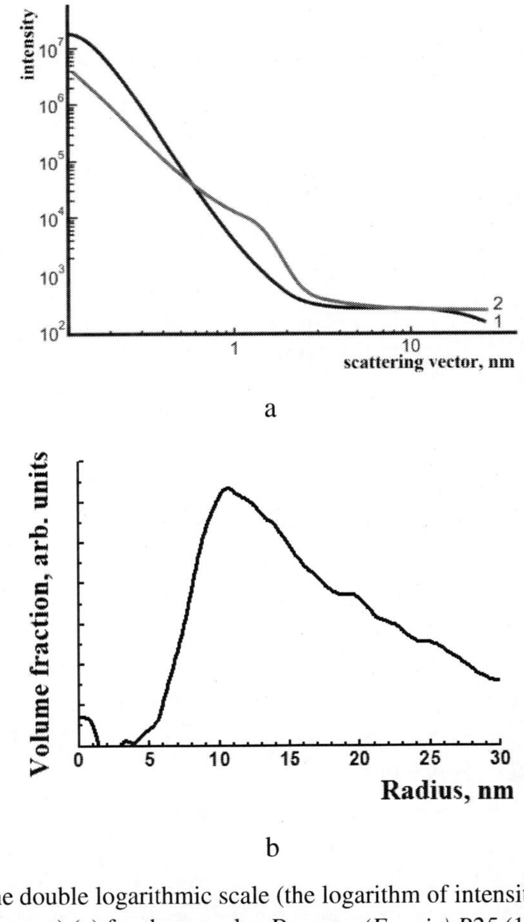

Figure 6. SAXS curve on the double logarithmic scale (the logarithm of intensity plotted versus the logarithm of the scattering vector) (a) for the samples *Degussa (Evonic) P25* (1) and *Hombikat UV100* (2), and nanoparticle size distribution for *Degussa (Evonic) P25* based on SAXS data *(b)*

Calculation of the diffraction pattern of *sample 1* with Debye formula (for model nanoparticle of about 5 nm in size: 12 translations along the crystallographic *a*- and *b*- axes, and 5 ones along the *c*-axis) results in the curve (Figure 5) repeating all the key features of the experimental curve (positions, relative intensities and half widths of reflections).

This implies that the almost isometric particles of anatase without a pronounced texture and specific defects present in the investigated samples.

According to SAXS analysis, the sample with *Degussa P25* (Figure 6a, *curve 1*) is characterized by a relatively broad particle size distribution with a maximum about 12 nm (Figure 6b).

Microstructures of *Hombikat UV100* (Figure 6a, *curve 2*) and *Hombifine N* radically differ from *Degussa P25* microstructure: *Hombikat UV100* and *Hombifine N* are polydisperse samples with a hierarchical system of fraction of large particles and independent fraction of fine particles with about 1.5 nm (*Hombikat UV100*) and 1.8 nm (*Hombifine N*) in radii (Table 2). For the self-made samples with anatase, the polydispersity is similar to *Hombikat UV100*. But as a result of averaging, a bimodal nanoparticle size distribution can be obtained: with primary (N_1=~10÷18 nm) and secondary (N_2=~30÷45 nm) nanoparticles. The presence of "large" (radius > 60 nm) and "small" (diameter 6.8 nm) inhomogeneities (Table 2) is observed for *sample 1*, particles of 8.8 nm in diameter being the almost homogeneous fraction. An interesting feature of *sample 1* is a weak periodicity with a period of between 12, and 14 nm and about 3 nm. It's not excluded that such periodicity is due to ordering oxygen vacancies.

4.2. Electron Microscopy

Technique

The microstructures of *Degussa P25*, *Hombifine N*, *Hombikat UV100*, and *sample 1* (Figure 7) were examined by scanning electron microscopy (SEM) on a JEOL JSM-7500F field-emission scanning electron microscope in the low-energy secondary electron mode which ensured the highest resolution (the resolution was 1 nm and 1.5 nm at primary beam energy of 20 kV and 2 kV, respectively). The powders to be studied were applied to a conductive adhesive carbon tape, which were then covered with a platinum film of 5-10 nm in thickness using magnetron sputtering (electric current, 30 μA; deposition time, 20 s; the distance to the target, 40 cm; pressure, 5 Pa). In order to avoid artifacts associated with the application of the platinum layer on the surface of particles, the preliminary experiments on its application were carried out on silicon single crystals at the same conditions (platinum particle size in the specified deposition mode was 4 nm). We also used gentle beam mode, intended to ensure high resolution at a low accelerating voltage.

Quantitative and qualitative electron probe microanalysis (EPMA) was performed on an Oxford Instruments INCA Penta FETx3 energy dispersive X-ray spectrometer system. The calibration (6 iterations) was carried out in accordance with two lines of K-series of cobalt standard sample. The quantitative analysis of elements with atomic number N ≥ 3 was performed using the INCA Penta FET program. Repeatability limit was 0.05%.

Electron micrographs and electron diffraction patterns of *samples 4* and *5* (Figures 8-10) were obtained on a JEOL JEM 2100 high-resolution transmission electron microscope (TEM)

(LaB_6 cathode as an electron source; accelerating voltage, 200 kV; minimum electron beam diameter in transmission mode, 20 nm; specimen tilts of ± 60° about the x axis and ± 25° about the y axis; point and line resolutions of 2.3 and 1.4 A, respectively; maximum accelerating voltage, 200 kV; direct magnification up to 1.5×10^6). The powder to be studied was immersed in heptane, which is nonreactive toward the powders under investigation and causes no agglomeration during drying. Selected area diffraction patterns were obtained using selector apertures. The minimum selected area size was 100 nm.

Investigation of morphology and electron diffraction analysis of *sample 1* (Figure 4d) were carried out on a JEM-2000FX transmission electron microscope (JEOL, Japan) at 150 kV. The JEM-2100 electron microscope (JEOL, Japan) was used to observe the microstructure of nanoparticles in high-resolution mode at the accelerating voltage of 200 kV. The samples to be studied were prepared by applying the TiO_2 diluted with heptane to a carbon thin film or perforated carbon film coated, in turn, with a carbon thin film [27].

Results

According to the SEM data, the commercial samples differ in particle sizes (Figure 7): ~10÷~40 nm for *Degussa P25* (Figure 7a); ~10÷~80 nm for *Hombifine N* (Figure 7b) and ~10÷~100 nm for *Hombikat UV100* (Figure 7c).

The particle sizes in the *sample 1* (Figure 7d) are ~10÷~45 nm. Micrographs demonstrate that *Hombifine N* (Figure 7b), *Hombikat UV100* and *sample 1* particles are characterized by agglomeration (aggregation).

a

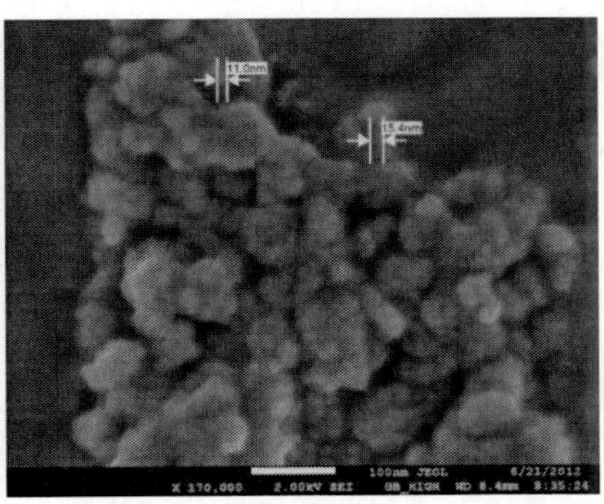

b

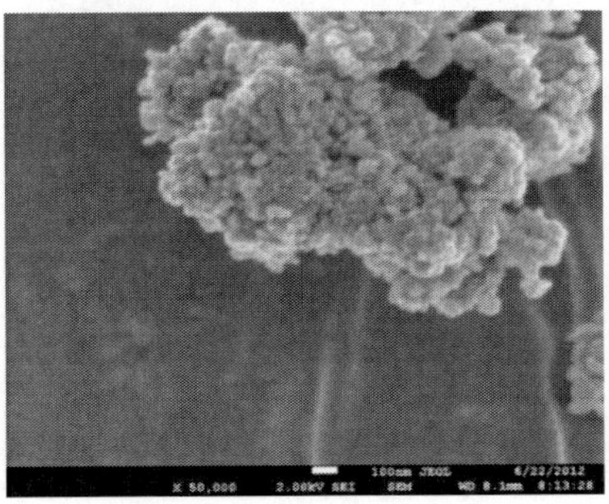

c

d

Figure 7. SEM images of *Degussa P25* (a), *Hombifine N* (b), *Hombikat UV100* (c), *sample 1* (Figure 4d) (d).

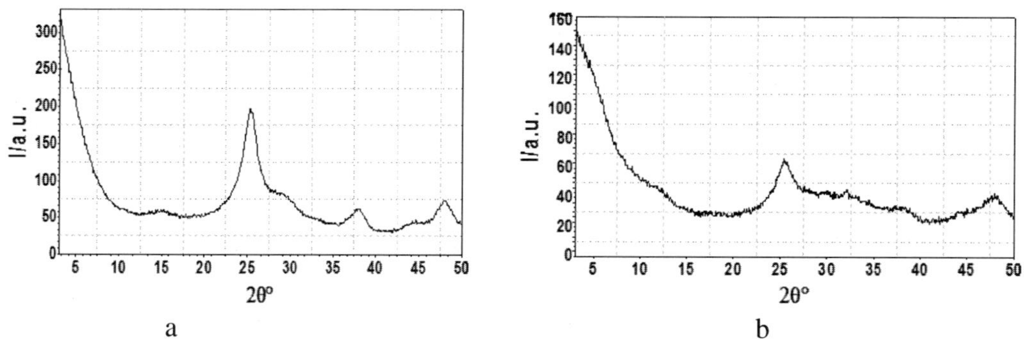

Figure 8. Diffraction patterns of *sample 4* (a) and *sample 5* (b).

According to the EPMA data, sulfur, titanium and oxygen contents in our samples vary in 2.17÷3.73 wt.%, 49.71÷53.14 wt.% and 38.58÷40.12 wt.% ranges, respectively; the particles have small variation in titanium/oxygen ratios.

Figure 8 shows the diffraction patterns of *samples 4* and *5* which have been studied by TEM (Figures 9, 10) [21].

Sample 4 contains anatase (D=45(2) Å) and β-TiO_2 (Figure 8a). It consists of agglomerates of fine nanoparticles about 10 nm in average size (Figure 9a). It seems likely that the material is amorphous and contains anatase particles. In addition, aggregates may contain platelets or flakes, which consist of fine grains, as evidenced by the large width of electron diffraction rings (Figure 9b). There are no peaks of β-TiO_2 because of the low content of this phase and the local character of the technique used.

Sample 5 (D=40(2)Å) contains anatase and a small amount of η-TiO_2 (Figure 8b). It consists of complex-shaped agglomerates of particles of different sizes (Figure 10a). The majority of the particles (average size is ~ 500 nm), which are predominantly faceted, are covered with an amorphous layer (Figure 10b).

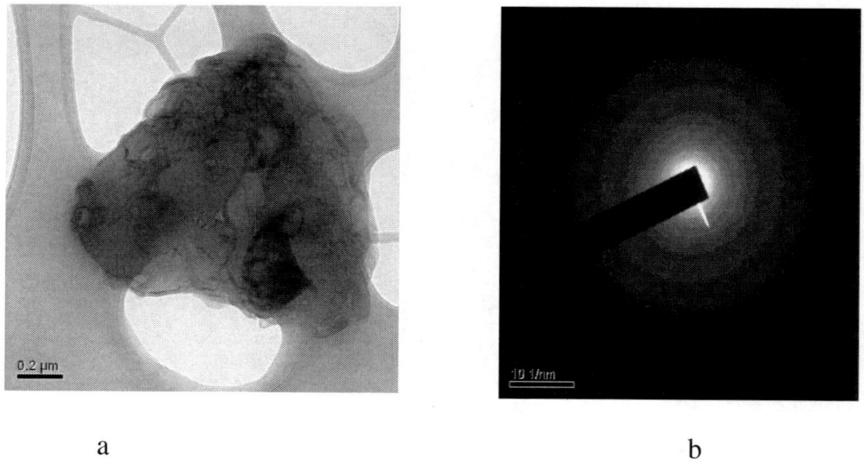

Figure 9. TEM micrograph (a) and selected-area diffraction pattern (b) of a particle agglomerate in *sample 4*.

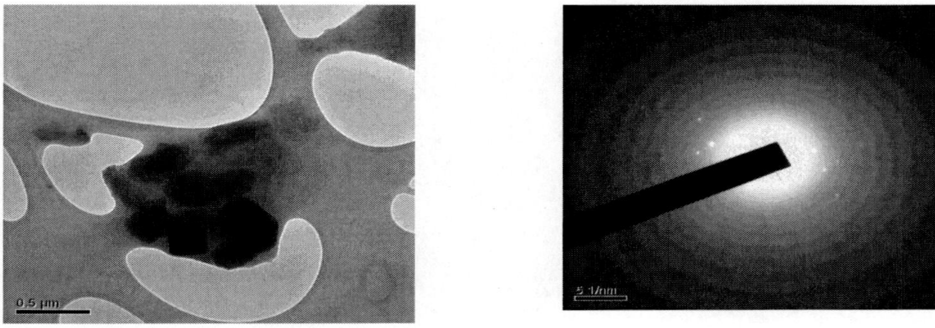

a b

Figure 10. TEM micrograph (a) and selected-area diffraction pattern (b) of a particle agglomerate in *sample 5*.

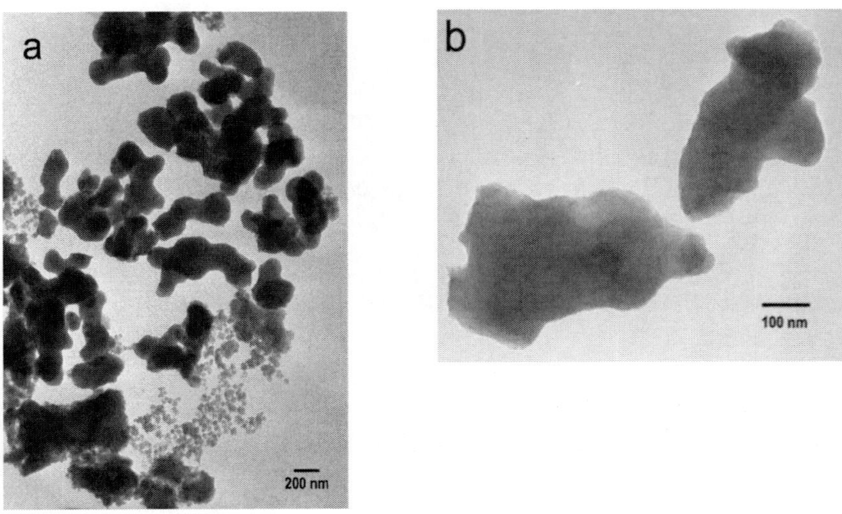

Figure 11. TEM micrographs of nano-objects (a) and microstructure (b) of *sample 1*.

A distinctive feature of samples with η-TiO_2 is the strong reflection with an interplanar spacing in the range of *d*~17÷~21Å, and less intense reflection with *d*~2.7÷~2.9Å. Other reflections (*d*~3.5, ~1.9, ~1.5Å) correspond to reflections of well-known TiO_2 polymorph with anatase structure [28].

High-resolution micrographs of almost single-phase *sample 1* (Figure 4d) shows that each particle consists of spherical particles (Figure 11).

General view of *sample 1* microstructure (Figure 11*b*) confirms its polydispersity: the particle size varies over quite a wide range (A=130÷1000 nm). According to the analysis of more than 100 particles, it was established that the small particles coalesce to form the large ones (average size, A_{av}=340 nm). TEM micrograph of nano-objects of *sample 1* (Figure 11a) shows that ~~the~~ small spherical particles about 9÷50 nm in size can both form the nanoparticles and exist separately. High-resolution micrograph of *sample 1* shows that the sizes of crystallites varying in the 3÷8 nm range (Figure 12).

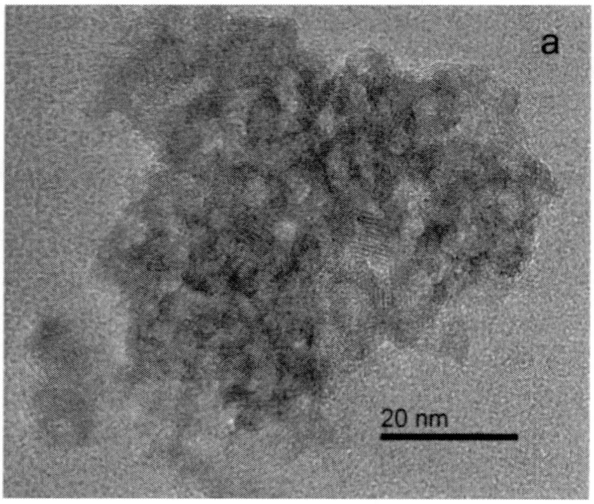

Figure 12. TEM micrograph of nanoparticles of *sample 1*.

Thus, nanoparticles of *sample 1* do not contain amorphous phase, the average sizes of the coherent scattering region (D=84(4) Å) and crystallites (L=30÷80 Å) being in the same range of values, which eliminates the defect structure of anatase.

4.3. Low Temperature Nitrogen Adsorption

Technique

The nitrogen sorption capacity at –196°C was measured volumetrically using a Micromeritics ASAP 2020 gas analyzer and represented in the form of sorption isotherms (adsorption and desorption). The specific surface area of the investigated samples was determined by the Brunauer–Emmett–Teller (BET) [29] and comparative [30] methods. The BET specific surface area was calculated using the formula: $S_{BET} = a_m \cdot N_A \cdot n_m$ (Equation 7), where a_m is the cross-section of the adsorbed molecule, m^2 ($0.162 \cdot 10^{-18}$ m^2 for nitrogen); N_A is the Avogadro number, mol^{-1}; n_m is the monolayer capacity, mol/g.

Monolayer capacity (n_m) was determined using the BET equation: $\frac{P/P_0}{n \cdot (1-P/P_0)} = \frac{1}{n_m \cdot C} + \frac{P/P_0 \cdot (C-1)}{n_m \cdot C}$ (Equation 8), where n is the sorption capacity measured for five to seven points in the 0.05–0.35 range of equilibrium relative nitrogen pressures; P_0 is the pressure of saturated nitrogen vapor at a temperature of –196°C, mm Hg; P is the equilibrium nitrogen pressure in the gas phase, mm Hg; C_{BET} is the energy constant of the BET equation.

When the specific surface area was determined with the comparative method, a graphic comparison of the sorption capacity values of the investigated sample and a nonporous reference sample was performed at the same values of equilibrium nitrogen pressure. The nonporous reference sample was ASX-1-4 from the GSO 9732-2010 kit.

The volumes of nanopores and ultrananopores, V_n and V_U, respectively (in this work, "nanopores" refers to pores with diameters of less than 100 nm; "ultrananopores" refers to pores with diameters of less than 2 nm [31, 32] that are filled via the volumetric mechanism [33] at the region of low nitrogen pressures), were calculated from Equations 9 and 10 [34, 35], respectively:

$V_n = n_{0.98}\dfrac{\rho_g}{\rho_l}$ (Equation 9), $V_U = n_{0.015}\dfrac{\rho_g}{\rho_l}$ (Equation 10), where $n_{0.98}$ is the sorption capacity, cm^3 (normal conditions)/g, at equilibrium nitrogen pressure $P/P_0 = 0.98$; $n_{0.015}$ is the sorption capacity, cm^3 (normal conditions)/g, at the equilibrium nitrogen pressure $P/P_0 = 0.015$; ρ_g is the density of gaseous nitrogen at normal conditions, 1.2504 g/mm^3 (normal conditions)/g; and ρ_l is the density of liquefied nitrogen at the temperature of measurement, 0.8086 g/cm^3.

The average sizes of particles (N, nm) were calculated from formula: $N = \dfrac{K_p}{\rho S}\cdot 10^9$ (Equation 11), where ρ is the X-ray density of titanium dioxide, g/m^3; S is the specific surface area determined by the comparative (S_{CM}, m^2/g) or BET (S_{BET}, m^2/g) methods; K_p is the shape factor of the particles (6 - for cubics and spheres, 4 - for cylinders and round rods, 2 - for plates) [36]. The sample density (the ρ value, g/cm^3) was calculated from the XRD data, or the literature data were used. The shape factor of the particles, K_p, was determined from the type of the hysteresis loop on the adsorption isotherm [36,37].

Results

Figure 13 shows the diffraction patterns of the samples for which the adsorption isotherm of nitrogen at -196 ° C were investigated; the results of a study of their microstructures are presented in Table 2.

Nitrogen adsorption isotherms for the investigated samples with anatase structure belong to type IV according to the IUPAC classification [38]. They are associated with the capillary condensation of nitrogen in the porous (interparticle) space of the materials. A sharp increase is observed in the initial regions of isotherms of *sample 7, sample 8,* and *Hombifine N*, which differ by the growing content of ultrananopores, 0.040–0.083 cm^3/g (Table 2); this is characteristic of type I isotherms, according to the IUPAC classification [38], due to the filling of ultrananopores via the volumetric mechanism in the region of low pressures of the adsorbed gas [33]. Hysteresis loops observed at the isotherms of *sample 8* and *Hombifine N* are type H3 ones; those of *sample 6* and *sample 7* are type H4 ones; and that of *Degussa P25* is type H1 one, according to the IUPAC classification [38]. Type H1 hysteresis loops are typical of globular models [37], and K_p=6 [36,37]. Hysteresis loops of types H3 and H4 are typical of nitrogen adsorption in gap-like pores between plane-parallel particles [37], and K_p=2 [36,37].

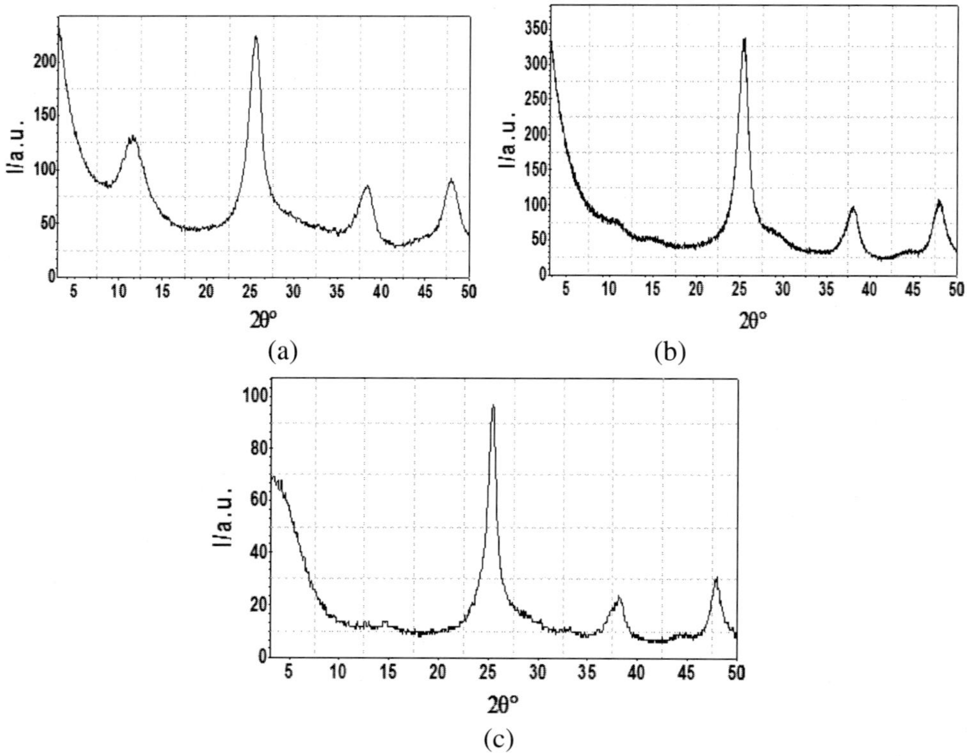

Figure 13. Diffraction patterns of *sample 6* (a), *sample 7* (b), *sample 8* (c).

The values of BET specific surface area for the studied samples, S_{BET}, calculated from Equation 7, where $S = S_{BET}$, are within the range of a few to few hundred m^2/g; the average sizes of particles, N_{BET}, range from few to hundreds of nanometers (Table 2). However, the presence of ultrananopores in the investigated sample (pores with diameters below 2 nm (Table 2) and filled according to a volumetric mechanism [33]) results in incorrect calculations according to BET Equation 7. It is known that the BET model is limited by a series of assumptions, particularly the layer-by-layer adsorption onto an energetically homogeneous surface, and others [29,37]. It should be also noted that for *sample 8* and *sample 9*, the values of the BET constant, C_{BET}, are not optimal [37], indicating that the determination of the monolayer capacity with the BET Equation 8 is not accurate [37]. The sample's specific surface area was also determined using the comparative method [30,37] based on the metrological approach, and the increments of the sorption capacity values of the investigated sample were analyzed relative to those for a nonporous reference material. A region of the equilibrium gas pressures, where the effect of the surface structure is weak, has been considered. These conditions are met when the statistical thickness of the adsorption film is above 1.5 monolayers [37,39]. The region of the equilibrium nitrogen pressures above 0.4 corresponds to these conditions. The specific surface areas of the investigated samples, obtained with the comparative method (S_{CM}), are within the range of few to hundreds of m^2/g (Table 2) and agree quite well with the BET results for *Degussa P25* (Table 2).

The specific surface areas according to BET for *samples 6–8* and *Hombifine N* strongly exceed those obtained with the comparative method (Table 2). The average size of nano-objects of *sample 1*, obtained according to the specific surface area value (S_{CM}, m^2/g) with the comparative method, was found to be N_{CM}=30 nm (S_{CM}=16.5 m^2/g) in the spherical approximation. Large differences between the values of the specific surface area determined by the BET method and the comparative method can be explained by the presence of a sufficiently large volume of micropores filled according to a volumetric mechanism. Therefore, to calculate the particle size, the specific surface area data obtained by the comparative method (S_{CM}) was taken into account because this method is more accurate for the determination of the values for the surfaces on which the adsorption of nitrogen occurs, as compared with BET one. As a result, the calculated values of the average sizes of particles for these samples according to BET are less than those calculated according to the comparative method, $N_{BET} < N_{CM}$.

The sizes of the particles of the investigated samples were compared to the data obtained by electron microscopy (SEM and TEM) and small angle X-ray scattering methods (SAXS) [21,40]. According to the TEM data obtained for *sample 7*, the size of titanium dioxide particles is 10÷15 nm, this agrees with the data calculated by the comparative method, N_{CM}= 13 nm, while, according to BET, the average size of particles is 5 nm, that is unrealistically small. According to the SAXS results for *samples 6* and *8*, these values are 16 and 10 nm for primary particles, respectively, and 37 nm for secondary particles; there is also better agreement between data calculated according to the comparative method (22 nm for both samples) as compared to the data calculated according to BET (13 and 3 nm, respectively) (Table 2) [41].

Thus, the samples with anatase obtained by us differ from the commercial samples *Degussa (Evonik) P25*, *Hombifine N* and *Hombikat UV100* by compositions, characteristics of the microstructure (the specific surface area; the volume of nano- and ultrananopor; the size of nanoparticles and aggregates) and substructure (the size of the coherent scattering regions and microstresses) (Table 2).

The investigated nano-objects are characterized by the $(D=L)\neq N$ or $D=L=N$ models (Figure 1). Nanoparticles with anatase can contain or not contain an amorphous phase of titanium dioxide. The small sizes of the coherent scattering regions (D~40÷50 Å) indicates the presence of the amorphous component of anatase.

4.4. Spectroscopy

Technique

IR-spectral analysis was performed using an EQ 5.5 IR-Fourier spectrophotometer.

Raman spectra were recorded on a Horiba-Jobin-Yvon T64000 spectrometer in reflection geometry using a 514.5-nm Ar laser (<0.2 mW on the sample surface). The size spot and spectral resolution were 1 μm and 1 cm^{-1}, respectively. In addition, the Raman spectra of some samples were analyzed on an RFS 100/S FT Raman spectrometer (Bruker) using a 1060-nm Nd laser (the power density on the sample surface did not exceed 50 W/cm^2) with a spectral resolution of no worse than 2 cm^{-1}.

Table 2. Characteristics of the investigated samples

Sample	Phase composition	*D*, nm	Adsorption method								Methods		
			Type of hysteresis loop (IUPAC)	V_{np}, cm^3/g	V_{unp}, cm^3/g	BET			Comparative method		SAXS	SEM	TEM
						S_{BET}, m^2/g	Constant C_{BET}	N_{BET}, nm	S_{CM}, m^2/g	N_{CM}, nm	N_{SAXS}, nm (N_1, N_2)*	N_{SEM}, nm	N_{TEM}, nm
Sample 1	Anatase + β-TiO_2 (traces)	8.4(4)	H3	0.048	0.013	36.5±0.4	-79	40	16.5		8.8** (6.8÷60)	~10÷~45	3÷8
Sample 2	Anatase + β-TiO_2	5.8(3)	H3	-	0.055	122.75	-54	12.2	11.4		-	-	-
Sample 6	Anatase + $TiO_2 \times H_2O$	4.6(2)	H4	0.044	0.011	40.3	134	13	23.6	22	(16,37)	-	-
Sample 7	Anatase + β-TiO_2	5.0(3)	H4	0.082	0.040	112.6	-101		38.7	13	-	-	10÷15
Sample 8	Anatase + β-TiO_2+η-TiO_2	8.0(4)	H3	0.118	0.052	171	40		23.2	22	(10,37)	-	-
Hombifine N	Anatase	8.3(4)	H3	0.329	0.083	312.8	111		116.4	9	1.8** (6÷40)	~10÷~80	-
Degussa P 25	Anatase (85%) + rutile (15%)	25.0(1)	H1	0.131	0.015	53.1	103	30	54.3	28	24	~10÷~40	-
Hombikat UV100	Anatase + HD	8.2(4)	H3	0.344	0.082	314	105		123	12	1.5** (6÷40)	~10÷~100	-

* The size of primary (N_1) and secondary (N_2) nanoparticles.

** The polydisperse sample.

X-ray photoelectron spectra (XPS) were recorded on an Axis Ultra DLD spectrometer (Kratos) using monochromatic AlK_α radiation and nonmonochromatic MgK_α one. All spectra were recorded at constant analyzer energy: 160 eV for the full-range (0–1200 eV) spectra and 20 eV for local spectra (Ti 2*p*, O 1*s*, C 1*s*). The spectrometer energy scale was previously calibrated with an error of ±0.03 eV based on the following standard peak positions: 83.96 eV (Au $4f_{5/2}$), 932.62 eV (Cu $2p_{3/2}$), and 368.21 eV (Ag). The spectra were approximated by the Gaussian–Lorenzian convolution with the simultaneous optimization of the background parameters.

Results

According to the factor group analysis, there are fifteen optical modes of anatase TiO_2 including A_{1g}, $2B_{1g}$ and $3E_g$, active in Raman spectroscopy, and A_{2u} and $2E_u$, infrared active [42].

The IR spectra of the investigated samples showed bands in the ranges 1620–1656, 1200–1207 and 1057–1072 cm^{-1} (Table 3), due to bending vibrations of surface Ti–O–H groups, which were probably hydrogen-bonded to H_2O molecules [43,44]. There were also bands in the 3255–3374 cm^{-1} range, due to O–H stretches, and at low frequencies (<700 cm^{-1}), due to Ti–O vibrations [43,45].

Table 3. Assigment of bands (cm^{-1}) in the IR spectrum of *sample 4*

Frequency, cm^{-1}	Assignment
3300	ν(OH) Ti–OH H_2O
1632	δ(OH) Ti–OH H_2O (hydrogen-bonded to the surface)
1200	δ(OH) Ti–OH H_2O (hydrogen-bonded to the surface)
1054 1131	δ(OH) Ti–OH H_2O (hydrogen-bonded to the surface), ν(SO)

The presence of sulfur inferred from the EPMA data leads us to assume that there are sulfate groups on the surface of the nanocrystallites in our samples. A free sulfate ion is known to have four vibrational modes [46]: ν_s(SO), δ_s(OSO), ν_{as}(SO), and δ_{as}(OSO) at 983, 450, 1105, and 611 cm^{-1}, respectively. Coordination may slightly shift and/or split these modes. Symmetry reduction may cause the former two bands to be IR-active. The bands of the sulfate ion are, however, difficult to identify with certainty in the spectra of our samples because they overlap with other vibrational modes.

According to [43,47,48], the Raman spectra of anatase exhibit six modes: A_{1g} (513 cm^{-1}), $2B_{1g}$ (399 and 519 cm^{-1}), and $3E_g$ (144, 197 and 639 cm^{-1}). The corresponding bands were found in the spectra of the anatase samples synthesized by us, *sample 9* (Figure 14), in particular, at about 160 (E_g), 520 (B_{1g}), and 645 (E_g) cm^{-1} [20].

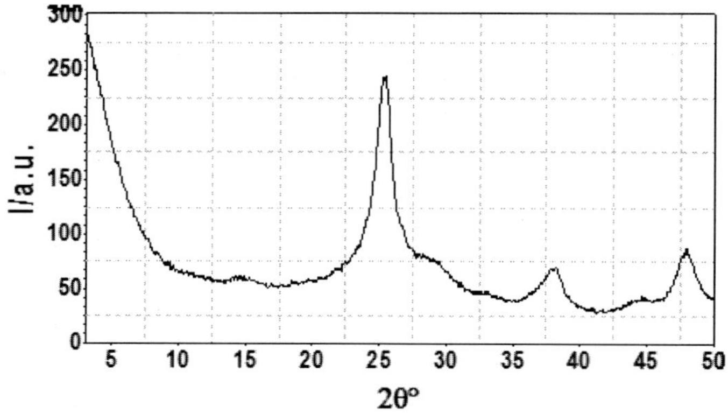

a

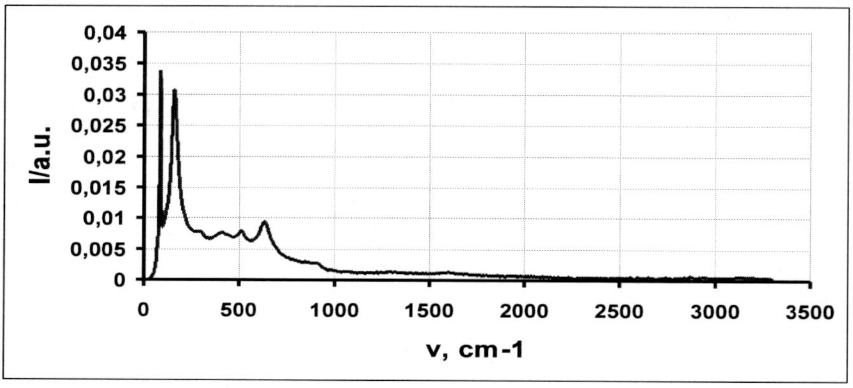

b

Figure 14. Diffraction pattern (a) and Raman spectrum (b) of *sample 9*.

In addition, the spectra of the samples contain bands at 297–314, 420–473, and 910–940 cm^{-1}. The origin of the bands at about 300 cm^{-1} remains unclear, although similar bands were observed in the spectra of titanate samples in [49]. The bands at about 450 cm^{-1} can be assigned to the Ti–O–Ti vibrations [50]. The occurrence of the bands at 910–940 cm^{-1}, which can be assigned to the vibrations of Ti–O terminal bonds [43,49], is apparently due to the small size of particles and, therefore, the large number of such bonds on their surfaces. In addition, the spectra of the anatase samples contain bands at about 1030 and 1600 cm^{-1}, which are caused by bending vibrations of the Ti–O–H groups [43,44].

According to the XPS data from the full-range spectra (Figure 15) and local (Ti 2*p*, O 1*s*, C 1*s*) spectra (Figure 16) [20], the surface of the *sample 8* contains (in large amounts) the main elements of the phases (titanium and oxygen), carbon (including hydrocarbon contaminations), and sulfur and nitrogen (impurities).

Three components are selected in the oxygen spectra which correspond to the Ti–O bonds (~530 eV), adsorbed OH^- and CO_3^{2-} particles (~532 eV), and adsorbed water or crystallohydrate water molecules (~534 eV) [51]. Since the carbon spectra indicate that the surface of powders contains a small amount of CO_3^{2-}, the component in the O 1*s* spectra (~532 eV) is mainly assigned to OH^-.

Thus, the surfaces of the nanoparticles obtained by us contain different quantities of H_2O, OH^- and SO_4^{2-} - groups, and, probably, titanates of various compositions. It may cause reducing specific surface area of nanoparticles (Table 2) and should affect their properties.

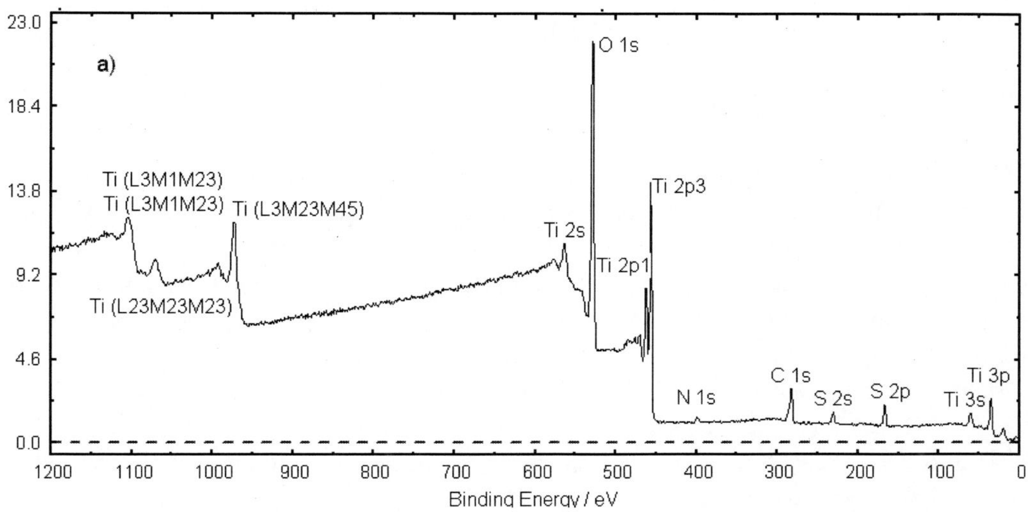

Figure 15. Full-range spectrum of *sample 8*.

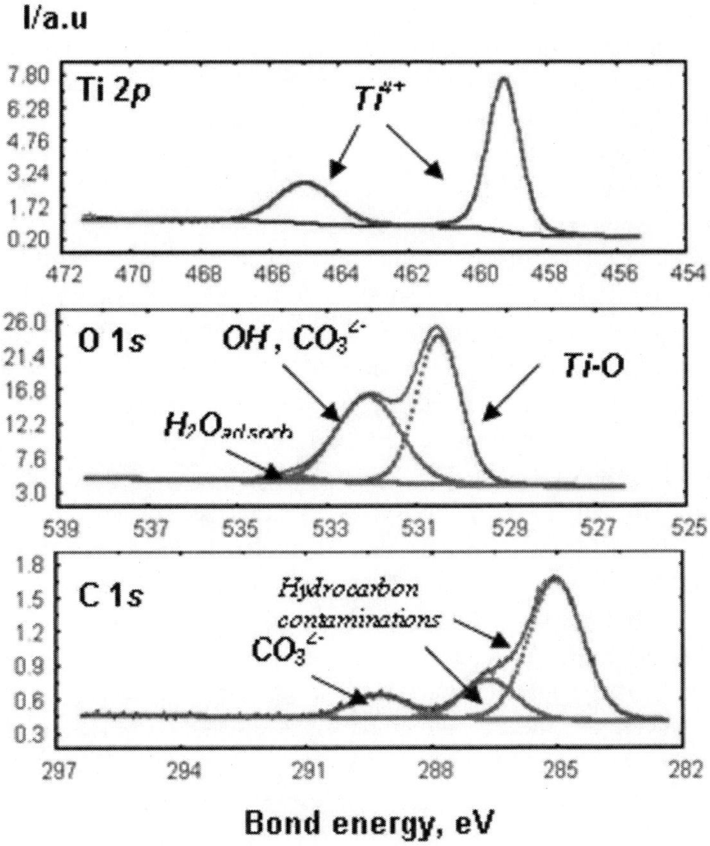

Figure 16. Local Ti 2*p*, O 1*s*, C 1*s* spectra; isolated components; and their interpretation for the *sample 8*.

5. PROPERTIES

5.1. Adsorption Properties

The purification, preservation and restoration of water resources are currently the most important issues. It applies to all types of water systems (natural, domestic and industrial). In recent years the requirements for drinking water have increased substantially. Nanosized titanium dioxide stands out among the inorganic sorbents: all its modifications show high adsorption capacity for As(III), As(V), U(VI), Bi(V) extracting from aqueous solutions [4,28,52,53] and favorable adsorption kinetics for removing low concentrations of other insoluble metals in aqueous systems (Cu(II) [54], As(V) [55,56,57], V(V) [58], Cu(II) and Mn(II) [59], Cd(II) [60]). These properties lead to water high purification from heavy metals (or) contribute to reduction in their concentration up to a few micrograms per liter.

It is known that the adsorption capacity of substances depends on *the surface properties*: chemical composition and physical condition, the nature of porosity and specific surface area. Dadachov showed that the adsorption properties of a large crystal are much worse compared with the adsorption of several small crystallites with the same surface [4,28]. Dadachov suggested that the generation of surface hydroxyl groups is the result of photo-decomposition of water molecules by nanosized semiconductors, in this case, by the nano-TiO_2 [28].

Composition of liquid phase, the solution from which the analytes are extracted (pH, ionic strength, the nature of the analytes and their concentration), as well as *conditions of the sorption process* (temperature, contact time, adsorbent weight, etc.) exercise significant influence over the sorption of metals. In research papers, devoted to the research of extraction of analytes from aqueous solutions using titanium dioxide as adsorbent, the emphasis is on the investigation of the influence on the sorption of the solution composition, temperature, and, to a lesser extent, the type of analyte, the nature of which is highly limited.

Study of the adsorption capacity of the samples with anatase structure was performed on synthetic mixtures (ICP-MS-68A standard, High-Purity Standards, USA; the solution contains 68 elements in a concentration of 10 mg/l each): 10 ml of standard solution with 1 ppm (C_i) of Al, As, Ba, Be, Bi, B, Ca, Cd, Ce, Co, Cr, Cs, Cu, Dy, Er, Eu, Ga, Gd, Ho, In, Fe, La, Pb, Li, Lu, Mg. Mn, Nd, Ni, P, K, Pr, Re, Rb, Sm, Sc, Se, Na, Sr, Tb, Tl, Th, Tm, U, V, Yb, Y, Zn, Sb, Ge, Hf, Mo, Nb, Si, Ta, Te, Sn, Ti, W, Zr, Ir, Os, Pd, Pt, Rh, Ru ions in 3% nitric acid was added to the specific weighed amount of sample. The solutions were shaken and left with the sorbent for a while, and then filtered through a 0.45-μm membrane filter and diluted with 3% nitric acid. The concentrations in the solutions after sorption (C_f) were measured by an inductively coupled plasma mass spectrometry (Perkin Elmer ELAN DRC-E, USA). The sorption degree (the R value, %) was evaluated as R=100-(C_f)/(C_i). Standard deviation for C_f and C_i determination was ± 2%.

We have examined adsorption capacity of the commercial samples with nano-TiO_2 (*Degussa P25, Hombifine N* and *Hombikat UV100*) (Figure 4 a-c), *sample 1* (Figure 4d), and *Hombikat UV100* and *sample 10* modified with H_2O_2 (Figure 17).

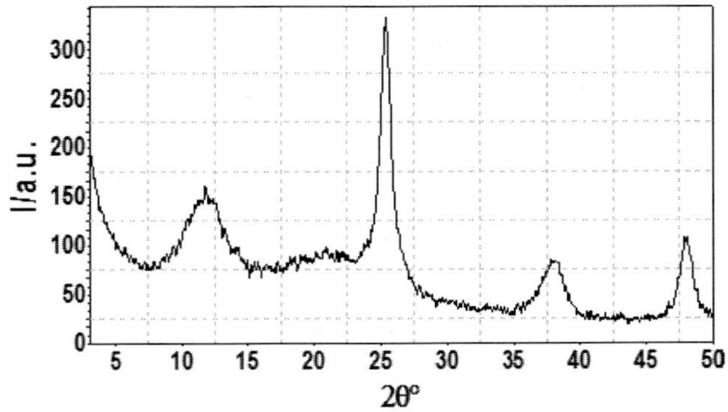

a

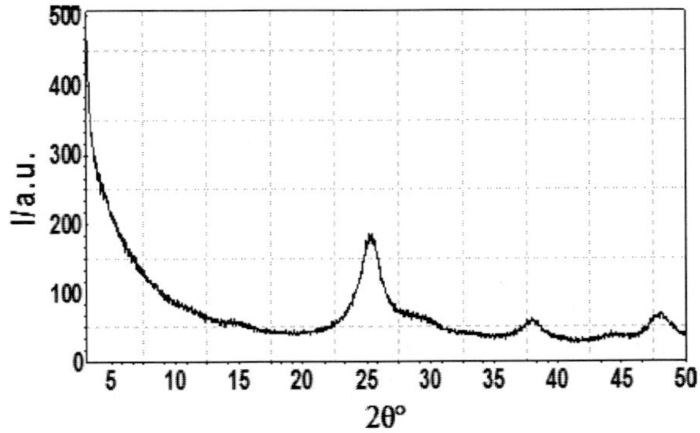

b

Figure 17. Diffraction patterns of samples modified with H_2O_2: *Hombikat UV100* (a) and *sample 10* (b).

It was found that the sorption degree (*R*,%) depends on the sorption conditions. The *R* value increases with the increasing of the sorption duration (τ_S) (Figure 18a), the sample weight (m) (Figure 18b) and the use of the membrane filter with 0.45 μm pores (Figure 18c).

It can be seen that the amount of Mo(VI) and W(VI) ions adsorbed by the TiO_2 nanoparticles from aqueous solutions first decreases and then increases with increasing the temperature from 4°C to 10°C and from 10°C to 85°C, respectively, using both filtration through a membrane filter and centrifugation; the latter method allows to decrease duration of residue removal (Figure 18 d) [61]. This behavior suggests exothermic nature of the process (in range of 4°C-10°C) which can be caused by physical modification of the adsorbents. Surfaces of metal oxides usually have positive charge at low pH (our case: pH=0.35, *i*=0.43(1), *i*-ionic strength). Mo(VI) and W(VI) ions in hydrated associates of different compositions probably have a negative charge (a similar to As(V) and Mo(VI) ions according to the literature data [56,57,62]). At low temperatures, they are adsorbed at the expense of forces of electrostatic character.

With the temperature increas from 4°C to 10°C, the attractive forces between TiO_2 and analytes weaken and adsorption decreases. Besides, while temperature increas, thickness of the boundary layer decreases, due to increasing tendency of an adsorbate to escape from the surface of the adsorbent to the solution, which results in decrease of adsorption as was recently reported in [63].

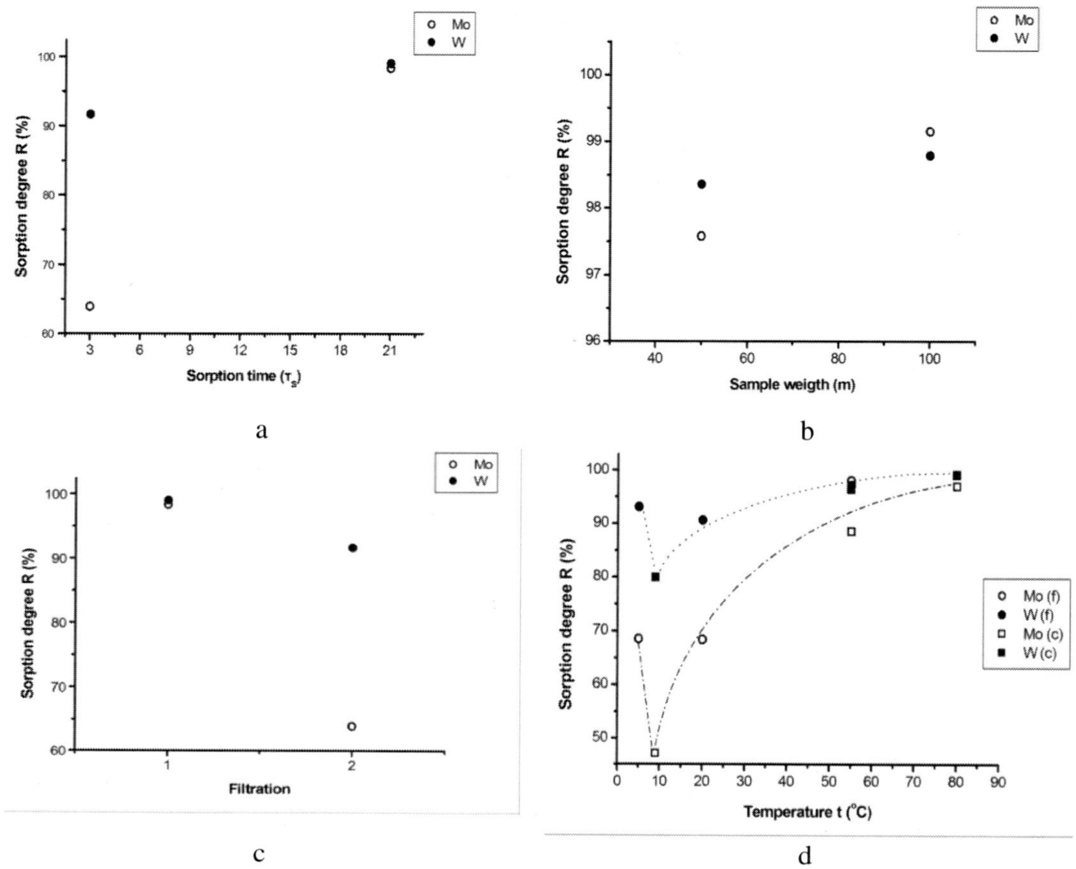

Figure 18. The sorption degree (R, %) vs.: sorption duration for *sample 1* (a), sample weight for *Hombikat UV100* (b), with (1) and without (2) filtration through a membrane filter for *sample 1* (c), sorption temperature (f- filtration through a membrane filter, c- centrifugation during 1 min) for *sample 1* (d).

It was found that in the temperature range from 10°C to 85°C, the process is endothermic, since the adsorption increases with increasing temperature (Figure 18d). Therefore, the rate of physical adsorption and the amount of ions adsorbed by this way decrease, meanwhile the degree of chemical adsorption increases. The increase of the sample dispersity under grinding can result in either increase (for Mo(VI)) or decrease (for W(VI)) of the sorption degree.

Correlations between the sorption degree and sample characteristics are less obvious, since it is difficult to choose samples differing by only one parameter.

Anatase modified with H_2O_2 (*sample 10*), which is a mixture of anatase and β-TiO_2 (Figure 16b), has the smallest value of the coherent scattering region size (D=40 Å) and the highest (comparing to other samples) sorption degree of W(VI) (Figures 18, 19).

Similar results were obtained earlier for As(III), As(V), Bi(V) and V(V) [4,28,51,55,56].

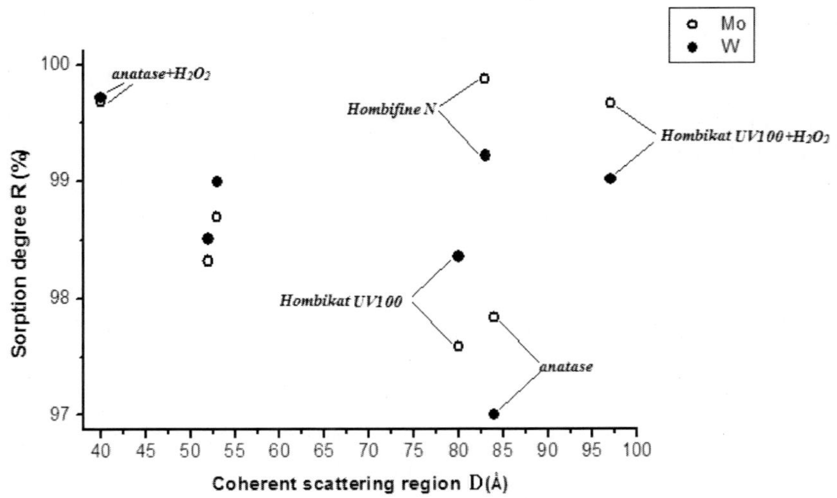

Figure 19. The sorption degree (*R*, %) of studied samples with anatase vs. the coherent scattering region size (*D*, Å).

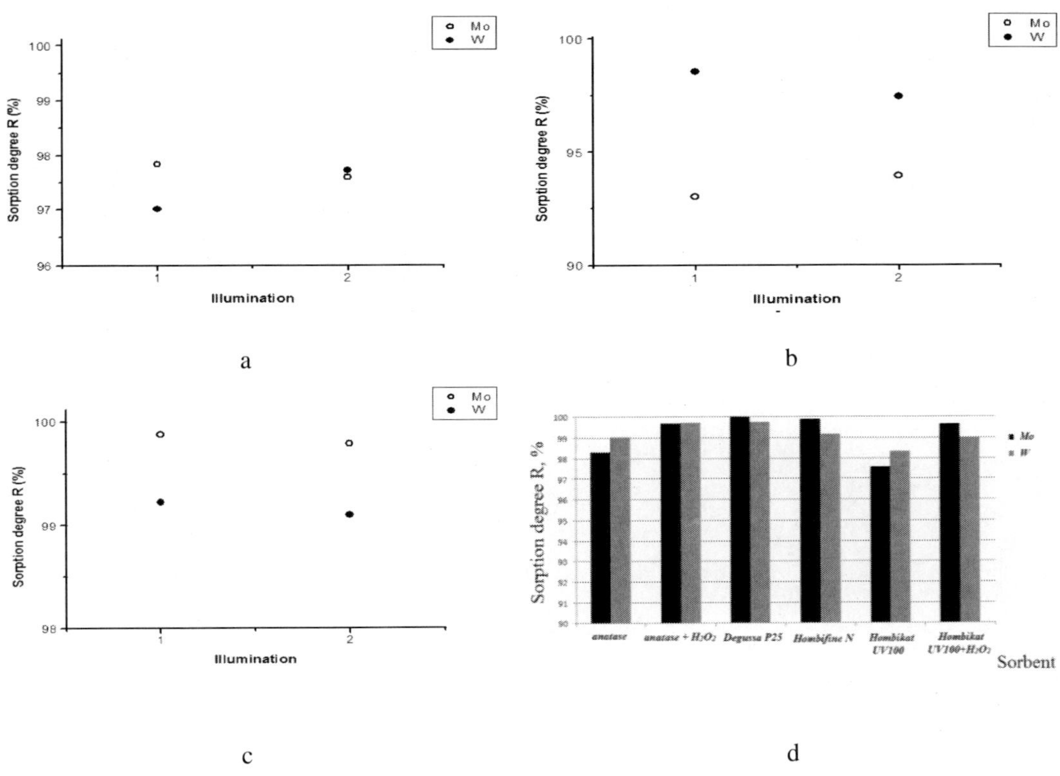

Figure 20. The sorption degree (*R*,%) with (1) and without (2) illumination for different sorbents: *sample 1* (a), *Degussa P25* (b), *Hombifine N* (c); the sorption degree (*R*,%) vs. type of adsorbents and analytes (Mo(VI) and W(VI)) (d).

However, there is no correlation between the coherent scattering region size, *D* (Å), and the sorption degree, *R* (%). The maximal recovery efficiency from aqueous solutions of

Mo(VI) ions has been obtained for *Hombifine N* with *D*=83 Å (Figure 19). *Hombikat UV100* has characteristics similar to *Hombifine N* except the specific surface area and composition of the sample. Nevertheless, *Hombikat UV100* (*D*=80Å) showed worse adsorption capacity as compared with *Hombifine N* (Figure 19). The results led to the conclusion that the maximal recovery efficiency depends on the concentration of adsorption centers – OH-groups – on the adsorbent surface as described previously by Hashimoto et al. [10] and Dadachov [28].

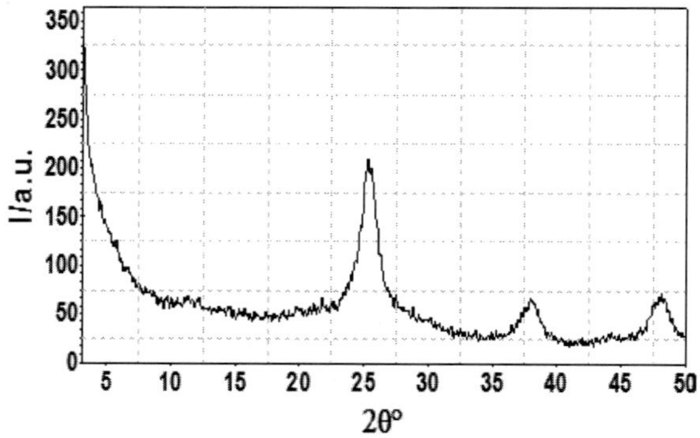

a

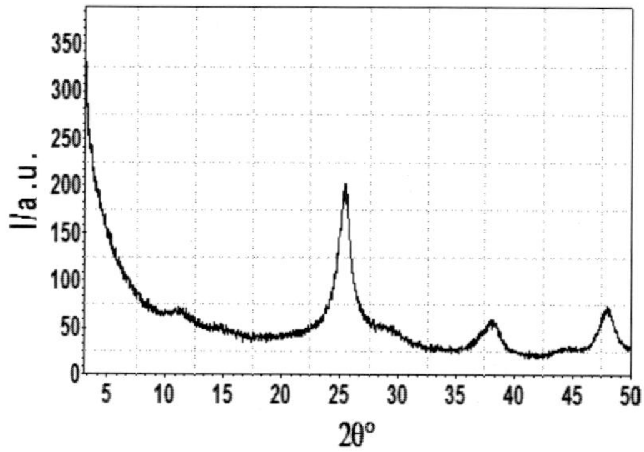

b

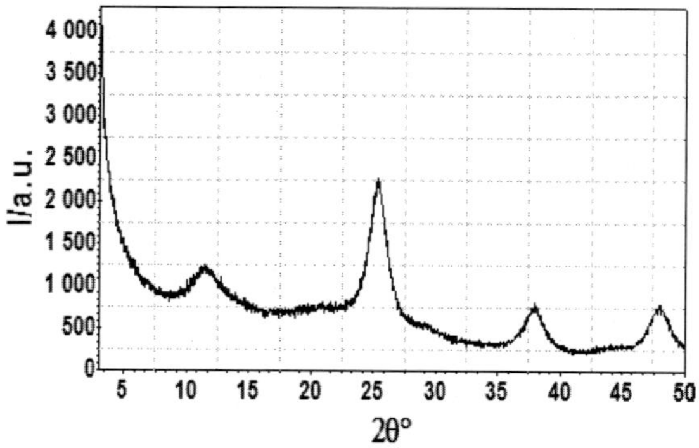

c

Figure 21. Diffraction patterns of *sample 11* (a), *sample 12* (b), *sample 13* (c).

Indeed, although the specific surface area of *Hombikat UV100* is higher than that for *Hombifine N* because of the presence of hydrous amorphous titanium dioxide (Figure 4c), the concentration of OH-groups on the surface of *Hombikat UV100* is lower. Only the sample containing pure hydrous amorphous titanium dioxide has the characteristics (S_{BET} = 317.5 m^2/g and S_C =46.0 m^2/g; V_U =0.099 cm^3/g; V_n =0.229 cm^3/g) and sorption degree (R_{Mo}=91.70% and R_W=97.72%) that are lower than for *Hombikat UV100* with R_{Mo}=97.50% and R_W=98.40%. However, *Hombikat UV100* modified with H_2O_2 by cold impregnation (*D*=97 Å) (Figure 17*a*) shows the highest sorption degree (R_{Mo}=99.70% and R_W=99.05%) and the highest (among other samples) *D* value (Figure 19). So, treatment of the reaction mixture with hydrogen peroxide led to decrease of the *D* value for anatase in contrast to the cold impregnation that increases this value. Our experience suggests that the additional treatment of samples with hydrogen peroxide is of particular value for increase of the sorption degree associated with appearance of the O-O-H-groups on the surface of particles.

On the other hand, the nature of analytes affects the sorption degree. The nanoparticle surfaces of the studied samples contain various cracks and pores which penetrate more or less deeply into the adsorbent body. Their existence, especially ultrananopores, which size is comparable to sizes of ions (r_{Mo}=0.69Å, r_W=0.60Å − ionic radius), considerably complicates the adsorption process. Large ions cannot penetrate into ultrananoporous of small sizes due to steric reasons; therefore, the ultraporous adsorbent possesses lower specific surface for such ions as compared to smaller adsorbed ions.

The observed correlations between the *R* value and the sample illumination depend on the kind of analytes and the characteristics of adsorbents (specific surface area, volume of nano- and ultrananopores, nanoparticles and coherent scattering region sizes) (Figure 20a-c).

We have reached the maximal recovery efficiency of Mo(VI) ions sorption from aqueous solutions for the commercial sample *Hombifine N* (*R*=99.90%; sorption conditions with the use of membrane filter with 0.45-μm pores: t=20°C, τ_S=21 h, m=50 mg), of Mo(VI) and W(VI) ions sorption for the commercial sample *Degussa P25* (R_{Mo}=99.99% and R_W=99.75%; sorption conditions with the use of centrifugation during 1 min: t=80°C, τ_S =3 h, m=50 mg), and for nanosized anatase modified with H_2O_2 (*sample 10*) (*R*=99.75%; standard sorption conditions with the use of membrane filter with 0.45-μm pores) (Figure 20d).

In general, the obtained experimental data shows the role of sorption conditions, sorbent characteristics and analyte nature in the process of Mo(VI) and W(VI) extraction from aqueous solutions (Figure 18d).

The adsorption capacity of commercial samples *Degussa P25, Hombifine N*, *Hombikat UV100,* and *sample 1*, *sample 11* (Figure 21a), *sample 12* (Figure 21b), as well as *Hombikat UV100* (Figure 16 a), *sample 10* (Figure 17b) and *sample 13* (Figure 21c) modified with H_2O_2 in the process of Nb(V) and Ta(V) extraction from aqueous solutions was investigated [64].

We have found correlations between the sorption degree (*R*, %) and *the main conditions of the sorption process* (Figure 22), which can be summarized as follows:

1. With increase of sorption time τ_C, the sorption degree increases for *Hombifine N* and *sample 1*, the *R*(Nb) value is almost unchanged and the *R*(Ta) value decreases for *Degussa P25* (Figure 22a);
2. With increase of adsorbent weight *m*, the sorption degree decreases for Ta(V) and is almost unchanged for Nb(V) (Figure 22b);

3. When using filtration of the aqueous solution through a membrane filter with 0.45-μm pores (Figure 22c), the *R*(Nb) and *R*(Ta) values decreases and increases, respectively, for *sample 1*; the *R*(Nb) is almost unchanged and the *R*(Ta) decreases for *Degussa P25* and *Hombifine N*;
4. The sorption degree is almost independent of illumination (the sorption was carried out under daylight fluorescent lamp and in the dark) (Figure 22d-f), that eliminates the effect of photocatalysis in the sorption process.

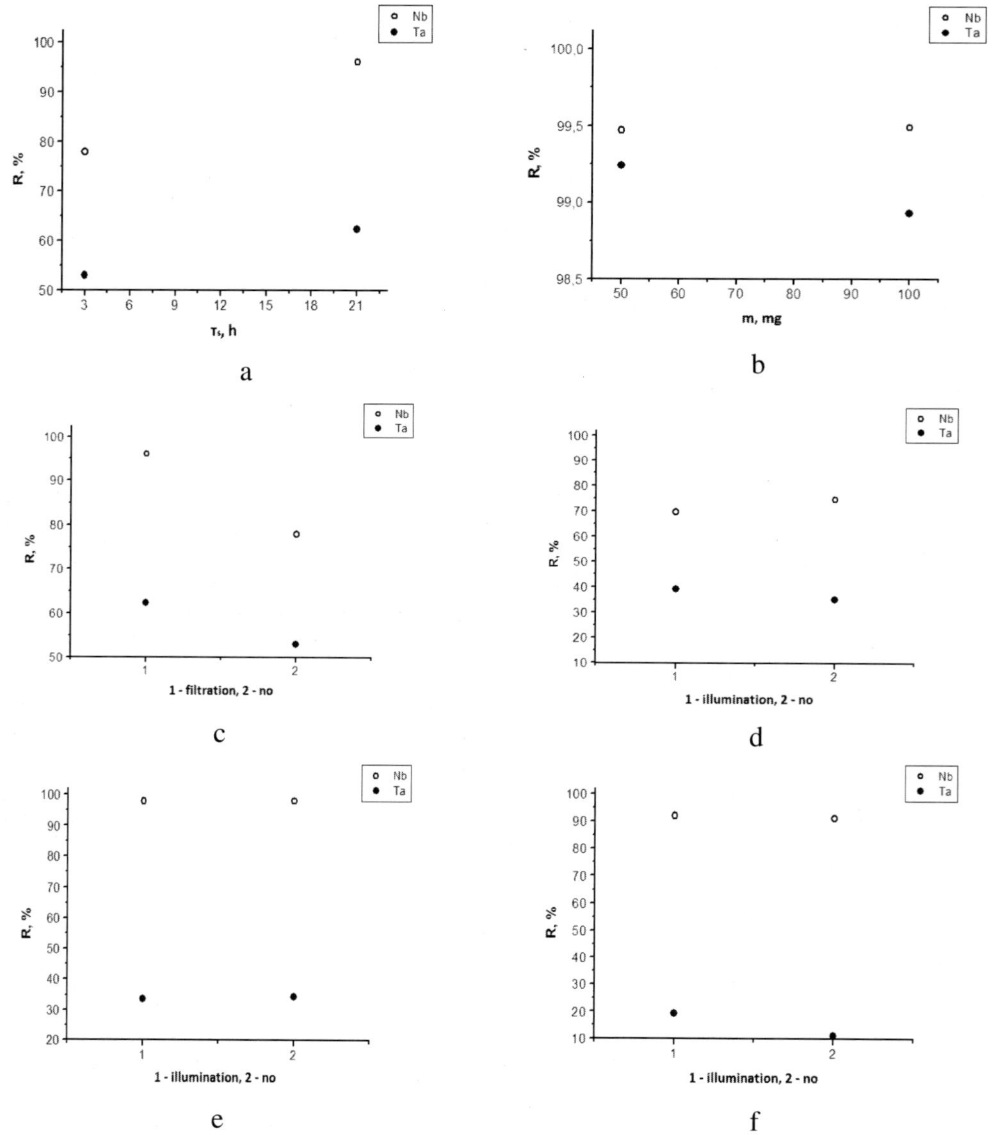

Figure 22. The sorption degree (*R*, %) vs.: sorption duration for *sample 1* (a), sample weight for *Hombikat UV100* (b), with (1) and without (2) filtration through a membrane filter for *sample 1* (c), illumination for *Degussa P25* (d), *Hombifine N* (e), *sample 1* (f).

As can be seen from Figure 22, for all the adsorbents studied the extraction of Nb(V) ions is higher as compared to Ta(V) ions under any sorbtion conditions. However, *R*(Ta)>*R*(Nb) at the temperature range of 4-10°C (Figure 23); the extraction of Nb(V) ions increases with increasing temperature.

Figure 23 shows that, when using both filtration through a membrane filter and centrifugation, the amount of adsorbed Nb(V) and Ta(V) ions first decreases (in the range of 4-10°C for Nb(V) and 4-20°C for Ta(V)), and then increases (in the range of 10-80°C for Nb(V) and 20-80°C for Ta(V)) with the temperature increase; centrifugation reduces the duration of the residue removal. This behavior indicates exothermic nature of the process in the temperature range of 4°C-10°C and 4°C-20°C for Nb and Ta, respectively, due to physical adsorption. It was found that at higher temperature (up to 80°C), the process is endothermic, since adsorption increases with increase of temperature (Figure 23). It also leads to increase of the chemisorption rate.

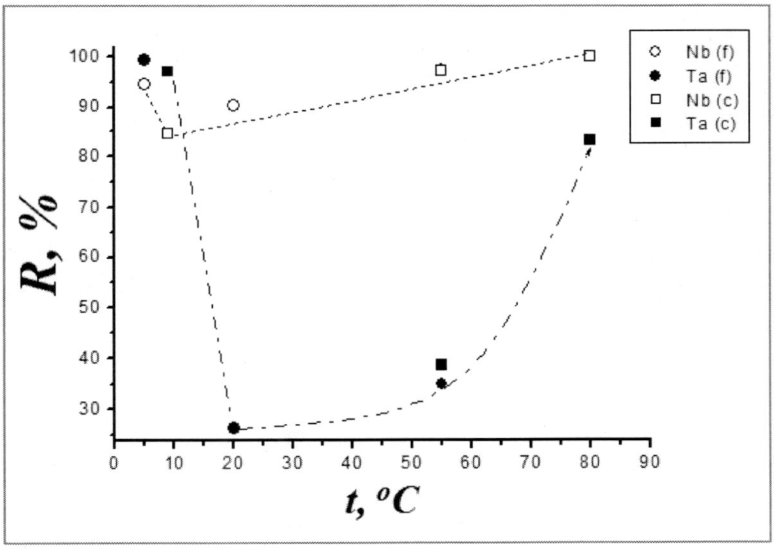

Figure 23. The sorption degree (R, %) vs. sorption temperature: f- filtration through a membrane filter, c- centrifugation during 1 min.

All the studied samples have different characteristics of their microstructure. Figure 24 shows the relationship between the *R* value and characteristics of samples (*N* - size of nanoparticles, *S* - specific surface area, *V* - volume of nano- and ultrananopores) under sorption conditions t = 20°C, τ_S = 21 h, m = 50 mg. Figure 24 shows that the sorption degree depends only slightly on *N, S,* and *V* values for *sample 1*, *Hombifine N,* and *Degussa P25:* the *R* value for Nb(V) ions is almost the same at intervals of N=12÷28 nm (Figure 24c), S_{CM}=54÷116 m^2/g (Figure 24d), V_U=0.016÷0.083 sm³/g (Figure 24e), V_n=160÷340 sm³/g (Figure 24f). The same tendency is observed for Ta(V) ions, the *R* values being smaller for them as compared with Nb(V) ions (Figure 24c-f). At almost equal *D* values for *samples 11* and *12* (*D* ~ 50 Å) as well as for *sample 1* and *Hombifine N* (*D* ~ 83 Å), the *R* values for Ta(V) ions are significantly different (*R*(Ta) = 27.79%, 92.81%, 89.83%, 33.65% for samples *11*, *12*, *1* and *Hombifine N*, respectively), although the sorption conditions for them are the same (standard conditions: t = 20 °C, τ_S = 3 h, m = 50 mg) (Figure 24a). The *R* value for

Nb(V) ions is almost unchanged in the range of D = 50÷85 Å: the highest value R(Nb) = 98.44% is observed for *sample 12* (Figure 24a). Therefore, the sorption degree of Nb(V) and Ta(V) depends on the composition of the samples (Table 2) and it is largely defined by *type of analyte* rather than by the characteristics of adsorbents, in particular, size of coherent scattering region [10,28]. It is assumed that we have not reached the high content of crystallites associated with the amount of reactive OH groups. Nevertheless, the maximum sorption degree was reached for: the Nb(V) and Ta(V) ions with R = 98.9% for *Hombifine N*; the Nb(V) and Ta(V) ions with R_{Nb} = 99.5% and R_{Ta} = 99.2% for *Hombikat UV100*; the Nb(V) and Ta(V) ions with R_{Nb} = 99.9% and R_{Ta} = 97.5% for *Degussa P25* (Figure 25).

The diffraction patterns of the samples modified with H_2O_2 (*Hombikat UV100* and *sample 10*, Figure 17; *sample 13*, Figure 21c) show that *sample 13* and *Hombikat UV100* contain anatase and HD. *Sample 10* contains β-TiO_2 which is formed from $H_2Ti_3O_7$ [65]. Introduction of hydrogen peroxide during the titanium dioxide synthesis process leads to a decrease of coherent scattering region size, and the treatment of titania powder with hydrogen peroxide leads to its increase (Figure 24b), adsorption capacity of *Hombikat UV100* and anatase (Figure 24) decreasing sharply, and increasing, respectively.

It should be noted that the sorption degree of Ta(V) ions becomes greater as compared to Nb(V) ions (under standard sorption conditions), regardless of the method of hydrogen peroxide introduction (Figure 24).

The Ta(V) and Nb(V) ions have the same ionic radii (r^{VI} = 0.64Å) and similar electronegativity values (χ_{Nb}=1.6, χ_{Ta}=1.5). However, Nb(V) ions are more active chemically, have a more pronounced acid character, and are easily reduced in aqueous solutions as compared to Ta(V) ions (the difference in chemical properties of the niobium and tantalum is due to structures of electron shells of their atoms: $4d^45s^1$ for Nb and $5d^36s^2$ for Ta) [66]. Indeed, the present work shows different nature of Nb(V) and Ta(V) ion extraction from aqueous media using adsorbents with anatase: the sorption degree of Nb(V) ions is much greater than that of Ta(V) ions due to chemisorption and is less, vice versa, due to physical sorption (Figure 23a,b), except of the samples with anatase modified with hydrogen peroxide (Figure 23b), for which strong chemical interaction between the adsorbent and Ta(V) is observed. The maximal Ta(V) ion extraction with R = 99.9% was found for the *sample 10*, which was modified with hydrogen peroxide during the synthesis process.

The dependence of adsorption on the molecular structure of the analyte is very complicated and it is quite difficult to establish any regularities. However, the obtained experimental data indicate that Nb(V) and Ta(V) form different configurations of niobium and tantalum in the solution. It is assumed that an oxo-group of a niobium-containing particle, which is absent in a tantalum-containing particle, interacts well with hydroxyl groups on titania surface by hydrogen bonds. In the presence of chelating peroxo groups on the surface of titanium dioxide, hydroxyl groups of tantalum-containing particles, which is absent in niobium-containing particles, promote formation of hydrogen bonds [67].

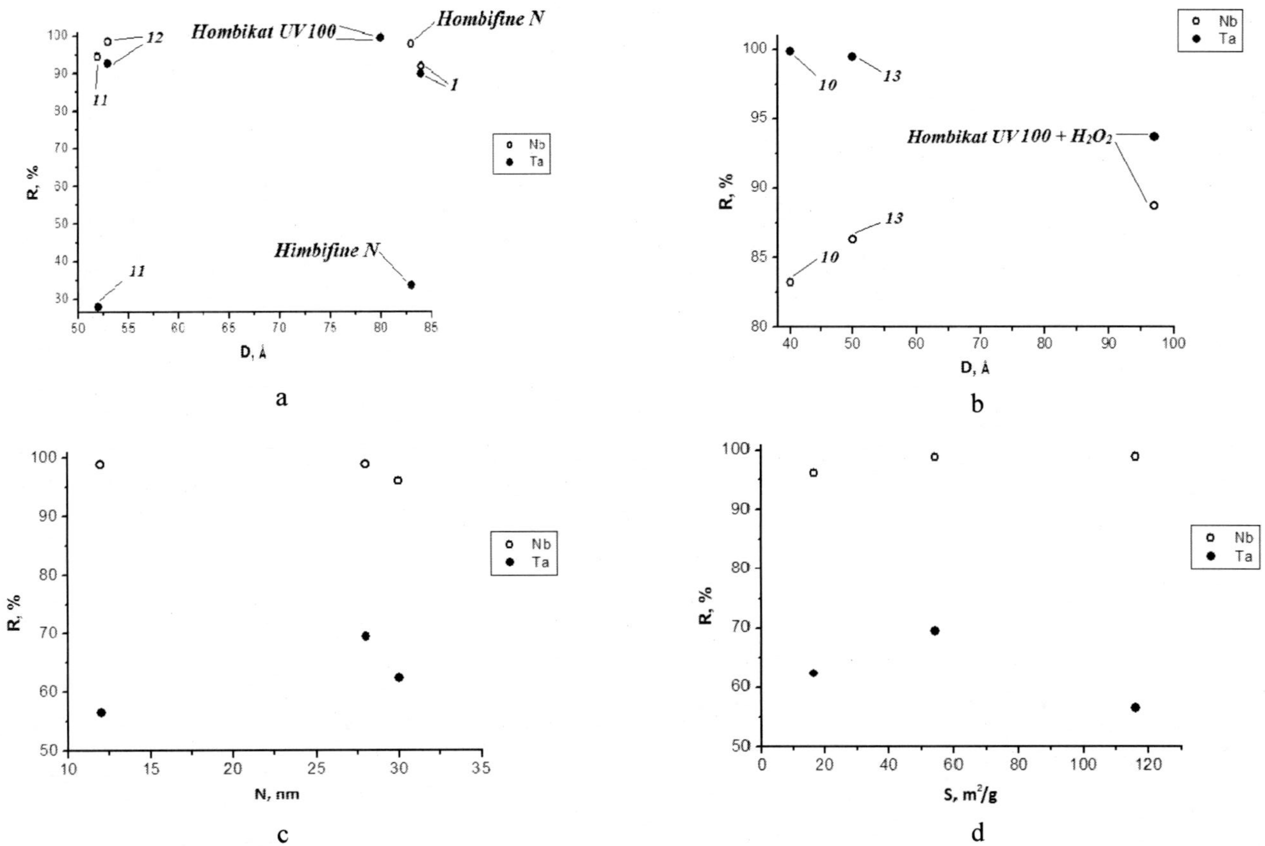

Figure 24 (Continued)

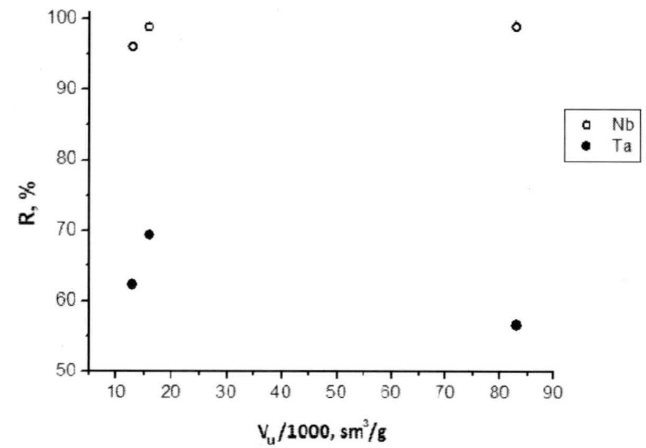

e

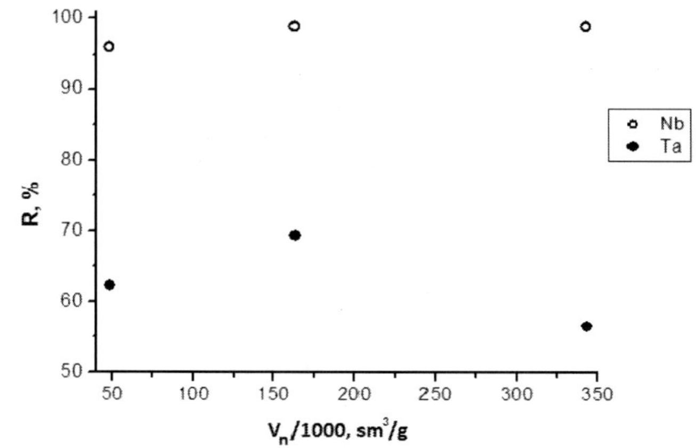

f

Figure 24. The sorption degree (*R*, %) vs.: size of coherent scattering region (*D*, Å) of samples unmodified (a) and modified with hydrogen peroxide (b): 1 –*sample 11*, 2- *sample 12*, 3-*sample 1*, 4-*sample 10*, 5-*sample 13*; the size of nanoparticles (*N*, nm) (c); the specific surface area (comparative method) (S_{CM}, m^2/g) (d); the volume of ultrananopores (V_U, sm^3/g) (e) and nanopores – (V_n, sm^3/g) (f) – for *sample 3*, *Hombifine N,* and *Degussa P25.*

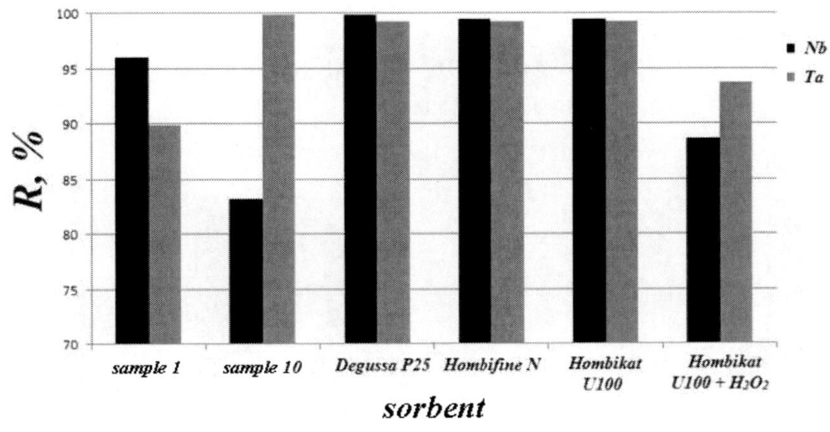

Figure 25. The sorption degree (*R*, %) of Nb(V) and Ta(V) ions vs. the type of analyte.

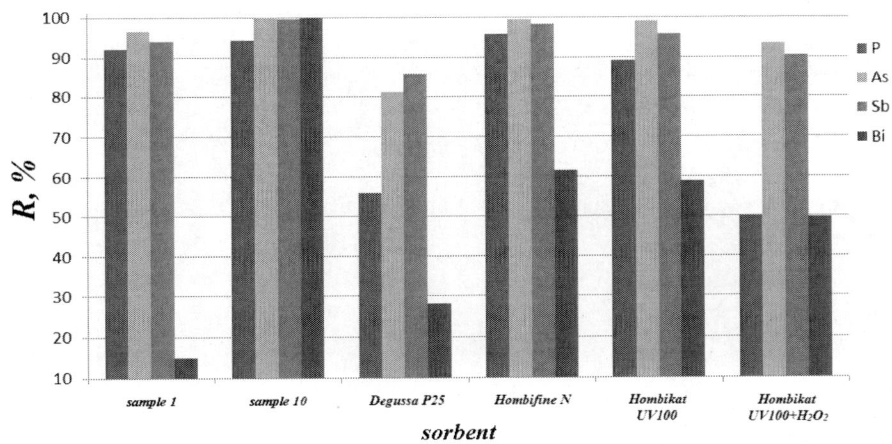

Figure 26. The sorption degree vs. the nature of titanium dioxide and the type of analyte - P(V), As(V), Sb(V), and Bi(V).

The maximum sorption degree was obtained for:

- P(V) ions with *R*=95.69%, As(V) with *R*=99.3%, Sb(V) with *R*=98.28%, and Bi(V) with *R* = 61.55% for *Hombifine N*; P(V), As(V), Sb(V) and Bi(V) ions with *R*=89.19%, 98.94%, 95.82% and 58.91%, respectively, for *Hombikat UV100*; P(V), As(V), Sb(V) and Bi(V) ions with *R* = 55.98%, 81.22%, 85.72% and 27.97%, respectively, for *Degussa P25* (Figure 26).
- Se(VI) ions with *R*=99.17% and Te(VI) ions with *R* = 98.61% for *Hombifine N*; Se(VI) and Te(VI) ions with *R*=97.55% and 99.14%, respectively, for *Hombikat UV100*; Se(VI) ions with *R*=65.67% and Te(VI) ions with *R* = 91.57% for *Degussa P25* (Figure 27).
- Ge(IV) and Sn(IV) ions with *R* = 56.19% and 90.04%, respectively, for *Hombifine N*; Ge(IV) and Sn(IV) ions with *R*=44.02% and 86.22%, respectively, for *Hombikat*

UV100; Ge(IV) ions with *R*=22.88% and Sn(IV) ions with *R* = 70.38% for *Degussa P25* (Figure 28).

- Mo(VI), Nb(V), Ta(V), As(V), Bi(V), and Sn(IV) ions up to the maximum allowable concentration in aqueous media.

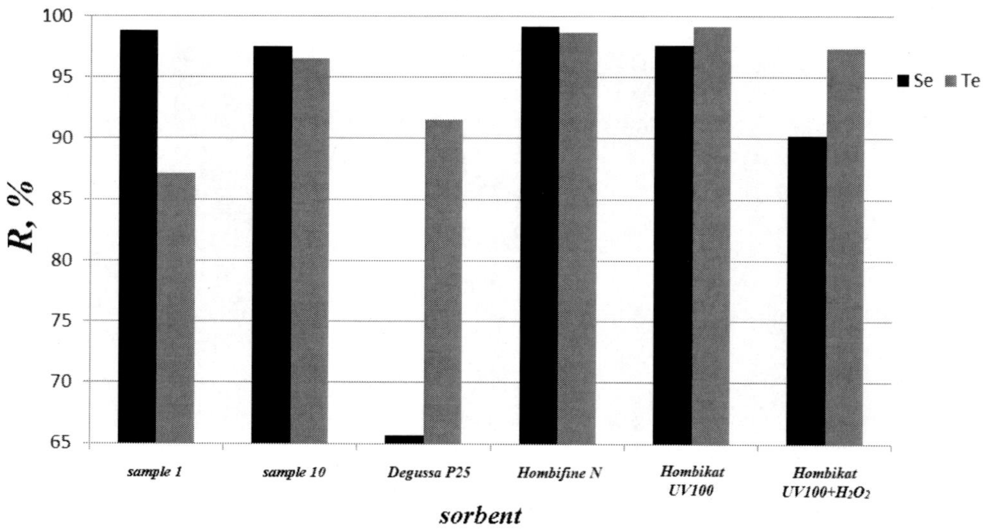

Figure 27. The sorption degree vs. the nature of titanium dioxide and the type of analyte – Se(VI) and Te(VI).

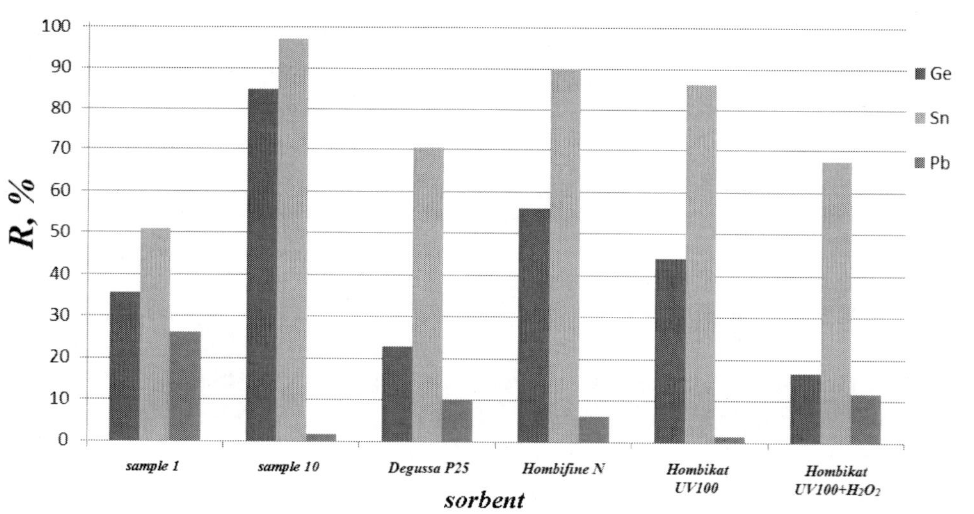

Figure 28. The sorption degree vs. the nature of titanium dioxide and the type of analyte – Ge(IV), Sn(IV), and Pb(IV).

5.2. Photocatalytic Properties of Anatase Modified with Manganese

Titanium dioxide is considered as the most effective photocatalyst for the decomposition of a wide range of organic and inorganic toxic substances [68], the photocatalytic properties of nanosized anatase being considered the best [7]. However, such a photocatalyst can work only under UV radiation with a wavelength < 386 nm, which is comparable with the band gap of anatase (3.2 eV corresponds to 387 nm) [69]. Decrease in TiO_2 band gap or appearance of energy levels of the impurities in it leads to photoabsorption shift toward the longer wavelength of the visible region, i.e., the maximum wavelength of the "working" spectral region increases, and the minimum quantum energy necessary for the photo-excitation decreases.

Table 3. Synthesis conditions of *samples 14-17* (*Series I*)

Sample	Synthesis Conditions
14	1) 7.35 g $TiOSO_4 \cdot H_2O \cdot H_2SO_4$ (Aldrich, 2012) were mixed with 40 mL H_2O; the mixture was heated up to t = 91 °C and stirred at t = 95–98 °C for 16 min and allowed to stay at t = 80±0.5 °C for 4 h; 2) 5 mL of 10M H_2O_2 and 20 mL of 0.05M $KMnO_4$ were added into the reaction mixture after 1 and 2.5 hours of thermostating, respectively; 3) In several days, the precipitate was separated by vacuum filtration, washed with water and acetone, dried at t ~ 100 °C for 1 h, and UV irradiated (125 W) for 1 h.
15	1) 0.24 g of commercial nano-TiO_2 (*Hombifine N* type) were dispersed in 24 mL of 0.05M $KMnO_4$ and stirred at t = 56–60 °C for 10 min; 2) The reaction mixture was UV irradiated (125 W, with stirring) for 1 h; 3) In several days, the precipitate was separated by vacuum filtration, washed with water and acetone, dried at t ~ 100 °C for 1 h, and UV irradiated (125 W) for 1 h.
16	1) 7.35 g $TiOSO_4 \cdot 2H_2O$ (USSR, 1990) were mixed with 40 mL H_2O; the mixture was heated up to t = 98 °C and stirred at t = 97–102 °C for 25 min; 10 mL of 10M H_2O_2 and 40 mL of 0.01M $KMnO_4$ were added into the reaction mixture after 2 min of stirring; 2) The resulting suspension was kept at t = 75±0.5 °C for 5 h; 3) In several days, the precipitate was separated by vacuum filtration, washed with water and acetone, dried at t ~ 100 °C for 1 h.
17	1) 3.85 g $TiOSO_4 \cdot H_2O \cdot H_2SO_4$ (Aldrich, 2012) were mixed with 20 mL H_2O, heated up to t = 88 °C and stirred at t = 90–94 °C for 4 min, while the reaction mixture became transparent (suspension conversed to colloidal solution); 2) The reaction mixture was stirred at 88–92°C for 2.5 min and quickly cooled down to 0°C in an ice bath; 3) 32 mL of 0.6M $Mn(CH_3CO_2)_2$ were mixed with 32 mL of 0.9M $KMnO_4$; the mixture was stirred and irradiated with a halogen lamp (250W) at t = 76 °C for 1 h up to the disappearance of the violet tinge of the liquid phase; 4) In several days, the precipitate was separated by vacuum filtration, washed with water and acetone, dried at t ~ 100 °C for 1 h, and UV irradiated (125 W) for 1 h;
	5) 0.282 g of the powder obtained in Step 4 was dispersed in titania sol obtained in Step 2; the suspension was kept at t = 80±0.5 °C for 7 h; 6) In several days, the precipitate was separated by vacuum filtration, washed with water and acetone, dried at t ~ 100 °C for 1 h.

For the preparation of photocatalysts based on titanium dioxide, which are active under visible light, photosensitization, in particular doping [70], or modifying [71] are preferably used. Doping is usually carried out at elevated temperatures; it can be avoided by modifying the titanium dioxide by other methods. In this case, organic dyes as well as colored inorganic and coordination compounds are used. The conditions for the preparation of modified samples affect the material properties.

Series I. The samples were synthesized using sulfate method [1]. The conditions of Mn(IV) modifying are given in Table 3.

Photocatalytic activity of the samples was studied in the model reactions of photocatalytic degradation of organic dye Methyl Orange (MeO) [72]: the initial concentration of the MeO aqueous solution, c_0 = 1.04 mmol/L; the weight of suspended photocatalyst was 20 mg per 50 mL of the reaction mixture. The ratio of the MeO current concentration (after the irradiation of mixture for τ minutes by halogen lamp, 250 W) to its initial concentration (the c/c_0 value) was monitored spectrophotometrically by measuring intensity of the MeO absorption band at 490÷510 nm.

The *sample 14* modified with manganese (IV) (Table 3) was taken as the initial one. The diffraction pattern of *sample 14* is shown in Figure 29a.

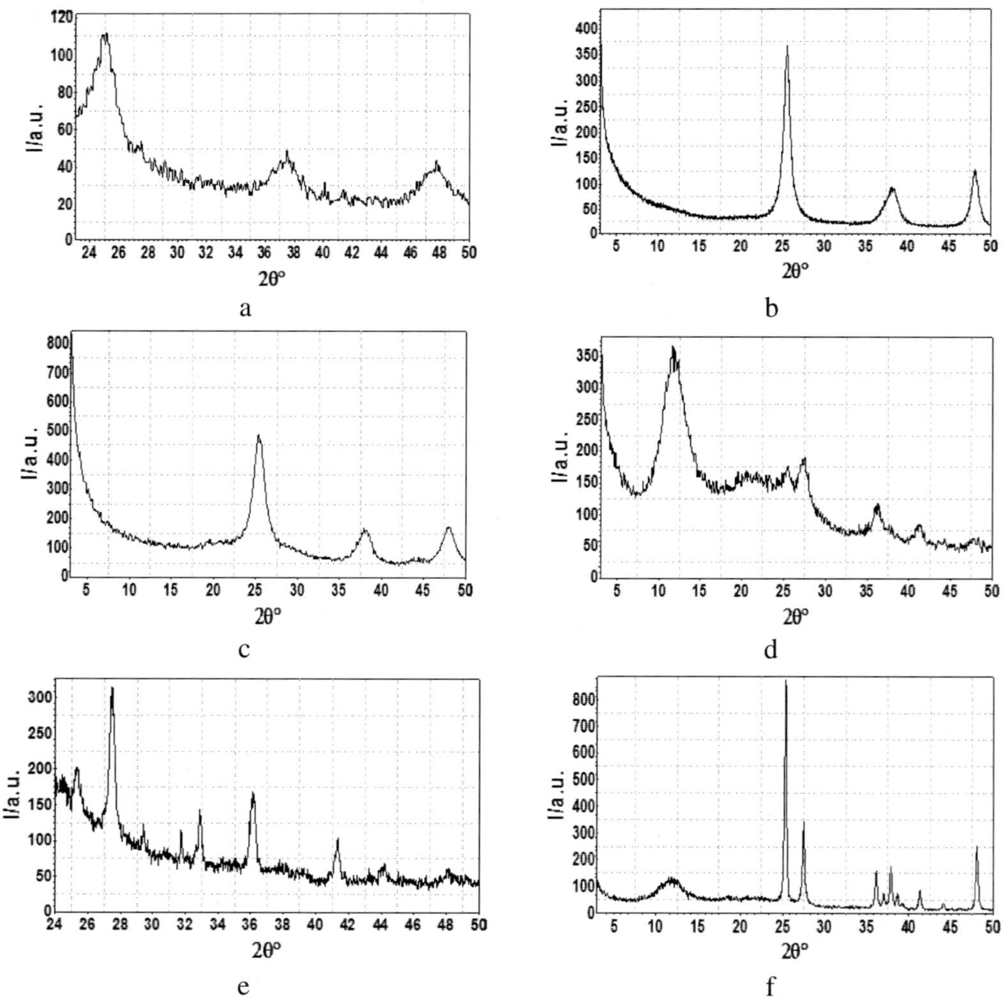

Figure 29. Diffraction patterns of *samples 14-18* (a-e) and $TiO_2(Mn^{IV})$ commercial photocatalyst (Aldrich, 2012, CAS № 13463-67-7) (f).

Sample 14 contains nano-anatase with coherent scattering region D=60(3) Å. Rutile content in the sample increases from ~61.5% to ~92% with increasing of duration of annealing at t = 400°C from 1 h to 7 h, respectively. It is accompanied by increase of the D value for anatase (D = 265(1) Å) and decrease of photocathalitic activity from k = 0.08685 min^{-1} (*sample 14*) to k = 0.00046 min^{-1}.

Samples 15 and 16 (Figure 29b,c) contain nanosized anatase with D=80(4) Å and D=55(3) Å, respectively. The photocatalytic properties of *sample 16* are comparable to those of *sample 14* and better as compared to *sample 15 - Hombifine* (Mn^{IV}).

Sample 17 represents an initial step of nanosized titania formation (Figure 29d): ~84% of hydrolytic titanium dioxide, ~6% of nano-anatase, and ~10% of nano-rutile. It should be noted that this sample already contains nano-rutile, the formation of which under given synthesis conditions is due to high concentration of MnO_2 in solution (Table 3), i.e., the introduction of Mn(IV) promoted the phase transition from anatase to rutile. Annealing of *sample 17* at $t \approx$ 400°C for 1 hour (*sample 18*) results in increase in the nano-rutile content up to ~65% (Figure 29e). Further annealing at the same temperature leads to 100% nano-rutile. In the same series of samples, increase of the coherent scattering region size is observed for nano-anatase: D=70(4) Å for *sample 17* and D=90(4) Å for *sample 18*. Photocatalytic properties of *sample 17* and *sample 18* were inferior to those of *sample 14* and *sample 16*, but significantly better than those of commercial sample TiO_2 containing 1% Mn (Sigma Aldrich, USA), which consists of ~75% rutile (D = 275(1) Å) and ~25% anatase (D = 275(1) Å).

Based on the obtained data, it is not excluded that the photocatalytic properties are correlated with the size of the coherent scattering region, and the maximum photocatalytic activity is achieved at D~55÷60 Å [73]. Modifying titanium dioxide with Mn(IV) leads to formation of localized impurity states in the band gap of titanium dioxide, hence, the decrease of sizes of TiO_2 nanocrystals results in an increase of concentration of impurity centers and increase of activity. According to [74,75], formation of the so-called phase junctures between anatase and rutile were considered to further promote the photocatalytic activity of TiO_2 by a synergy effect. In our case, judging from the decrease of the photocatalytic activity of the annealed samples containing a mixture of rutile and anatase, compared with synthesized samples without rutile the phase junctures were not formed.

Series II. Table 4 shows the synthesis conditions of four titanium dioxide samples modified with manganese. The main difference in the synthesis conditions is the sequence of the introduction of oxidant (potassium permanganate) and reductant (hydrate of ammonia, which instantly form the ammonium cations in the reaction medium with pH<1) into the titanium-containing reaction mixture.

Photocatalytic activity of the samples was studied in the model reactions of photocatalytic degradation of organic dye Methyl Orange (MeO). A powdered sample (weight content, 0.5 g/L) was put into aqueous solution of the dye (initial concentration, 0.05 mmol/L). The obtained suspension (volume, 50 mL) was irradiated with visible light (halogen lamp; irradiation at wavelengths λ > 450 nm; nominal power, 60 W) in an open container (irradiation area, 0.07 m^2) with continuous stirring using a magnetic stirrer. Aliquots (approximately 3 mL) were taken from the reaction mixture; the solid phase was separated by centrifugation if necessary. MeO concentration during photoreaction (the c/c_0 value) was monitored by measuring the change in intensity of the absorption band of the chromaphoric group at λ = 455÷470 nm.

Figure 30 shows the diffraction patterns of the synthesized *samples 19-22*.

Table 4. Synthesis conditions of *samples 19-22* modified with manganese (*Series II*)

Sample	Synthesis conditions
19	1) 5 g $TiOSO_4 \cdot 2H_2O$ were mixed with 0.1 g $KMnO_4$ and 50 mL of 4.4% H_2SO_4; 2) The mixture was heated up to t = 90 °C and stirred at t = 90–95 °C for 40 min and then at t = 80±0.5 °C for 220 min; 3) 20 mL of 2.5% $NH_3 \cdot H_2O$ were added into the suspension obtained; 4) The suspension was kept at t = 80±0.5 °C for 40 min; 5) 30 mL of 1.8% HCl were added into the suspension; 6) In several days, the precipitate was separated by vacuum filtration, washed with water and acetone, dried at room temperature in air.
20	1) 5 g $TiOSO_4 \cdot 2H_2O$ were mixed with 50 mL of 4.4% H_2SO_4; 2) The mixture was heated up to t = 90 °C and stirred at t = 90–95 °C for 40 min and then at t = 80±0.5 °C for 220 min; 3) 20 mL of 2.5% $NH_3 \cdot H_2O$ were added into the suspension obtained; 4) The suspension was kept at t = 80±0.5 °C for 40 min; 5) 10 mL of 0.06 M $KMnO_4$ were added into the reaction mixture; 6) 30 mL of 1.8% HCl were added into the suspension obtained; 7) In several days, the precipitate was separated by vacuum filtration, washed with water and acetone, dried at room temperature in air.
21	1) 5 g $TiOSO_4 \cdot 2H_2O$ (Sigma Aldrich, USA) were mixed with 0.1 g $KMnO_4$ and 42 mL of 1.3 M $(NH_4)_2SO_4$ (in the form of H_2SO_4 and $NH_3 \cdot H_2O$ solutions); 2) The mixture was heated up to t = 90 °C and stirred at t = 90–95 °C for 40 min and then at t = 80±0.5 °C for 220 min; 3) The suspension was kept at t = 80±0.5 °C for 40 min; 4) 30 mL of 1.8% HCl were added into the suspension obtained; 5) In several days, the precipitate was separated by vacuum filtration, washed with water and acetone, dried at room temperature in air.
22	1) 5 g $TiOSO_4 \cdot 2H_2O$ (Sigma Aldrich, USA) were mixed with 42 mL of 1.3 M $(NH_4)_2SO_4$ (in the form of H_2SO_4 and $NH_3 \cdot H_2O$ solutions); 2) The mixture was heated up to t = 90 °C and stirred at t = 90–95 °C for 40 min and then at t = 80±0.5 °C for 220 min; 3) The suspension was kept at t = 80±0.5 °C for 40 min; 4) 10 mL of 0.06 M $KMnO_4$ were added into the reaction mixture; 5) 30 mL of 1.8% HCl were added into the suspension obtained; 6. In several days, the precipitate was separated by vacuum filtration, washed with water and acetone, dried at room temperature in air.

Samples 19-21 (Figures 30 a-c) have only one phase with anatase structure (sp.gr. $I4_1/amd$, a~3.785Å, c~9.50Å; JCPDS № 89-4921). *Sample 22* contains the phase with a characteristic diffraction reflections at 2θ ~ 5° and ~33° along with nano-anatase (2θ~25.5°, ~38°, ~48°, ~54°, ~62°) (Figure 30d). According to [28], these peaks belong to η-TiO_2. Thus, the introduction into the system of $(NH_4)_2SO_4$ aqueous solution (in the form of mixture of H_2SO_4 and $NH_3 \cdot H_2O$ solutions) along with titanyl sulphate at the first step promotes formation of a mixture of phases with anatase and η-TiO_2 structures.

In addition, the phases differ in sizes of coherent scattering regions: D=74.2(4)Å, 61.6(4)Å and 62.6(4)Å for *samples 19, 20* and *21*, respectively (Figure 30), and D=44.2(3)Å for *sample 22*. Significantly lower value for coherent scattering region for *sample 22* can be explained by a larger content of the amorphous component, characteristic for η-TiO_2 phase referred in [21].

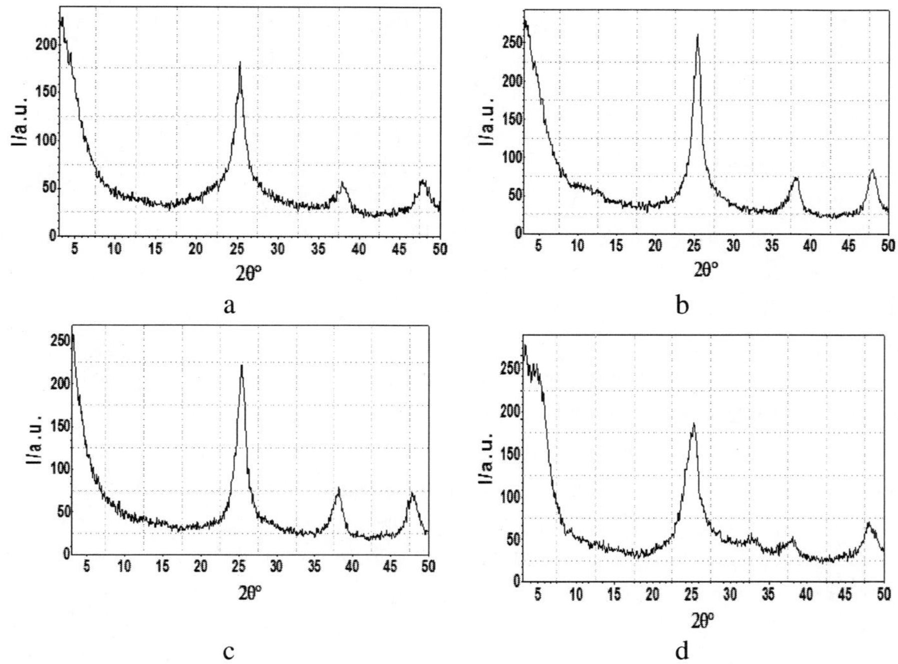

Figure 30. Diffraction patterns of samples 19-22 (a-d).

Figure 31 shows kinetic curves of Methyl Orange photodecomposition in the presence of investigated samples modified with manganese and commercial sample *Hombifine N* containing the nano-anatase as a reference sample.

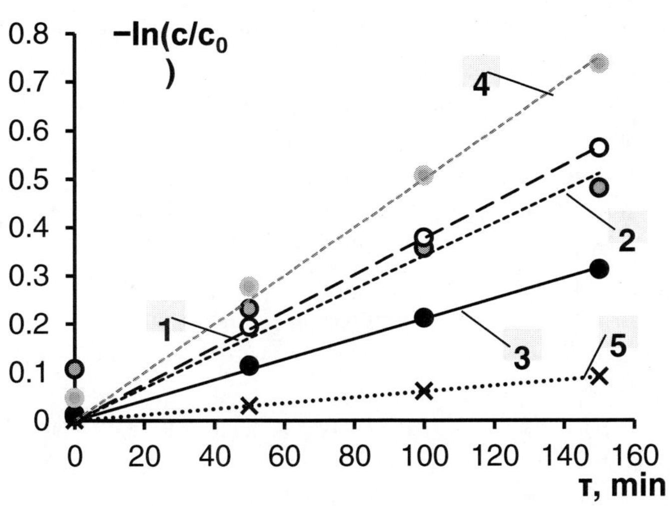

Figure 31. Interpolated values of the negative natural logarithm of the relative change in MeO concentration ($-\ln(c/c_0)$) vs. duration (τ) of irradiation of the photoreaction mixture in the presence of: 1 – *sample 19* ($k = 0.0037$ min^{-1}, k - reaction rate constant); 2 – *sample 20* ($k = 0.0025$ min^{-1}); 3 – *sample 21* ($k = 0.0020$ min^{-1}); 4 – *sample 22* ($k = 0.0046$ min^{-1}); 5 – *Hombifine N* ($k = 0.0006$ min^{-1}).

Photocatalytic activity of the modified samples is greater than that of the unmodified *Hombifine N*; *sample 22* shows the greatest photocatalytic activity in the visible wavelength range.

According to the EPMA data, the average manganese content is different in the samples under investigation: *samples 22* and *21* contain the lowest (0.30 at.%) and the highest (1.13 at.%) manganese content, respectively; *samples 19* (0.80 at.%) and *20* (0.55 at.%) are characterized by the intermediate values. The high manganese content in the samples is observed in the case of introduction of manganese source in the first stage of the synthesis process (*samples 19* and *21*) and the low content - in subsequent stages (*samples 20* and *22*) (Table 4).

As is known, the phase composition and crystallite size of TiO_2 can greatly affect its photocatalytic activity. Preparation of mixed-phase was thought to be an effective approach for the improvement of the photocatalytic activity of TiO_2 [74,76,77]. According to [9,78], the objects containing three phases of titanium dioxide (amorphous, anatase and rutile) simultaneously show enhanced photocatalytic activity as compared to the samples consisting of one phase. The explanation of this effect is given in terms of increase in space charge potential, which is created on a contact of two phases, and by the presence of localized electronic states of the amorphous phase. A similar situation was observed for *sample 22* (amorphous, anatase and η-TiO_2) exhibiting greater photocatalytic activity as compared to other samples (Figure 31). Synergy effect [74,79,80] was regarded as a possible reason for the enhancement of the photocatalytic activity in the mixed-phase *sample 22*.

5.3. Catalytic Properties of PdCl2-CuCl2/TiO2 System

Titanium dioxide with different crystal structures is widely used as supports for photocatalysts of many reactions [81]. There is a great practical interest in catalysts that enable room-temperature oxidation of carbon monoxide with atmospheric oxygen: $CO + 0.5O_2 = CO_2$. Such catalysts can be used as fillers in respirators and gas mask filter cartridges for personnel protection under hazardous workplace conditions and in emergency situations (plant fires and volcanic eruptions) and in devices (air conditioning systems) for CO removal from the air. The activity and stability of catalysts can be improved by optimizing the support.

Supported titania-based metal complex catalysts for low-temperature oxidation of carbon monoxide were prepared by cold impregnation. Palladium and copper chlorides were deposited from an aqueous solution at room temperature. The carrier particle size was <0.063 mm in all experiments [40].

The catalytic activity of a $PdCl_2$-$CuCl_2$/γ-Al_2O_3 standard catalyst and $PdCl_2$-$CuCl_2$/TiO_2 catalysts was assessed under comparable conditions: catalyst turnover rate, 12 000 h^{-1} (or 24 000 h^{-1}); initial CO concentration, 100 mg/m^3 (or 500 mg/m^3); gas–air mixture humidity and temperature, 9.3 g/m^3 and 25°C; test time, 120 min. The initial carbon monoxide concentration and the CO concentration at the catalytic reactor outlet was measured by a PKG-4-SO-MK-S sensor gas analyzer fitted with an electrochemical sensor (Technical Specification TU 4215-004-29359805-03, AOOT Praktik NTs, Russia). The initial gas-air mixture humidity was monitored with an IVTM-7-03-03-01 humidity meter (Technical Specification TU 4311-001-29359805-01, AOOT Praktik NTs, Russia). The flow rate of the gas–air mixture was measured by a GSB-400 flowmeter. Catalyst activity was inferred from the ratio of the CO conversion on the $PdCl_2$-$CuCl_2/TiO_2$ – CAT/TiO_2 catalysts to that on a standard catalyst, $PdCl_2$-$CuCl_2/\gamma$-Al_2O_3 – CAT/Al_2O_3, under comparable conditions: $X_{REL} = (X_{CAT/TiO2}/X_{CAT/Al2O3}) \times 100\%$.

All of the CAT/TiO_2 catalysts tested showed lower activity (X_{MAX} = 32.5%) in comparison with the standard catalyst used, CAT/Al_2O_3 [82]. One possible reason for this is that one should optimize the conditions for the deposition of the active component onto the surface of the TiO_2 carrier, like in the case of γ-Al_2O_3 [82], and the characteristics of the carrier. The CO conversion to CO_2 on the CAT/TiO_2 catalysts is comparable to that on gold-containing catalysts supported on mesoporous TiO_2 [83]. The highest activity (X_{REL} = 32.5%) was achieved with the CAT/TiO_2 catalysts supported on the samples containing the β-TiO_2 impurity phase (X_{REL}=30.5%) [40]. Catalysts supported on the impurity-free samples showed very low activity. We failed to find any correlation between catalytic activity of the investigated samples and their characteristics (microparticle and crystallite sizes, specific surface area), but the influence of impurity phases is well seen.

5.4. The Interaction between Linear DNA Molecules and Nanosized Titanium Dioxide with Anatase Structure

Investigating biological activity of nanostructured materials is of great importance with respect to safety and the prospects for their practical application. When TiO_2 interacts with DNA molecules *in vivo*, their double chains break down [84,85]. The mechanism of such damage, however, remains poorly studied, and the experimental data are contradictory. Anatase can thus degrade the DNA in mouse liver cells after introducing the suspension into the peritoneal cavity [86]. However, titanium dioxide neither blocks the polymerase chain reaction [87] nor affects the viability of E. coli from the TiO_2-catalyzed formation of organic compound derivatives [88], but it is toxic to daphnids [89]. There is still no clear answer as to the biological safety of nanoscale titanium dioxide, and it is therefore recommended that more detailed systematic studies be performed [90].

The interaction between double-stranded DNA fragments and titanium dioxide is estimated from changes in the length of the fragments and the concentration of DNA in buffers with varying ionic strength after treatment with a suspension of TiO_2. Suspensions of titanium dioxide samples with the structures of anatase in deionised water at a concentration of 20 mg/mL were prepared in advance. The DNA product was prepared by treating DNA Lambda phage with the enzyme Hind III and the hydrolyzing of DNA in strictly defined areas, resulting in a set of fragments made up of 125 to 23130 nucleotide pairs [91]. The resulting fragmented DNA was diluted with TE buffer (10 mM Tris-HCl and 1 mM EDTA; pH 7.8) to a concentration of about 20 μg/mL. The concentration of DNA was measured fluorimetrically using a Qubit DNA fluorimeter (Invitrogen) and a standard set of reagents.

Low-salt TE buffer and a standard hybridization SSC-buffer (20X stock with a composition of 3 M NaCl and 300 mM of sodium citrate; pH 7.8) were used in our experiments. Two groups of experiments on the interaction between TiO_2 and DNA in visible light were performed: one in light (daylight fluorescent lamp; surface illumination, 100 lux; exposure, 1 h) and one in the dark, using buffers with varying ionic strength. The method for sample preparation is described in Table 5.

Prepared working solutions were divided into aliquots to minimize errors of dosing. The mixing of solutions was performed by smoothly rocking the tubes to avoid the mechanical rupture of DNA molecules. After incubating the DNA in the presence of titanium dioxide, samples were centrifuged at a relative centrifugal acceleration of 18000×*g* for 5 min to separate the particles of TiO_2 and DNA. The DNA remaining in solution was analyzed by microfluidic electrophoresis on an Experion Bio-Rad automated analyzer (United States) using the standard reagents in the Experion DNA 12K Analysis Kit according to the recommendations of the manufacturer. This allowed us to determine with a high degree of accuracy the concentration and length of the DNA fragments in solution. DNA of known concentration with molecules 50 and 17 000 nucleotide pairs in length (internal standard reference) was added into each analyzed sample as well. To increase our experimental accuracy, an analysis of the external standard (a mixture of DNA molecules of known length and concentration that served as a marker of molecular mass) was performed in parallel. The results from our analysis are displayed in the form of the distribution of DNA molecules in the sample according to length (electrophoregrams) [92].

Table 5. Features of preparing the experiment on interaction between DNA and titanium dioxide with anatase structure

Experiment	Sample preparation
1	**Reference sample:** Solution of DNA, 100 μL; TE buffer, 50 μL; deionized H_2O, 50 μL. Without subsequent processing by TiO_2. Mixture was subjected to illumination of 100 Lux for 1 h in a transparent 1.5 mL Eppendorf-type test tube; volume of solution, 200 μL.
2	**Anatase* + TE:** Solution of DNA, 100 μL; suspension of TiO_2 with the structure of anatase, 50 μL; TE buffer, 50 μL. Mixture was kept in the dark for 1 h in a transparent 1.5 mL Eppendorf-type test tube.
3	**Anatase* + TE + *hυ*:** Solution of DNA, 100 μL; suspension of TiO_2 with the structure of anatase, 50 μL; TE buffer, 50 μL. Mixture was subjected to illumination of 100 Lux for 1 h in a transparent 1.5 mL Eppendorf-type test tube.
4	**Anatase* + 2XSSC:** Solution of DNA, 100 μL; suspension of TiO_2 with the structure of anatase, 50 μL; 20X SSC buffer, 10 μL (to a final concentration of 1X); deionized H_2O, 40 μL. Mixture was kept in the dark for 1 h in a 1.5 mL Eppendorf-type test tube.
5	**Anatase* + 2XSSC + *hυ*:** Solution of DNA, 100 μL; suspension of TiO_2 with the structure of anatase, 50 μL; 20X SSC buffer, 10 μL (to a final concentration of 1X); deionized H_2O, 40 μL. Mixture was subjected to illumination of 100 Lux for 1 h in a transparent 1.5 mL Eppendorf-type test tube.

* The sample with anatase and η-TiO_2 impurity (*D*=50 nm).

After centrifuging and the separation of titanium dioxide, part of DNA was adsorbed on the solid phase and removed, upon which its total concentration in the solution fell. The lower amount of residual DNA and the qualitative change in its composition are also associated with chemical degradation catalyzed by TiO_2. Studies have shown a relatively high efficiency of adsorption for DNA on titanium dioxide particles. Exposure to visible light raises the adsorption capacity, apparently as a result of the photoenhanced adsorption of DNA on the surface of the particles by analogy with other photoactive materials [91] and the photo degradation of DNA molecules in the presence of titanium dioxide [93,94].

The increased ionic strength weakens the interaction between DNA and anatase due to the size and number of micropores on the surface of the particles and, presumably, a different mechanism of interaction based on the partial penetration of DNA molecules in the pore of clusters. Such a mechanism for the adsorption of DNA is characteristic of mesoporous silica [95]. This property of anatase can be used in developing biosensors by using it as a substrate (a carrier of DNA in molecular biology).

After interaction with titanium dioxide, the number of truly deterministic DNA fragments in solution and the emergence of additional fragments that are not characteristic of the DNA initial solution declines due to discontinuities in the molecules of DNA, and only in certain places (otherwise, there would be a fuzzy continuous band of degraded DNA). This indicates preferential interaction between the TiO_2 particles and molecules of DNA with a specific sequence.

Photosensitization of the adsorption and degradation of DNA molecules under the action of both ultraviolet radiation and visible light is characteristic of titanium dioxide with the anatase structure. Titanium dioxide with the structure of anatase is very active toward DNA in buffers with low ionic strength. The strengthening of the adsorption and degradation of DNA by titanium dioxide with the structures of anatase under visible light can be used in developing photosensitive testing systems, and in the separation and purification of substances in bio-chemistry [92].

Conclusion

The results of the synthesis by sulfate method of samples with nano-anatase both nominally pure and modified with hydrogen peroxide and manganese are reported. The samples were single-phase or contained other polymorphs of titanium dioxide with the rutile, η-TiO_2 and β-TiO_2 structures as impurities. Characteristics of the investigated samples (composition of samples and surfaces of nanoparticles; sizes of specific surface areas, coherent scattering regions, nanoparticles, aggregates and agglomerates; volumes of nano- and ultrananopores) were compared with those of the known commercial samples (*Aeroxide Degussa (Evonik) P25*, *Hombifine N* and *Hombikat UV100*, Sachleben Chemie GmbH), that allowed to find their similarities and differences. Investigation and analysis of properties of the synthesized and commercial samples indicates the positive influence of modifying by H_2O_2 on their adsorption properties for extracting of a wide range of M(IV)-M(VI) ions from aqueous solutions. In the case of the presence of two polymorphs in the sample, the synergy effect may occur for the catalytic and photocatalytic properties.

Photosensitization of the adsorption and degradation of DNA molecules under the action of both ultraviolet radiation and visible light is characteristic of titanium dioxide with the anatase structure. Titanium dioxide with the structure of anatase is very active toward DNA in buffers with low ionic strength. The strengthening of the adsorption and degradation of DNA by titanium dioxide with the structures of anatase under visible light can be used in developing photosensitive testing systems, and in the separation and purification of substances in bio-chemistry.

ACKNOWLEDGMENTS

The authors acknowledge the Russian Foundation for Basic Research (project no. 13-03-00367) for supporting this work.

REFERENCES

[1] Eliseev, A.A.; Lukashin, A.V. Functional matherials; FIZMATLIT: Moscow, 2010, 452 p.
[2] Kormann, C.; Bahnemann, D.W.; Hoffmann, M.R. *J. Phys. Chem.* 1988, 92, 5196–5201.
[3] Lin, H.; Huang, C.P.; Li, W.; Ni, C.; Shah, S. I.; Tseng, Y.H. *Applied Catalysis B: Environmenta,* 2006, 68, 1-11.
[4] Dadachov, M. US Pat. Appl. Publ. US 2006/0144793, 2006.
[5] Grätzel, M. *J. Photochem. Photobiol. C: Photochem. Rev.,* 2003, 4, 145-153.
[6] Carp, O.; Huisman, C.L.; Reller, A. *Progr. in Solid State Chem.,* 2004, 32, 33-177.
[7] Matsumoto, Y.; Unal, U.; Tanaka, N.; Kudo, A.; Kato, H. *J. Solid State Chem.*, 2004, 177, 4205-4212.
[8] Artemiev, Y.M.; Ryabchuk, V.K. An introduction to the heterogeneous photocatalysis; St. Petersburg, 1999, 304 p.
[9] Ismagilov, Z.R.; Tsikoza, L.T.; Shikina, N.V.; Zaritova, V.Ph.; Zinoviev, V.V.; Zagrebel'nii, S.N. *Russian Chemical Reviews.*, 2009, 9, 943-955.
[10] Hashimoto, K.; Irie, H.; Fujishima, A. *AAPPS Bulletin*, 2007, 17, 12-29.
[11] Malekshahi Byranvand, M.; Nemati Kharat, A.; Fatholahi, L.; Malekshahi Beiranvand, Z. *J. of Nanostructures*, 2013, 3, 1-9.
[12] Gupta, Sh.M.; Tripathi, M. *Central European Journal of Chemistry*, 2012, 10, 279-294.
[13] Hayashi, H.; Torii, K. *J. of Materials Chemistry*, 2002, 12, 3671-3676.
[14] Khataee, A.R.; Aleboyeh, H.; Aleboyeh, A. *J. of Experimental Nanoscience*, 2009, 4, 121-137.
[15] Kolen'ko, Y.V.; Burukhin, A.A.; Churagulov, B.R.; Oleynikov, N.N. *J. Inorg. Chem. (Russia),* 2004, 40, 942-949.
[16] Kolen'ko, Y.V.; Kovnir, K.A.; Gavrilov, A.I.; Garshev, A.V.; Frntti, J.; Lebedev, O.I.; Churagulov, B.R.; Van Tendeloo, G.; Yoshimura, M. *J.Phys.Chem. B. (Russia),* 2006, 110, 4030-4038.

[17] Maksimov, V.D.; Shaporev, A.C.; Ivanov, V.K.; Churagulov, B.R.; Tretyakov, Y.D. *J.Chemical Technology (Russia)*, 2009, 10, 70-75.

[18] Kolen'ko, Y.V.; Burukhin, A.A.; Churagulov, B.R.; Oleynikov, N.N. *Mater. Lett.*, 2003, 57, 1124-1129.

[19] Xia, B.; Weibin, L.; Zhang, B.; Xie, Y. *J. of Materials Science*, 1999, 34, 3505-3511.

[20] Kuz'micheva, G.M.; Savinkina, E.V.; Obolenskaya, L.N.; Belogorokhova, L.I.; Mavrin, B.N.; Chernobrovkin, M.G.; Belogorokhov, A.I. *Crystallography Reports,* 2010, 55, 913-918.

[21] Savinkina, E.V.; Kuz'micheva, G.M.; Tabachkov, N.Y.; Obolenskaya, L.N.; Demina, P.A.; Yakovenko, A.G. *Inorganic Materials,* 2011, 47, 489-494.

[22] Jurov, V.V.; Ivanov, S.A. *Crystallography Reports,* 1997, 42, 239-243.

[23] Chernyshov, A.A.; Veligzhanin, A.A.; Zubavichus, Y.V. *Nucl. Instr. Meth. Phys. Res. A.,* 2009, 603, 95-98.

[24] Trofimova, N.N.; Veligjanin, A.A.; Murzin, V.Y.; Chernyshov, A.A.; Khramov, E.V.; Zabluda, V.N.; Edel'mal, I.S.; Slovokhotov, Y.L.; Zubavichus, Y.V. *Nanotechnologies in Russia*, 2013, 8, 108-113.

[25] Hammersley, A. P.; Svensson, S.O.; Hanfland, M.; Fitch, A. N.; Hausermann, D. *High Press. Res.* 1996, 14, 235-248.

[26] Svergun, D.I. *J. Appl. Crystallogr.,* 1992, 25, 495-503.

[27] Nokolaychik, V.I.; Kovalchuk, M.G.; Zaporojez, M.A. *Bulletin of the Russian Academy of Sciences: Physics,* 2011, 9, 1332-1334.

[28] Dadachov M. US Pat. Appl. Publ. US 2006/0171877, 2006.

[29] Brunauer, S.; Emmett, P.H.; Teller, E. J. *Amer. Chem. Soc.,* 1938, 60, 309.

[30] Zhagraphskaya, R.V.; Karnaukhov, A.P; Fenelonov, V.B. *Kinetics and Catalysis,* 1976, 17, 739.

[31] Fomkin, A.A.; Shkolin, A.V. *Doklady Akademii Nauk (Russia),* 2008, 423, 80-82.

[32] Fomkin, A.A. Physical Chemistry of Surfaces and protection materials; Nauka: Moscow, 2009, 133 p.

[33] Dubinin, M.M. Basic problems of the theory of physical adsorption; Nauka: Moscow, 1970, pp.251-259.

[34] ISO 15901-2 Pore size distribution and porosity of solid materials by mercury porosimetry and gas adsorption - Part 2: Analysis of mesopores and macropores by gas adsorption. ISO: Geneva, 30, 2006.

[35] ISO 15901-3:2007. Pore size distribution and porosity of solid materials by mercury porosimetry and gas adsorption - Part 3: Analysis of micropores by gas adsorption ISO: Geneva, 27, 2007.

[36] Diz'ko, V.A.; Karnaukhov, A.P.; Tarasova, D.V. Physical chemical principles of the synthesis of oxide catalysts; Novosibirsk, Russia, 1978, 378 p.

[37] Greg, S.; Sing, K. Adsorption, specific surface area, porosity; Mir: Moscow, 1984, 306 p.

[38] Sing, K.S.W.; Everett, D.H.; Haul, R.A.W.; Moscou, L.; Pierotti, R.A.; Rouquerol, J.; Siemieniewska, T. *Pure and Appl. Chem*., 1985, 57, 603.

[39] Partuka, S.; Rouquerol, J. *J. Colloid. Interface Sci.,*1979, 68, 21.

[40] Kuz'micheva, G.M.; Savinkina, E.V.; Chernobrovkin, M.G.; Titov, D.N.; Demina, P.A.; Obolenskaya, L.N.; Bruk, L.G.; Yakovenko, A.G. *Inorganic Materials (Russia),* 2011, 47, 838-844.

[41] Zhilkina, A.V.; Gordienko, A.A.; Prokudina, N.A.; Trusov, L.I.; Kuz'micheva, G.M.; Dulina, N.A.; Savinkina, E.V. *Russian Journal of Physical Chemistry*, 2013, 87, 685-691.

[42] Kment, S.; Kmentova, H.; Kluson, P.; Krysa, J.; Hubicka, Z.; Cirkva, V.; Gregora, I.; Solcova, O.; Jastrabik, L. *Journal of Colloid and Interface Science*, 2010, 348, 198-205.

[43] Quian, L.; Du, L.; Yang, Sh.Y.; Jin, Z.Sh. *J. Molecular Structure.*, 2005, 749, 103-107.

[44] Bezrodna, T.; Puchkovska, G.; Shymanovska, V.; Baran, J.; Ratajczak, H. *J. Molecular Structure*, 2004, 700, 175-181.

[45] Babkov, L.M.; Bezrodnaya, T.V.; Puchkovskaya, G.A.; Uspenskii, K.E.; Shimanovskaya, V.V. *Proceedings of Saratov Universit, Physical Series, Russia,* 2007, 7, 44-49.

[46] Hakamoto, K. IR and Raman spectra of inorganic coordination compounds; Mir: Moscow, 1991, 536 p.

[47] Hearne, G.R.; Zhao, J.; Dawe, A.M.; Pischedda, V.; Maaza, M.; Nieuwoudt, M. K.; Kibasomba, P.; Nemraoui, O.; Comins, J. D. *Phys. Rev.,* 2004, 70, 134102-134112.

[48] *Ohta*, S.S.; *Kamei, S.*; Hanakawa M.; Kurita, S. *J. Phys. Chem. Solids*, 2001, 62, 717-721.

[49] Viana, B.C.; Ferreira, O.P.; Souza Filho, A.G.; Filho, J.M.; Alves, O.L. *J. Braz. Chem. Soc.*, 2009, 20, 167-175.

[50] Dobal, P.S.; Katiyar, R.S.; Jiang, Y.; Guo, R.; Bhalla, A.S. *J. of Physics and Chemistry of Solids,* 2000, 61, 1805-1808.

[51] Wagner, C.D.; Naumkin, A.V.; Kraut-Vass, A.; Allison, J.W.; Powell, C.J.; Rumble, J.R.Jr. NIST XPS database. Version 3.4.

[52] Meng, X.; Dadachov, M.; Korfiatis, G.P.; Christodoulatos C. US Pat. Appl. Publ. US 2003/0155302, 2003.

[53] Kuz'micheva, G.M.; Natykan, A.A.; Savinkina, E.V.; Obolenskaya, L.N.; Domoroshchina, E.N.; Yakovenko, A.G. RU Patent 2463252, 2012.

[54] Georgaka, A.; Spanos, N. *Global NEST Journal*, 2010, 12, 239-247.

[55] Valencia-Trejo, E.; Villicaña-Méndez, M.; Alfaro-Cuevas-Villanueva, R.; Garnica-Romo, M.G.; Cortés-Martínez, R. *J. Appl.Sci. in Environ. Sanit,* 2010, 5, 171-184.

[56] Deedar, N.; Irfan, A.; Ishtiaq, Q. *J. of Environmental Sciences*, 2009, 21, 402-408.

[57] Xiao, Y.; Ling, J.; Qian, Sh.; Lin, A.; Zheng, W.; Xu, W.; Luo, Yu.; Zhang, M. *Water Environ. Res.,* 2007, 79, 1015-1022.

[58] Tella, E.; Panagiotou, G.D.; Petsi, T.; Bourikas, K.; Kordulis, C.; Lycourghiotis, A. *Global NEST Journal*, 2010, 12, 231-238.

[59] Bleam, W.F.; McBride, M.B. *Journal of Colloid and Interface Science*, 1986, 110, 335-346.

[60] Gao, Y.; Wahi, R.; Kan, A.T.; Falkner, J.C.; Colvin, V.L.; Tomson, M.B. *Langmuir*, 2004, 20, 9585-9593.

[61] Demina, P.A.; Kuz'michev, A.N.; Tsybinsky, A.M.; Obolenskaya, L.N.; Kuz'micheva, G.M.; Domoroshchina, E. N.; Savinkina, E.V. *Appl Nanoscience*, 2013. DOI 10.1007/s13204-013-02799.

[62] Von Tytko, K.H.; Baethe, G.; Hirschfeld, E.R.; Mehmke, K.; Stellhorn, D. *Z.anorg.allg.Chem.,* 1983, 5083, 43-66.

[63] *Igwe, J. C.; Abia, A.A. Global J. of Environ. Res.*, 2007, 1, 22-26.

[64] Demina, P.A.; Zubinskii, A.M.; Kuz'micheva, G.M.; Obolenskaya, L.N.; Savinkina, E.V.; Prokudina, N.A. *Crystallography Reports*, 2014, 59, 473-479.
[65] Feist, T.P.; Davies, P. K. *J. of Solid State Chemistry*, 1992, 101, 275-295.
[66] Gibalo, I.M. Analytical chemistry of niobium and tantalum; Nauka: Moscow, 2012, 177 p.
[67] Ivanenko V.I. Ph. Sc. Dissertation; Apatity, Russia, 2011, 40 p.
[68] Chen, X.; Mao, S.S. *Chem. Rev.*, 2007, 107, 2891-2959.
[69] Hashimoto, K.; Irie, H.; Fujishima, A. *AAPPS Bulletin*, 2007, 17, 12-28.
[70] Banerjee, A.N. *Nanotechnology, Science and Applications*, 2011, 4, 35-65.
[71] Pei, D.; Luan, J. *International Journal of Photoenergy*, 2012, 2012, 13.
[72] Savinkina, E.V.; Kuzmicheva, G.M.; Obolenskaya, L.N. *Int. J. Energy Environ.*, 2012, 6, 268-275.
[73] Obolenskaya, L.N.; Domoroshchina, E.N.; Savinkina, E.V.; Kuz'micheva, G.M. *Fundamental research (Russia),* 2013, 1, 796-801.
[74] Zhang, J.; Xu, Q.; Feng, Z.C.; Li, M.J.; Li, C. *Angew. Chem. Int. Ed.,* 2008, 47, 1766-1769.
[75] Yu, F.; Wang, Y.; Tang, H.; Zhang, Y.; Jin, Ch.; Liu, X.; Li, M.; Wang, J. *Catalysis Communications,* 2014, 51, 46-52.
[76] Jang, H.D.; Kim, S.K.; Kim, S.J. *J. Nanopart. Res.,* 2001, 3, 141-148.
[77] Zhang, J.; Xu, Q.; Feng, Z.C.; Li, M.J.; Li, C. *Angew. Chem. Int. Ed.*, 2008, 47, 1766-1769.
[78] Gusev, A.I. Nanomaterials, nanostructures, nanotechnology; FIZMATLIT: Moscow, 2009, 416 p.
[79] Scanlon, D.O.; Dunnill, C.W.; Buckeridge, J.; Shevlin, S.A.; Logsdail, A.J.; Woodley, S.M.; Catlow, C.R.A.; Powell, M.J.; Palgrave, R.G.; Parkin, I.P.; Watson, G.W.; Keal, T.W.; Sherwood, P.; Walsh, A.; Sokol, A.A. *Nature Mater.*, 2013, 12, 798-801.
[80] Kho, Y.K.; Iwase, A.; Teoh, W.Y.; Madler, L.; Kudo, A.; Amal, R. *J. Phys. Chem. C*, 2010, 114, 2821-2829.
[81] Parmon, V.N. Photocatalytic conversion of solar energy; Novosibirsk, Russia, 1991, pp.7-17.
[82] Bruk, L.G.; Budyka, A.A.; Vorojtsov, G.N.; Golub, Y.M.; Kaliya, O.L.; Kulikov, N.K.; Lujkov, Y.M.; Oshanina, I.V.; Syrychko, V.V.; Temkin, O.N.; Shelyapin, I.P.; Shepelev, A.D. Patent RU 2267354 C1.
[83] Zhang, D.; Qi, L.; Ma, J.; Cheng, H. *J.Mater.Chem.,* 2002, 12, 3677-3680.
[84] Serpone, N.; Salinaro, A.; Emeline. A. *Proc. SPIE*, 2001, 4258, 86-98.
[85] Vevers, W.F.; Jha, A.N. *Ecotoxicology*, 2008, 17, 410-420.
[86] Li, N.; Ma, L.; Wang, J.; Zheng, L.; Liu, *J.;* Duan, Y.; Liu, H.; Zhao, X.; *Wang*, S.; *Wang*, H.; Hong, F.; Xie, Y. *Nanoscale Res. Lett.*, 2010, 5, 108-115.
[87] Khaliq, A.R.; Sonawane, P.J.; Sasi, B.K.; Sahu, B.S.; Pradeep, T.; Das, S.K.; Mahapatra, N.R. *Nanotechnology*, 2010, 21, 1-11.
[88] Ibrahim, R.; Shi, M.; Wu, C. *J. Experimental Microbiology and Immunology*, 2004, 12, 67-71.
[89] Bang, S.H.; Le, T.H.; Lee, S.K.; Kim, P.; Kim, J.S.; Min, J. *Environmental Health Toxicology*, 2011, 26, 1-6.
[90] Tran, D.T.; Salmon, R. *Australasian Journal of Dermatology,* 2011, 52, 1-6.
[91] Kochevar, I.E.; Garcia, C.; Geacintov, N. *Photochem. Photobiol.,* 1998, 78, 692-698.

[92] Kutsev, M.G.; Kuz'micheva, G.M.; Obolenskaya, L.N.; Savinkina, E.V. *Russian Journal of Physical Chemistry A*, 2012, 86, 1697-1701.
[93] Hirakawa, K.; Mori, M.; Yoshida, M.; Oikawa, S.; Kawanishi, S. *Free Radic Res.,* 2004, 38, 439-447.
[94] Terence, K.J.; Robin, M.A. *Progress in Reaction Kinetics and Mechanism*, 2007, 32, 219-229.
[95] Lakshmi, R.; Baskar, V.; Ranga, U. *Anal. Biochem.,* 1999, 272, 109-112.

In: Titanium Dioxide
Editor: Jerri Brown

ISBN: 978-1-63321-391-3

Chapter 10

EFFECTS OF TITANIUM DIOXIDE NANOPARTICLES ON THE INFLAMMATORY PROCESS

***Elizabeth Huerta-García*[1,†], *Angélica Montiel-Dávalos*[2,†], *María del Pilar Ramos-Godínez*[3], *Ernesto Alfaro-Moreno*[2,‡§] *and Rebeca López-Marure*[1,§]**

[1]Laboratorio de Biología Celular, Departamento de Fisiología, Instituto Nacional de Cardiología "Ignacio Chávez", México

[2]Laboratorio de Toxicología Ambiental, Subdirección de Investigación Básica, Instituto Nacional de Cancerología, México

[3]Laboratorio de Microscopía Electrónica, Subdirección de Patología, Instituto Nacional de Cancerología, México

ABSTRACT

Titanium dioxide nanoparticles (TiO_2 NPs) are manufactured in large quantities worldwide for use in a wide range of applications. Recently, the IARC included TiO_2 as a type-2 carcinogenic compound. Additionally, at the nanoscale size, several toxic effects have been reported. Humans may be exposed to TiO_2 NPs through three main pathways: respiratory (airborne nanoparticles), dermal (cosmetics) and oral (food) routes.

After nanoparticles are absorbed, they may be distributed by the blood and lymphatic system to reach different targets in the body. TiO_2 NPs can reach bone, kidney, pancreas, liver, heart, and the nervous system, where they are retained and can become toxic. Although the toxic mechanisms are not well understood, *in vitro* experimental studies have reported cell membrane damage, production of reactive oxygen species (ROS) that may lead to oxidation of biomolecules (i.e., lipids, DNA, and proteins), genotoxicity, and

---

† These authors contributed equally to this work.

‡§ Corresponding author: Dr. Ernesto Alfaro-Moreno. Laboratorio de Toxicología Ambiental, Sudbirección de Investigación Básica, Instituto Nacional de Cancerología. Av. San Fernando 22, Tlalpan. CP 14080, México D.F. e-mail: ealfaro.incan@gmail.com.

§ Corresponding author: Dr. Rebeca López-Marure. Departamento de Biología Celular, Instituto Nacional de Cardiología "Ignacio Chávez". Juan Badiano No. 1, Colonia Sección 16, Tlalpan, C. P. 14080, México D. F., MEXICO. Tel: (52-55) 55 73 29 11 ext. 1337 Fax: (52-55) 55 73 09 26 e-mail: rlmarure@yahoo.com.mx.

pro-inflammatory effects. We previously showed that TiO_2 NPs exerted different effects on different cell types and induced a strong inflammatory response in human endothelial cells (HUVEC), mediating increases in cellular adhesion (endothelial-monocyte) and adhesion molecule expression. TiO_2 NPs also induce oxidative stress, NF-κB translocation, and cell death, indicating a strong toxic effect.

The aim of this chapter is to present a broad description of the main toxic effects of TiO_2 NPs related to inflammatory responses in various cells types and target organs.

INTRODUCTION

The rapid expansion of nanotechnology has generated a large variety of nanoparticles (NPs) with new electronic, mechanical, and physical-chemical properties. Currently, many NPs are used in the cosmetic, pharmaceutical, and alimentary industries. Their small size provides them with special physical-chemical properties such as high surface area, high surface charge, and high photocatalytic capacity, in comparison with the same materials with a larger size (non-nanometric).

NPs can reach a wide variety of targets within the human body by entering or diffusing through the pores of cellular membranes, causing unknown health-risks to humans and other species [1, 2].

Titanium dioxide (TiO_2) has been considered a non-toxic agent in the past and was used as an innocuous control in particle toxicology studies [3]. Once the size of the TiO_2 was reduced to nanoscale size, it has been observed to cause several toxic effects in both *in vivo* and *in vitro* studies [4]. TiO_2 NPs can enter into the body by three main pathways: 1) digestive, 2) dermal, and 3) inhalation, the latter being the primary route of exposure in humans. The deposition and toxicological effects of TiO_2 NPs in the respiratory tract are associated with their size, aggregation and agglomeration [5, 6].

The proportion of NPs exhaled or deposited in the respiratory tract, as well as in the nasopharyngeal, tracheobronchial and alveolar regions, depends on their size. For example, NPs of 1 nm in size are deposited at a rate of 90% in the nasopharyngeal region, while only 10% of the particles of this size are deposited in the tracheobronchial region. Furthermore, NPs of 5 nm are deposited in equal proportions in the three regions, and NPs of 20 nm are primarily deposited in the alveolar region, with only approximately 15% deposition in the tracheobronchial and thoracic regions [5, 6].

After NPs are deposited in the alveolar region, alveolar macrophages remove insoluble NPs by phagocytosis [7, 8]. Nevertheless, several studies have suggested that individual, non-artificial and non-aggregated NPs are not efficiently phagocytized by macrophages; therefore, NPs can accumulate in the alveoli and may persistently interact with epithelial cells [9, 10, 11]. The contact of NPs with the alveolar cells triggers the release of proinflammatory factors, which may induce endothelial activation [12, 13, 14]. If NPs cross the biological barrier between the alveoli and the blood vessels, the endothelium is activated by its direct contact with NPs [15, 16, 17] (Figure 1). This large accumulation of NPs can cause inflammatory responses that may lead to the development of lung diseases.

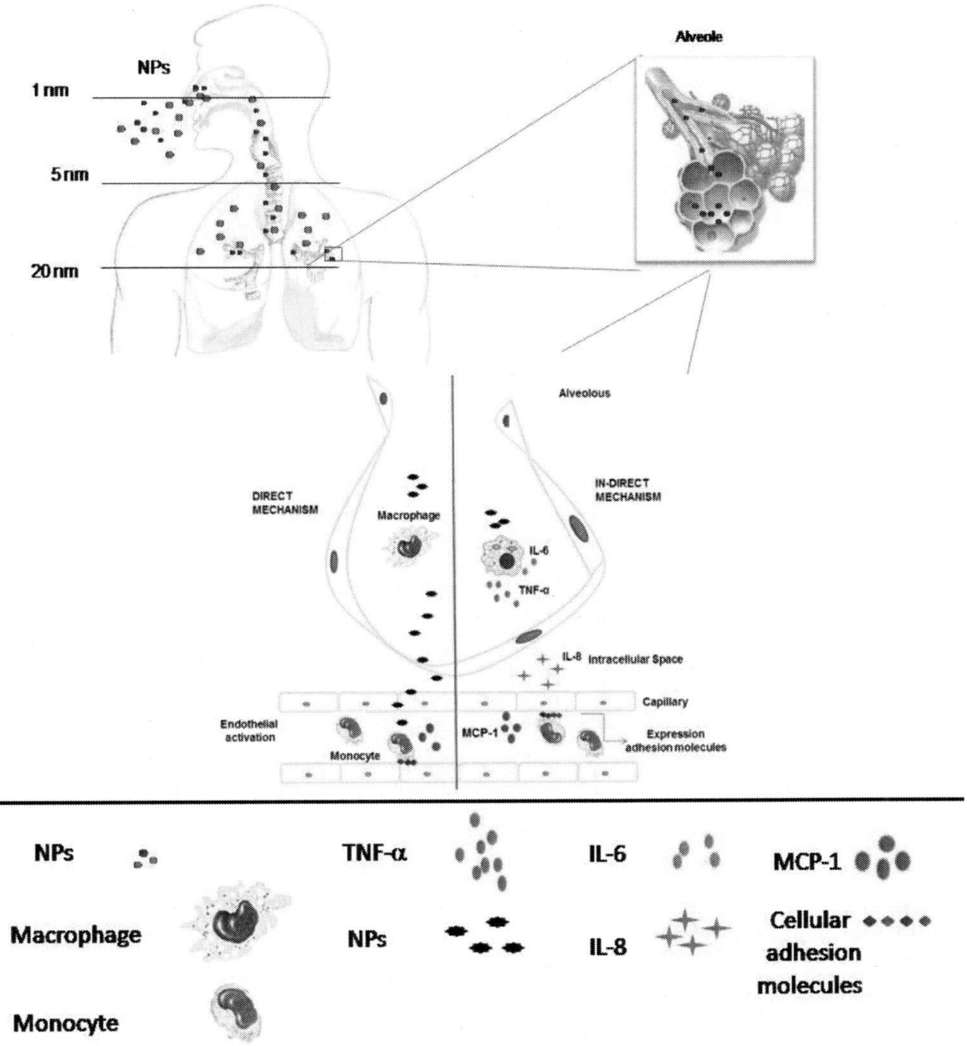

Figure 1. Schematic of nanoparticle inhalation, accumulation, and translocation to the different regions of the respiratory system and blood vessels. Nanoparticles (NPs) can activate the endothelium by two different pathways, via direct contact or via the secretion of proinflammatory factors. If NPs translocate from the alveoli to the blood vessel without provoking any response from the alveolar macrophages or epithelial cells, the endothelial cells may be activated, express adhesion molecules, and produce ROS and NO. If the NPs cause the activation of alveolar macrophages or epithelial cells, causing the release of cytokines and interleukins involved in the inflammatory response, the activation of the endothelium may be induced.

1. Internalization

The extremely small size of NPs allows them to cross a variety of physical barriers and translocate to tissues and organs that are far from the primary site of contact [17]. Smaller particles translocate more easily than larger particles. NPs are able to cross biological membranes and, therefore, can be internalized into cells, tissues, and organs. It has been shown that TiO_2 NPs of 20, 30, and 50 nm tend to accumulate in the cytoplasm [18, 19, 20].

There is also evidence that silica NPs (SiO_2-NPs) may be internalized in the neuronal stem, neurons, microglia and astrocytes [21], suggesting that different NPs can be internalized independently of their composition or the cell type with which they are interacting. It has been reported that TiO_2 NPs induce conformational changes in tubulin protein, altering the cytoskeleton [22]. Moreover, the size and charge of different NPs has been associated with variations in the membrane potential, leading to changes in cell permeability. Currently, several studies have described how different NPs are internalized into the cell [23]. The mechanisms that have been associated with this process are endocytosis, phagocytosis, and macropinocytosis [24].

Endocytosis is a process that ensures the entry of extracellular macromolecules through the cellular membrane into the cytoplasm, mediating a variety of mechanisms that can be characterized as either clathrin-dependent or independent [25]. Due to the nanometric size of NPs, cells can take up NPs through these pathways [26]. Clathrin is a cytosolic protein that is synthesized by free ribosomes in the cytosol, and its function is initiated by the recognition of loading receptors by the adapter protein (AP) in the membrane of the Golgi complex. AP is responsible for binding structures with 3 heavy and light chains of clathrin, also called trisqueliones. The membrane-bounded trisqueliones cause invagination and give rise to vesicles [27, 28]. Other proteins are involved in clathrin-independent-endocytosis, such as rafts and caveolins [29, 30, 31]. Caveolins are a family of integral membrane proteins, which are the main component of caveolae (invaginations of the membrane). Caveolins are less specific in the transport of molecules, in comparison with clathrins. Their main function is to transport material from the outside to the inside of the cell, mediating a process called transcytosis, which is often observed in the endothelial cells of blood vessels [32, 33]. Lipid rafts are molecules organized into domains in the membrane, formed by sphingolipids and cholesterol. Their function is to regulate specific lipid-protein and protein-protein interactions that are involved in the activation of signal transduction, cell adhesion, membrane trafficking and molecular sorting [34, 35, 36].

Phagocytosis is the other mechanism of NP internalization [37]. In this process, the cells generate membrane and cytosol projections called pseudopods that surround NPs. Once NPs are surrounded, the pseudopods fuse to form a vesicle around the particle called a phagocytic vesicle or phagosome [38]. Physiologically, the materials taken up within vesicles are digested by enzymes released from lysosomes [39]. In the case of mineral nanoparticles, digestion is not possible, and the persistence of NPs may trigger various cytotoxic effects [40, 41, 42]. The last mechanism of NP internalization is macropinocytosis, this mechanism is a type of endocytosis that involves the formation of membrane projections (lamellipodia) supported by a network of actin filaments forming micropinosomes [43]. In a study performed with Chinese hamster ovary cells (CHO), magnetic NPs coated with a polyacrylic were internalized by macropinocytosis and clathrin-dependent endocytosis [44]. Similarly, in one study using SiO_2-NPs, macropinocytosis was the internalization pathway in NCI-H292 cells [45]. In contrast, during the endocytosis of gold nanoparticles (Au-NPs) in human HepG2 cells, there was a reduction in the expression and degradation of clathrin, and no changes in caveolin expression were observed. In another study, the change of the positive- to negative-charge of the Au-NPs decreased phagocytosis in HepG2 cells, whereas there was no change in phagocytosis in regard to charge in 264 killer cells [46, 47]. Moreover, zinc oxide nanoparticles (ZnO-NPs) induced IL-8 mRNA and protein expression and promoted internalization by endocytosis because the use of an inhibitor of phagocytosis, cytochalasin B, decreased the production of IL-8 and the internalization of these NPs [48].

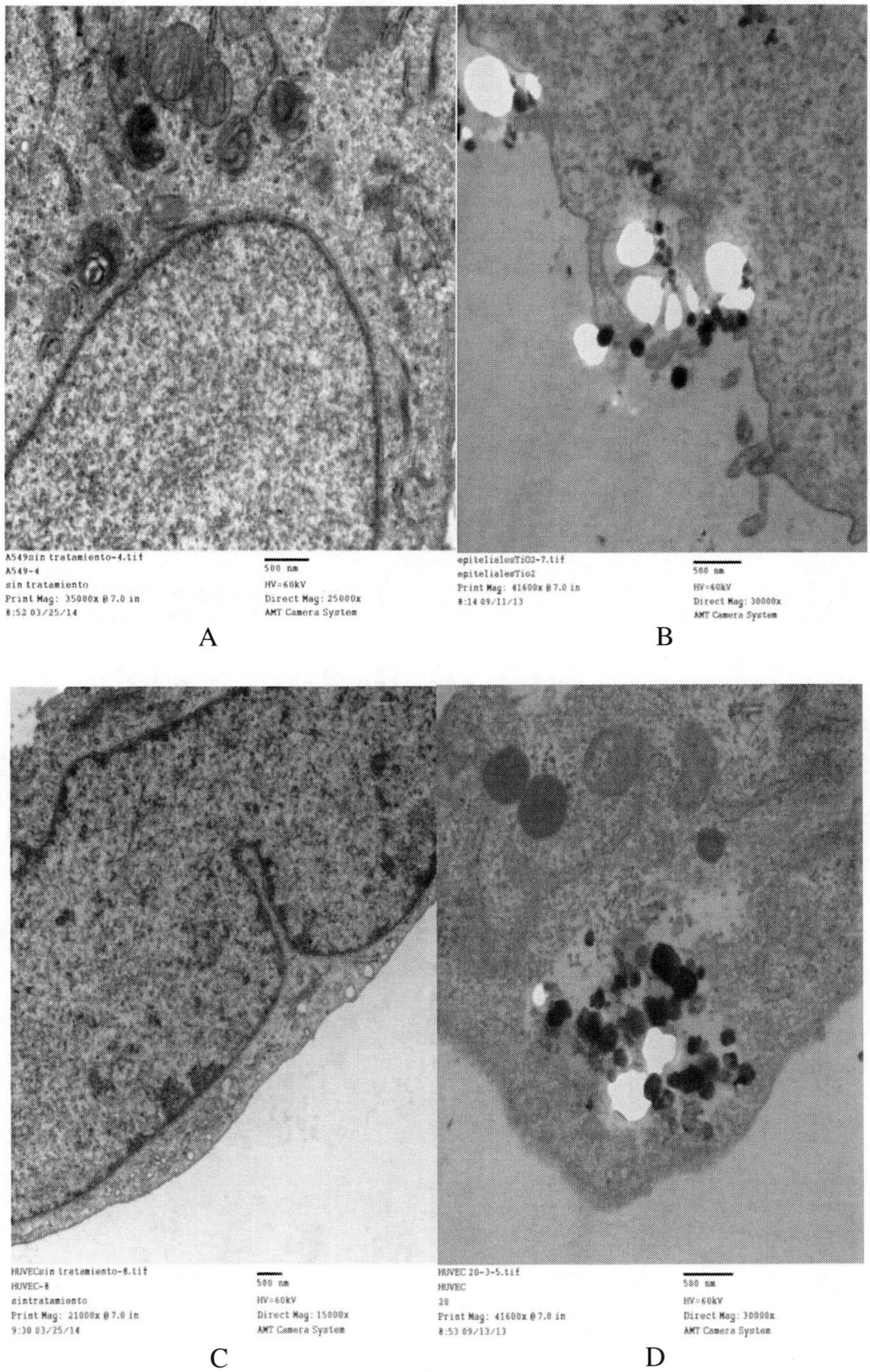

Figure 2. (Continued)

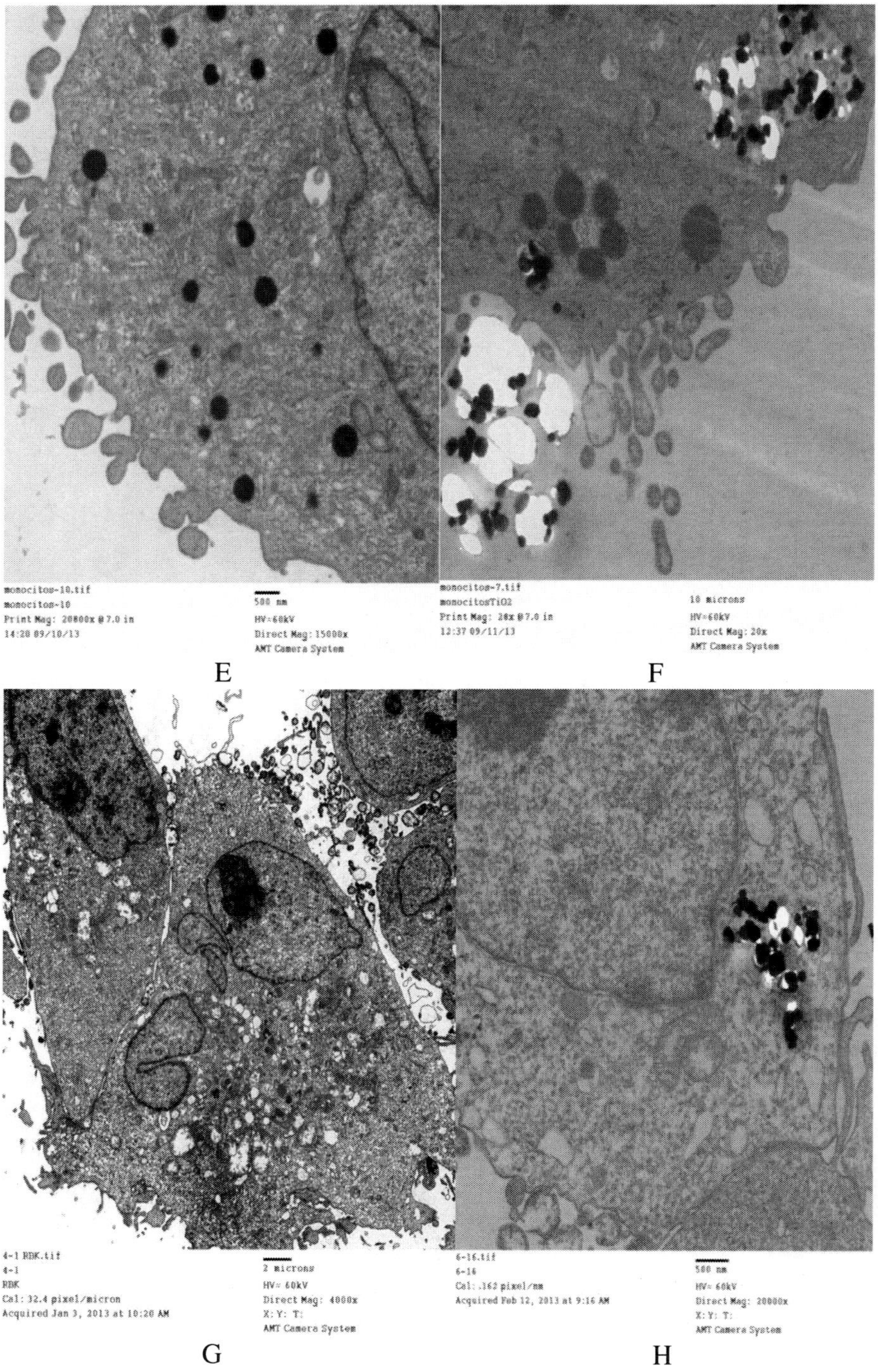

Figure 2. Transmission Electron Micrography (TEM) of different human cell types unexposed or exposed to 20 μg/cm^2 of TiO_2 NPs. A-B, pneumocytes (A549); C-D, human umbilical vein endothelial cells (HUVEC); E-F, monocytes (U937); G-H, glial cells (U373). Despite the differences within the cell types, all the exposed cultures show internalization of the TiO_2 NPs. In figures 2B and 2F, phagocytosis internalization is observed, while in figure 2D, pinocytosis is the process of internalization observed.

There are very few reports describing the mechanisms involved in the internalization of TiO_2 NPs. In L929 fibroblasts exposed to TiO_2 NPs for 24 h, it was observed that their internalization involved a combination of actin-dependent uptake of large agglomerates, and non actin-dependent uptake of small agglomerates. These results revealed that agglomerates of all NP types were attached to the cellular membrane, internalized and confined inside cytoplasmic vesicles [49]. In a previous report, we demonstrated that glial cells from human (U373) and rat (C6) internalized TiO_2 NPs of approximately 50 nm in size. We also showed the internalization of TiO_2 NPs in other cells involved in the inflammatory process, such as human umbilical vein endothelial cells (HUVEC), a human leukemic monocyte lymphoma cell line (U937), and adenocarcinomic human alveolar basal epithelial cells (A549), analyzed by transmission electronic microscopy (TEM) (Figure 2).

TiO_2 NPs were internalized mediated by a phagocytic process, where the formation of phagosomes was observed. In conclusion, these results indicated that TiO_2 NPs can be internalized into cells independently of the cell type, wherever the phagocytotic process is involved.

2. Inflammatory Response

After NPs are taken up by cells, different types of toxic effects are observed. Inflammation is the main process involved after exposure to TiO_2 NPs [41, 50]. Inflammation is a biological response that results in response to harmful stimuli, causing damage to cells and tissues [51, 52], and is mainly a protective action to eliminate pathogens and to repair damaged tissue. In the inflammatory process, there is participation of polymorphonuclear cells (neutrophils, eosinophils, basophils), and mononuclear cells (monocytes and macrophages). Their main function is to engulf dead cells and to act as antigen presenting cells. They also release proinflammatory factors such as cytokines, which are responsible for proinflammatory events in the brain and other systems [53, 54, 55, 56]. Other important events in the inflammatory process are the synthesis and release of chemical mediators such as endothelin-1, nitric oxide (NO), and angiotensin, among others.

The inflammatory response is classified as acute or chronic, depending on if the proinflammatory stimulus is short or sustained, respectively [57, 58]. The onset of acute inflammation is very quick, and its duration is relatively short. It is characterized by local vasodilation, increased capillary permeability, and leukocyte infiltration. Migration of leukocytes from microcirculation to the inflammatory focus to remove the insult occurs under the influence of chemotactic factors. During the processes of chemotaxis and phagocytosis, there is the release of active metabolites and proteases, which can cause tissue damage [54, 59, 60]. In contrast, chronic inflammation persists for weeks and even months and is characterized by the infiltration of macrophages, lymphocytes and plasma cells into the inflamed lesion. Chronic inflammation is produced from acute inflammation, triggering diseases such as atherosclerosis, diabetes, metabolic syndrome and Alzheimer's disease, among many others [55, 61, 62, 63].

It has been shown that the interaction between NPs and cells induces a strong inflammatory response. Several investigations have reported that the size of NPs is crucial in the production of toxic effects. Garnett and Kallinteri [64] found that NPs with a size < 300 nm may translocate into the cells, whereas NPs of 70 nm can even enter the nucleus of cells, and cause even more damage [65, 66, 67]. Some NPs induce toxicity by inducing oxidative stress, production of proteins involved in the inflammatory response [68, 69], DNA damage [66], structural damage to the cell nucleus and mitochondria and interference with their activity, inhibition of cell growth [9, 65], and even cell death [66].

2.1. Oxidative Stress

Although the mechanisms of toxicity are not clear, the main mechanism is the induction of oxidative stress via the production of free radicals [70, 71, 72]. Oxidative stress is caused by an imbalance between the production of reactive oxygen species (ROS), and the cellular antioxidant systems [70]. Abnormalities in this normal redox state can cause toxic effects through the production of peroxides and free radicals that damage cell components, including proteins, lipids, and DNA.

ROS, or free radicals, are molecules that have an unpaired electron in their molecular structure, leading to a configuration that generates high instability [73, 74]. Superoxide anion (O^{-2}), hydroxyl radical (HO^{-1}), hydrogen peroxide (H_2O_2), NO, peroxyl ($ROO^{\bullet}$), thiol radicals ($RS^{\bullet}$) $O_2^{-\bullet}$, $NO^{\bullet}$, and peroxynitrite ($ONOO^{-}$), are the most common ROS in cells [74, 75, 76, 77, 78]. These species are generated by different pathways: mitochondrial electron transport, the Fenton and Haber Weiss reaction [79, 80, 81], and the activity of xanthine oxidase, among others. Oxidation reactions are important for cellular metabolic processes; however, these reactions often become toxic. Antioxidants are defense mechanisms that serve to counteract the negative effects of ROS. They donate an electron to free radicals and turn it into a weak free radical. There are two types of antioxidant systems: enzymatic and non-enzymatic. The enzymatic system includes enzymes such as catalase (CAT), superoxide dismutase (SOD), and glutathione peroxidase (GPx); and the non-enzymatic system includes vitamins E and C, carotenoids, flavonoids, and NO [82].

NO is synthesized by NO synthase from L-arginine and molecular oxygen. NO regulates vascular tone, increases blood flow, and decreases the adhesion of platelets and leukocytes to the endothelium. NO synthesized by endothelial cells reduces the tone of the smooth muscle cell [83]. When these protective systems fail to decrease ROS production, lipid peroxidation, protein oxidation and DNA damage occur.

Numerous *in vitro* and *in vivo* studies have shown the ability of different types and sizes of NPs to induce oxidative stress and increase ROS production [84]. These events have been associated with inflammation and genotoxicity, mediating the upregulation of redox-sensitive transcription factors such as nuclear transcription factor kappa B (NF-κB), and activator protein-1 (AP-1). Upregulation of the kinases involved in inflammation, such as mitogen-activated protein kinases (MAPK), has also been observed [85, 86]. Other molecules involved with oxidative stress are low density lipoproteins (LDL), which are a complex of lipids (cholesterol, phospholipids and triglycerides) and proteins (apolipoprotein B-100). Oxidized LDL (oxLDL) are a risk factor for the development of atherosclerosis because they induce dysfunction of the vascular endothelium by activation of their receptor, LOX-1 (lectin-like oxLDL receptor-1) [87, 88]. LOX-1 regulates the expression of metalloproteinases (MMPs) and cellular adhesion molecules, induces apoptosis and reduces eNOS and NO production

[89, 67, 90, 91]. Activation of LOX-1 is regulated by NF-κB and AP-1, which activate various inflammatory molecules such as TNF-α, angiotensin II, and oxLDL [92,]. Furthermore, macrophages with a high accumulation of oxLDL may be transformed to foam cells, favoring the development and progression of atherosclerotic plaque [93].

2.2. Oxidative Stress and TiO_2 NPs

The toxicity of TiO_2 NPs is related to increased oxidative stress. It has been demonstrated that intranasal administration of TiO_2 NPs in mice for 90 days induced an increase in p38, c-jun kinase, NF-κB, Nrf-2 and heme oxygenase activation, caused increased ROS production in the brain, and consequently, oxidation of lipids, proteins and DNA [94]. TiO_2 NPs also induce the production of free radicals and other products of oxidation such as O^{-2}, H_2O_2, malondialdehyde (MDA), carbonyl, and 8-OHgd; and the reduction of CAT, SOD, glutathione-S-transferases (GST), and other non-enzymatic antioxidants (ascorbic acid, glutathione, and thiol) in mouse heart and rat synovial cells RSC-364 [95, 96, 97]. In addition, vascular endothelial cells exposed to TiO_2 NPs increased oxidative stress, MAPK activation (JNK, ERK and p38), vascular cell adhesion molecule 1 (VCAM-1) expression, and MCP-1 mRNA. We have shown that TiO_2 NPs cause oxidative stress, mediating an increase in ROS and NO production in HUVEC, in A549 cells, and in U373 and C6 cells [98, 99, 100].

There are few reports discussing the effects of TiO_2 NPs on LDL. An *in vivo* study showed that serum collected from rats exposed to TiO_2 NPs caused increases in the concentration of LDL [101]. HUVEC exposed to TiO_2 NPs for 24 h showed increased expression of the LOX-1 receptor, indicating that endothelial cells activated by TiO_2 NPs are capable of increasing oxLDL. This is a crucial event favoring the development of cardiovascular diseases such as atherosclerosis, as well as other inflammatory diseases [98].

All of the described effects are not exclusive to TiO_2 NPs and are shared by other NPs. For instance, nickel (Ni-NPs), silver (Ag-NPs), copper (CuO-NPs), and cerium (CeO_2-NPs) nanoparticles induce depletion of glutathione, SOD, decreases in NO and DNA damage in different cell lines such as MCF-7, dermal fibroblasts, A549, and human neuroblastome IMR32 cells [102, 103, 104, 105]. Additionally, zinc nanoparticles (ZnO-NPs) cause increases in ROS and the mitochondrial membrane potential, as well as the release of lactate dehydrogenase in primary astrocytes. Moreover, phosphorylation of JNK, p38 and ERK proteins was found. All these effects were related to decreases in cell viability and apoptosis [102].

2.3. NF-κB and TiO_2 NPs

As it was previously mentioned, NF-κB activation is an important event during the inflammatory response and plays an important role in processes such as cell adhesion, proliferation, differentiation, apoptosis, and oncogenesis [106, 107, 108, 109]. NF-κB is a dimer of ubiquitous expression. In the majority of cellular types, NF-κB dimers are predominantly retained in the cytosol forming complexes with specific inhibitors called IκB protein family [110, 111, 112, 113]. In this context, NF-κB complexes are inactive; however, in response to different stimuli, NF-κB dimers are released from the cytosol and translocate to the nucleus, where the transcription of specific genes is regulated [114], maintaining the homeostasis of the endothelial cells in a normal context. However, NF-κB activation by inflammatory stimuli has been implicated in various diseases such as: rheumatoid arthritis, atherosclerosis, multiple sclerosis, asthma, cancer, etc. [115, 116, 117]. The signaling pathways for NF-κB activation are extremely complex, and they can vary according to the agent, cell type and specific conditions [118, 119].

Previous studies have shown that NPs activate NF-κB signaling. Previously, we have showed that TiO_2 NPs induced a strong pro-inflammatory response in HUVEC, mediating ROS production increases, NF-κB translocation and IκB-α degradation, and also cellular death [99]. Inhalation of TiO_2 NPs induced an inflammatory response through activation of NF-κB in epithelial cells of rat trachea [120, 121]. TiO_2 NPs are also capable of inducing pro-inflammatory events in a mouse embryonic fibroblast cell line (NIH/3T3) cells, which were transfected with the Toll-like receptor 4 (TLR4) gene, and NF-κB reporter gene [121].

2.4. Adhesion Molecules and TiO_2 NPs

NF-κB also regulates the expression of adhesion molecules on the surface of endothelial cells, such as E- and P- selectins, VCAM-1, intercellular adhesion molecule-1 (ICAM-1), and platelet endothelial cell adhesion molecule-1 (PECAM-1), during inflammatory responses [122, 123, 124]. Adhesion molecules are membrane receptors involved in various functions related to cell trafficking, interactions between cells, and cellular adhesion to the extracellular matrix. Additionally, they are involved in diverse processes such as growth, differentiation, migration, and apoptosis [125].

In cases of atherosclerosis, the expression of adhesion molecules is increased in the endothelium [126, 127, 128]. Some studies have shown that TiO_2 NPs induced increases in E- and P-selectins [98, 129], ICAM-1, VCAM-1, and PECAM-1 in HUVEC [98]. These increases correlated with the adhesion of monocytic cells (U937) to HUVEC previously exposed to TiO_2 NPs [98, 100]. In another study, we demonstrated that A549 cells exposed to TiO_2 NPs induced the activation of HUVEC, increasing the expression of adhesion molecules in a co-culture model [99]. Increases in the levels of mRNA for VCAM-1 and MCP-1 have also been reported for vascular endothelial cells [130].

These results are consistent with those observed with other types of NPs. For example, silica and zinc oxide NPs induced an increase in mRNA and proteins of ICAM-1 and VCAM-1 in the cell line EA.hy926, and in human pulmonary artery endothelial cells (HPAEC) [131, 132].

2.5. Cytokines and TiO_2 NPs

Another event regulated by the activation of NF-κB and p38 protein is the secretion of cytokines, which is a main event during inflammation [133, 134]. Cytokines are soluble extracellular polypeptides or glycoproteins that are produced by various types of cells.

Cytokines bind to specific receptors, triggering intracellular signals that regulate genetic transcription. Thus, cytokines regulate the activity, differentiation, proliferation and survival of immune cells, and also regulate the production and activity of other cytokines, which may increase (pro-inflammatory Th1) or attenuate (anti-inflammatory Th2) inflammatory responses [134, 135, 136]. Among the pro-inflammatory cytokines are the interleukins (ILs) 1, 2, 6, 7, and tumor necrosis factor (TNFα); in contrast, some of the anti-inflammatory cytokines are IL-4, IL-10, IL-13, and transforming growth factor beta (TGF-β) [135, 136, 137]. The exposure to different types of particles is related to a complex network of pro- and anti-inflammatory cytokines, but the pattern of cytokine regulation induced by NPs is not well understood.

Among the studies describing the cytokine induction by NP, in a report, TiO_2 ultrafine particles induced the activation of rat alveolar macrophages (NR8383), which was related to increases in TNF-α mRNA [138]. *In vitro*, TiO_2 nanocrystals induced the secretion of IL-8 in human bronchial and alveolar epithelial cells (NHBE, A549 and BEAS-2B). Similar results have been reported for neutrophils exposed to TiO_2 NPs [134, 139]. Other NPs, such as SiO-NPs, induced increased IL-1β and IL-6 expression in peripheral blood mononuclear cells [140].

2.6. Growth Factors and TiO_2 NPs

In addition to cytokines, growth factors also participate in the inflammation process. Growth factors are proteins, hormones, or neurotransmitters that play a central role in intercellular communication [141]. Growth factors stimulate cell proliferation through cell cycle regulation and, also participate in different processes such as cell survival, migration, differentiation and even apoptosis [142]. Growth factors act by binding to cellular receptors in the cellular membrane, triggering an intracellular signal, in which a specific set of different protein kinases participate. This process is mediated by the phosphorylation of certain proteins, activating a signaling cascade that ends with the transcription of one or more genes (signal transduction). However, the over-expression of some growth factors has been associated with the development of diseases; for example, VEGF promotes tumor angiogenesis [143, 144], while PDGF promotes the progression of tumor growth. PDGF plays an important role in migration and proliferation of smooth muscle cells, an event associated with atherosclerotic plaque formation [145, 146].

Particulate matter exposure increases growth factors such as derived platelet derived growth factor (PDGF), fibroblast growth factor (FGF), vascular endothelial growth factor (VEGF), granulocyte-colony stimulating factor (G-CSF), granulocyte-macrophage-colony stimulating factor (GM-CSF), and epidermal growth factor-1 (EGF-1) [147].

Among NPs capable of inducing the expression of growth factors, are Ag-NPs, which induced VEGF expression in BALBc female mice [148]. In another study similar results were obtained with Au-NPs, which inhibit VEGF expression and decreased proliferation of retinal pigment epithelial cells [149]. Regarding TiO_2 NPs, increases in VEGF in serum from rats orally exposed for 5 days have been reported [150]. HUVEC exposed to different concentrations of TiO_2 NPs presented high increases in VEGF expression [98]. Other studies have reported that oral administration of TiO_2 NPs in mice increased TGF-β expression [152], whereas no effect on TGF-β expression was observed in fibroblasts [153].

In a study with Ni-NPs in mesothelial cells derived from rat pleura (NRM2), these produced a higher activity of PDGF, an increase in ROS generation and prolonged activation of ERK [154].

2.7. Tissue Factor (TF) and TiO_2 NPs

TF is one of the factors that are produced to prevent the loss of homeostasis when the endothelium has been severely damaged. TF is of great interest due to its role as a trigger of the coagulation cascade [155, 156, 157, 158]. TF is a membrane glycoprotein produced in various cell types (fibroblasts, monocytes) and in the wall of blood vessels. Under physiological conditions, TF is absent in endothelial cells, but when damage occurs in the blood vessel, TF is released into the blood and is also expressed in endothelial cells and monocytes, triggering the coagulation cascade and the inflammatory response [156, 157].

Intravenous administration of SiO_2-NPs in Sprague Dawley rats for 7 days increased TF expression [159]. In an *in vitro* study, an increase in TF expression was observed in HUVEC exposed to TiO_2 NPs for 24 h [98].

2.8. Cell Death and TiO_2 NPs

Different particles and nanoparticles are capable of inducing cell death by apoptosis and necrosis [160, 161]. Cellular death is triggered when cells are not able to repair severe damage, which can be related to strong pro-inflammatory events, oxidative stress, and DNA damage, among others. When the damage is irreversible, there is a set of events that activates death signals. Apoptosis and necrosis are the primary and most common pathways of cell death [162, 163, 164]. Apoptosis is defined as a set of biochemical reactions occurring in a regulated way, and it occurs in response to various physiological or pathological events [165, 166]; whereas necrosis is defined as a pathological death in tissues caused by an acute and irreversible injury that cannot be repaired by mechanisms of adaptation and resistance.

Several studies have reported toxic effects of TiO_2 NPs that lead to death cell. In leukemia cells (L1210), human bronchial epithelial (BEAS-2B) and mouse epidermal cells (JB6), TiO_2 NPs induced apoptosis and were linked to DNA fragmentation and caspase-3 and -8 activation [167, 168, 169]. TiO_2 NPs with a size of 15 nm induced Bax activation in bronchial epithelial cells [171]. In HUVEC and U937 cells, TiO_2 NPs induced cell death by apoptosis and necrosis [98, 170].

In human keratinocytes (HaCaT), increases in oxidative stress and apoptosis were observed after 48 h of TiO_2 NPs exposure [171]. In an *in vivo* study in which mice were continuously exposed by intragastric administration to different concentrations (2.5, 5, 10 mg/kg) of TiO_2 NPs for 90 days, significant increases in apoptosis were observed in the spleen and thymus associated with TiO_2 NPs accumulation and lead to histopathological changes [172]. In another model, TiO_2 NPs produced a significant reduction in renal glomerulus numbers, apoptosis, infiltration of inflammatory cells, necrosis or disorganization of renal tubules in mouse kidney tissue [173].

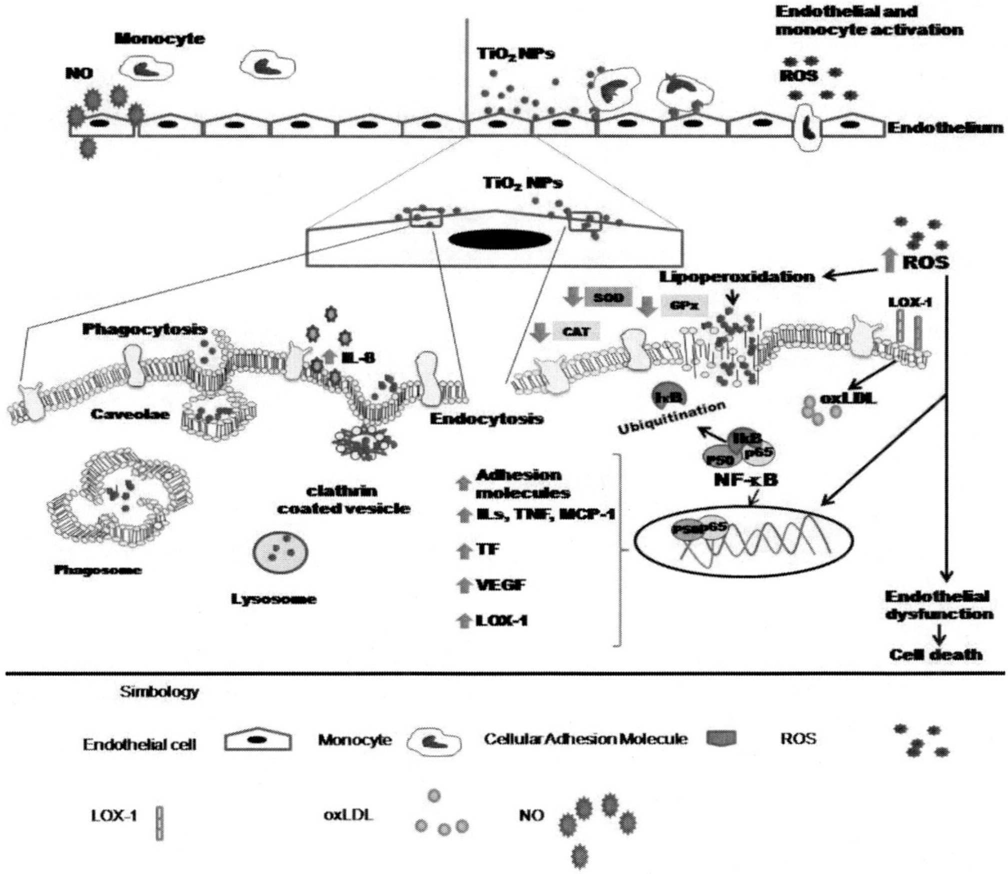

Figure 3. Schematic of the possible interaction of TiO_2 NPs with endothelial cells to induce an inflammatory response. NPs interact with the cellular membrane of monocytes and endothelial cells and are internalized by endocytosis, phagocytosis and a clathrin-dependent mechanism. Evidence collected in monocytes and endothelial cells indicates that after cellular activation, there are increases in ROS, which can induce lipoperoxidation, oxidation of LDL, and a decrease in the expression of antioxidant enzymes. Other consequences of ROS increases include the phosphorylation of IκB-α, inducing the translocation and activation of NF-κB. When NF-κB is activated, it may induce the expression of adhesion molecules as well as several pro-inflammatory molecules. Depending on the dosage of TiO_2 NPs, all these events may produce severe damage to the cells, causing death, mainly by apoptosis.

Conclusion

TiO_2 NPs are widely used for various purposes and can be found in toothpaste, sun-blocks, and paints, among many other products. Human exposure to TiO_2 NPs is not well documented, but considering the large quantities of TiO_2 NPs produced worldwide, the number of people working in the manufacturing of different products containing TiO_2 NPs, and the millions of people consuming those products, it is important to have a clear view of the potential toxic effects of these particles. As it has been emphasized in the present chapter, the physical-chemical properties of nanoparticles provide new toxicological characteristics to the TiO_2 NPs. The internalization process of TiO_2 NPs appears to occur via phagocytosis and pinocytosis. Internalization processes have been documented in different cell types by us and other authors, noting that the internalization of these particles is independent of cell types.

The inflammatory response triggered as a consequence of TiO_2 NPs exposure is related to oxidative stress (ROS and NO production), which is related to the activation and translocation of NF-κB and the further expression of adhesion molecules and many proinflammatory factors. The expression of the early and late adhesion molecules are key processes in the recruitment of inflammatory cells and therefore are among the main events that allow us to understand the inflammatory effects induced by TiO_2 NPs. In addition to all of the previously described effects, it is also known that TiO_2 NPs are capable of inducing cellular death, and the evidence suggests that despite the observation of necrosis, apoptosis appears to be the main cell death process induced by these particles (Figure 3).

REFERENCES

[1] Li JJ, Muralikrishnan S, Ng CT, Yung LY, Bay BH. 2010. Nanoparticles-induced pulmonary toxicity. *Exp. Biol. Med.* 9:1025-33.

[2] Salata O. 2004. Applications of nanoparticles in biology and medicine. *J. Nanobiotechnology*. 1:3-6.

[3] Matej Skocaj, Metka Filipic, Jana Petkovic, Sasa Novak. 2011. Titanium dioxide in our everyday life; is it safe? *Radiol. Oncol.* 4: 227-247.

[4] Ostiguy, C. 2006. *Nanoparticles-Current Knowledge about Occupational Health and Safety Risks and Prevention Measures.*

[5] Oberdörster G, Oberdörster E, Oberdörster JJ. 2005. Nanotoxicology: an emerging discipline evolving from studies of ultrafine particles. *Environ. Health Perspect.* 123(7): 823-39.

[6] Wang J, Zhou G, Chen C, Yu H, Wang, Ma Y, Jia G, Gai Y, Li B, Sun J, Li Y, Jiao F, Zhano Y, Chai Z. 2007b. Acute toxicity and biodistribution of different sized titanium dioxide particles in mice after oral administration. *Toxicol. Lett.* 168(2): 176-185.

[7] Baisch BL, Corson NM, Wade-Mercer P, Gelein R, Kennell AJ, Oberdörster G, Elder A. 2014. Equivalent titanium dioxide nanoparticle deposition by intratracheal instillation and whole body inhalation: the effect of dose rate on acute respiratory tract inflammation. *Part Fibre Toxicol.* 1-16.

[8] Oberdörster G, Ferin J, Lehnert BE. 1994. Correlation between particle size, in vivo particle persistence, and lung injury. *Environ. Health Perspect.* 5:173-79.

[9] Adamcakova-Dodd A, Stebounova LV, Kim JS, Vorrink SU, Ault AP, O Shaughnessy PT, Grassian VH, Thorne PS. 2014. Toxicity assessment of zinc oxide nanoparticles using sub-acute and sub-chronic murine inhalation models. *Part. Fibre Toxicol.* 1:15.

[10] Eleonore Fröhlich and Sharareh Salar-Behzadi. 2013. Toxicological Assessment of Inhaled Nanoparticles:Role of in Vivo, ex Vivo, in Vitro, and in Silico Studies *Int. J. Mol. Sci.* 4795 -4822.

[11] Blum JL, Rosenblum LK, Grunig G, Beasley MB, Xiong JQ, Zelikoff JT.2014. Short-term inhalation of cadmium oxide nanoparticles alters pulmonary dynamics associated with lung injury, inflammation, and repair in a mouse model. *Inhal. Toxicol.* 48-58.

[12] Maniatis NA, Kotanidou A, Catravas JD, Orfanos SE. 2008. Endothelial pathomechanisms in acute lung injury. Vascular Pharmacology. 149:119-133.

[13] Crandall ED, Matthay MA, Garfinkel S. 2001. Alveolar epithelial transport: basic science to clinical medicine. American Journal of Respiratory nd Critical Care Medicine. 163 (4): 1021-1029.

[14] Bengalli R, Mantecca P, Camatini M, Gualtieri M. 2013. Effect of nanoparticles and environmental particles on a co-cultures model of the air-blood barrier. Biomed Res Int. 1-8.

[15] Oberdörster G, Sharp Z, Atudorei V, Elder A, Gelein R, Kreyling W, Cox C. 2004. Translocation of inhaled ultrafine particles to the brain. *Inhal. Toxicol.* 16(6-7): 437-45.

[16] L E. Yu, Lin-Yue Lanry Yung, Choon-Nam Ong, Yueh-Li Tan, Kumar Suresh Balasubramaniam, Deny Hartono, Guanghou Shui, Markus R. Wenk and Dr Wei-Yi Ong. 2007. Translocation and effects of gold nanoparticles after inhalation exposure in rats. *Nanotoxicology*. 235-242.

[17] Béatrice L'Azou, Joana Jorly, Dinhill On, Elisabeth Sellier, Frédéric Moisan, Jocelyne Fleury-Feith, Jean Cambar, Patrick Brochard and Céline Ohayon-Courtès. 2008. In vitro effects of nanoparticles on renal cells. *Particle and Fibre Toxicology*.5:22.

[18] Stearns RC, Paulauskis JD, Godleski JJ. 2001. Endocytosis of ultrafine particles by A549 cells. *Am. J. Respir. Cell Mol. Biol.* 2:108-15.

[19] Mühlfeld C, Mayhew TM, Gehr P, Rothen-Rutishauser B. 2007. A novel quantitative method for analyzing the distributions of nanoparticles between different tissue and intracellular compartments. *J. Aerosol. Med.* 20(4): 395-407.

[20] Rothen-Rutishauser B, Mühlfeld C, Blank F, Musso C, Gehr P. 2007. Translocation of particles and inflammatory responses after exposure to fine particles and nanoparticles in an epithelial airway model. *Part. Fibre Toxicol.* 25:4-9.

[21] Izak-Nau E, Kenesei K, Murali K, Voetz M, Eiden S, Puntes VF, Duschl A, Madarász E. 2013. Interaction of differently functionalized fluorescent silica nanoparticles with neural stem- and tissue-type cells. *Nanotoxicology.* 29:45-57.

[22] Verma A, Stellacci F. 2010 Effect of surface properties on nanoparticle-cell interactions. *Small.* 12-21.

[23] Busch W, 2011. Internalization of engineered nanoparticles into mammalian cells in vitro: influence of cell type and particle properties. *J. Nanoparticle Res.* 13(1):293–310.

[24] Gheshlaghi ZN, Riazi GH, Ahmadian S, Ghafari M, Mahinpour R. 2008. Toxicity and interaction of titanium dioxide nanoparticles with microtubule protein. *Acta Biochim. Biophys. Sin.* (Shanghai). 40(9):777-82.

[25] Popova N.V, Deyev I.E., Petrenko A.G. 2013. Clathrin-Mediated. Endocytosis and Adaptor Proteins. *Acta Naturae.* 18: 62-73.

[26] Conner SD, Schmid SL. 2003. Regulated portals of entry into the cell. *Nature.* 37-44.

[27] FranÇois Aguet, Costin N. Antonescu, Marcel Mettlen, Sandra L. Schmid, and Gaudenz Danuser. 2013. Advances in Analysis of Low Signal-to-Noise Images Link Dynamin and AP2 to the Functions of an *Endocytic Checkpoint. Dev. Cell.* 26: 279–291.

[28] Bonifacino J. S. and B. S. Glick. 2004. The mechanisms of vesicle budding and fusion. *Cell.* 116: 2:153–166.

[29] Grant BD, Donaldson JG. 2009. Pathways and mechanisms of endocytic recycling. *Nat. Rev. Mol. Cell Biol.* 10: 597–608.

[30] Mayor S, Pagano RE. 2007. Pathways of clathrin-independent endocytosis. *Nat. Rev. Mol. Cell Biol.* 8:603–12.

[31] Sandvig K, Torgersen ML, Raa HA, van Deurs B. 2008. Clathrin-independent endocytosis: from nonexisting to an extreme degree of complexity. *Histochem. Cell Biol.* 129:267–76.

[32] Rothberg KG, Heuser JE, Donzell WC, Ying YS, Glenney JR, Anderson RG. 1992. Caveolin, a protein component of caveolae membrane coats. *Cell.* 4:673-82.

[33] Pavlides S, Gutierrez-Pajares JL, Iturrieta J, Lisanti MP, Frank PG. 2014. Endothelial caveolin-1 plays a major role in the development of atherosclerosis. *Cell Tissue Res.* 356(1): 147-157.

[34] Lingwood D, Simons K. 2007. Detergent resistance as a tool in membrane research. *Nat Protoc.* 2(9): 2159-65.

[35] Simons K, Gerl MJ. 2010. Revitalizing membrane rafts: new tools and insights. *Nat. Rev. Mol. Cell Biol.* 11(10): 688-99.

[36] Simons K, Toomre D. 2000. Lipid rafts and signal transduction. *Nat. Rev. Mol. Cell Biol.* 1(1):31-9.

[37] Kettler K, Veltman K, van de Meent D, van Wezel A, Hendriks AJ. 2014. Cellular uptake of nanoparticles as determined by particle properties, experimental conditions, and cell type. *Environ. Toxicol. Chem.* 481-92.

[38] Gehr P, Clift MJ, Brandenberger C, Lehmann A, Herzog F, Rothen-Rutishauser B. 2011. Endocytosis of environmental and engineered micro- and nanosized particles. *Physiol.* 1159-74.

[39] Sansonetti PJ. 2000. Phagocytosis, a cell biology view. *J. Cell Sci.* 19: 3355-3368.

[40] Chen Q, Xue Y, Sun J. 2013. Kupffer cell-mediated hepatic injury induced by silica nanoparticles in vitro and in vivo. *Int. J. Nanomedicine.* 1129-40.

[41] Lewinski N, Colvin V y Drezek R. 2008. Citoxicity of nanoparticles. *Small.* 4 (1): 26-49.

[42] Aillon K, Xie Y, Gendy N, Berkland C y Forrest M. 2009. Effects of nanomaterial physicochemical properties on in vivo toxicity. *Advanced Drug Delivery Reviewa.* 61:457-466.

[43] F. Sallusto, M. Cella, C. Danieli, A. Lanzavecchia. 1995. Dendritic cells use macropinocytosis and the mannose receptor to concentrate macromolecules in the major histocompatibility complex class II compartment: downregulation by cytokines and bacterial products *J. Exp. Med.* 182:389–400.

[44] Bregar VB, Lojk J, Sustar V, Veranič P, Pavlin M. 2013. Visualization of internalization of functionalized cobalt ferrite nanoparticles and their intracellular fate. *Int. J. Nanomedicine.* 8:919-31.

[45] Vranic S, Boggetto N, Contremoulins V, Mornet S, Reinhardt N, Marano F, Baeza-Squiban A, Boland S. 2013. Deciphering the mechanisms of cellular uptake of engineered nanoparticles by accurate evaluation of internalization using imaging flow cytometry. *Part. Fibre Toxicol.* 2-16.

[46] Liu X, Huang N, Li H, Jin Q, Ji J. 2013. Surface and size effects on cell interaction of gold nanoparticles with both phagocytic and nonphagocytic cells. *Langmuir.* 29:9138-48.

[47] Fraga S, Faria H, Soares ME, Duarte JA, Soares L, Pereira E, Costa-Pereira C, Teixeira JP, de Lourdes Bastos M, Carmo H. 2013. Influence of the surface coating on the cytotoxicity, genotoxicity and uptake of gold nanoparticles in human HepG2 cells. *J. Appl. Toxicol.* 10:1111-9.

[48] Yan Z, Xu L, Han J, Wu YJ, Wang W, Yao W, Wu W. 2014. Transcriptional and posttranscriptional regulation and endocytosis were involved in zinc oxide nanoparticle-induced interleukin-8 overexpression in human bronchial epithelial cells. *Cell Biol. Toxicol.* 30(2): 79-88.

[49] Allouni ZE, Høl PJ, Cauqui MA, Gjerdet NR, Cimpan MR. 2012. Role of physicochemical characteristics in the uptake of TiO2 nanoparticles by fibroblasts. *Toxicol. In Vitro*. 3:469-79.

[50] Luo M, Shen C, Feltis BN, Martin LL, Hughes AE, Wright PF, Turney TW. 2014. Reducing ZnO Nanoparticles citoxicity bay surface modification. Nanoescale.. (epub ahead on print)

[51] Ferrero-Miliani, L Nielsen, O.H Andersen, P.S Girardin, S.E. 2007. Chronic inflammation: importance of NOD2 and NALP3 in interleukin-1beta generation. *Clin. Exp. Immunol.* 147:227–235.

[52] Medzhitov, R. 2008. Origin and physiological roles of inflammation. *Nature.* 454:428–435.

[53] Gądek-Michalska A, Tadeusz J, Rachwalska P, Bugajski J. 2013. Cytokines, prostaglandins and nitric oxide in the regulation of stress-response systems. *Pharmacol. Rep*. 65(6):1655-62.

[54] Rita de Cássia da Silveira e Sá, Luciana Nalone Andrade, Rafael dos Reis Barreto de Oliveira, Damião Pergentino de Sousa.2014. A Review on Anti-Inflammatory Activity of Phenylpropanoids Found in Essential Oils. *Molecules*.19:1459-1480.

[55] Serhan C.N, Savill J. 2005. Resolution of inflammation: the beginning programs the end. *Nature Immunology*. 6:1191–1197.

[56] Summers C, Rankin SM, Condliffe AM, Singh N, Peters AM, Chilvers ER. 2013. Neutrophil kinetics in health and disease. *Trends Immunol.* 31:318–324.

[57] Nathan C. 2006. Neutrophils and immunity: challenges and opportunities. *Nat. Rev. Immunol*. 6:173–182.

[58] Abbas, A.B, Lichtman, A.H. 2009. Innate Immunity. In *Basic Immunology, Functions and Disorders of the Immune System*, 3rd ed.58

[59] de Rooij J. Adhesion in vascular biology: Mechanics control dynamics. 2014. *Cell Adh Migr.* 14:8(2). [Epub ahead of print].

[60] Gilroy D.W.2004. The endogenous control of acute inflammation–from onset to resolution. *Drug Discovery Today: Therapeutic Strategies.* 1:313.

[61] Savill J, Dransfield I, Gregory C, Haslett C. 2002. A blast from the past: clearance of apoptotic cells regulates immune responses. *Nature Reviews Immunology.* 2:965–975.

[62] Chapman KE, Coutinho AE, Zhang Z, Kipari T, Savill JS, Seckl JR. 2013. Changing glucocorticoid action: 11β-hydroxysteroid dehydrogenase type 1 in acute and chronic inflammation. *J. Steroid Biochem. Mol. Biol.*137: 82-92.

[63] Hart BA, Chalan P, Koopman G, Boots AM. 2013. Chronic autoimmune-mediated inflammation: a senescent immune response to injury. *Drug Discov. Today*. 18(7-8): 372-9.

[64] Garnett M, Kallinteri P. 2006. Nanomedicines and nanotoxicology: some physiological principles. *Occup. Med.* 56:307-311.

[65] Chen M, von Mikecz A. 2005. Formation of nucleoplasmic protein aggregates impairs nuclear function in response to SiO2 nanoparticles. *Experiment Cell Res.* 305:51-62.

[66] Geiser M, Rothen-Rutishauser B, Kapp N, Schürch S, Kreyling W, Schulz H, Semmler M, Im Hof V, Heyder J, Gehr P. 2005. Ultrafine particles cross cellular membranes by nonphagocytic mechanisms in lungs and in cultured cells. *Environ. Health Perspect.* 113(11):1555-60.

[67] Li N, Sioutas C, Cho A, Schmitz D, Misra C, Sempf J, Wang M, Oberley T, Froines J, Nel A. 2003. Ultrafine particulate pollutants induce oxidative stress andmitochondrial damage. *Environ. Health Perspect.* 111(4): 455-460.

[68] Oberdörster G, Oberdörster E, Oberdörster J. 2005b. Nanotoxicology: an emerging discipline from studies of ultrafine particles. *Environ. Health Perspect.* 113(7): 823-839.

[69] Xia, T., Kovochich, M., Brant, J., Hotze, M., Sempf, J., Oberley, T., Sioutas, C., Yeh, J.I., Wiesner, M.R. y Nel, A.E. 2006. Comparison of the abilities of ambient and manufactured nanoparticles to induce cellular toxicity according to an oxidative stress paradigm. *Nano letters*. 6: 1794-1807.

[70] Stone, V. y Donaldson, K. Nanotoxicology.2006. Signs of stress. *Nature Nanotechnology*. 1: 23-24.

[71] Lanone, S. and Boczkowski, J. 2006. Biomedical applications and potential health risks of nanomaterials: molecular mechanisms. *Current Molecular Medicine.* 6: 651-663.

[72] Kaysen GA, Eiserich JP. 2004. The role of oxidative stress-altered lipoprotein structure and function and microinflammation on cardiovascular risk in patients with minor renal dysfunction. *J. Am. Soc. Nephrol.* 3:538-48.

[73] Hanserberg W. 2002. Biología de las especies de oxigeno reactivas. *Mensaje Bioquimico UNAM*. 26: 19-54.

[74] Jimenez I, Speisky C, Noran C. 2000. Radicales libres y antioxidantes en la prevención de enfermedades: II mecanismo de defensa antioxidantes. ReuChilNutr.

[75] Ferreira R. 1988. *Estrés oxidativo y antioxidativo: de las ciencias básicas a la medicina aplicada.* Buenos Aires: Hospital Militar Central.

[76] Dröge W. 2002a. Aging-related changes in the thiol/disulfide redox state: implications for the use of thiol antioxidants. *Exp. Gerontol.* December 37(12): 1333-45.

[77] Dröge W. 2002b. The plasma redox state and ageing. *Ageing Res. Rev.* 1(2):257-78.

[78] Dröge W. 2002c. Free radicals in the physiological control of cell function. *Physiol. Rev.* 82(1):47-95.

[79] Bretón-Romero R, Lamas S. 2014. Hydrogen peroxide signaling in vascular endothelial cells. *Redox. Biol.* 2:529-534.

[80] Matés J, Pérez-Gómez C, Nuñez de Castro I. 1999. Antioxidant enzymes and human diseases. *Clin. Biochem.* 32: 595-603.

[81] Urrutia PJ, Mena NP, Núñez MT. 2014. The interplay between iron accumulation, mitochondrial dysfunction, and inflammation during the execution step of neurodegenerative disorders. *Front Pharmacol.* 10: 5:38.

[82] Helen R. Griffiths, Irundika H.K. Dias, Rachel S. Willetts, and Andrew Devitt. 2014. Redox regulation of protein damage plasma. *Redox. Biol.* 2: 430–435.

[83] Vanhoutte PM. 2003. Endothelial control of vasomotor function from health to coronary disease. *Circ. J.* 67: 572- 575.

[84] Cook-Mills JM, Deem L. 2005. Active participation of endothelial cells in inflammation. *J. Leukoc. Biol.* 77:487- 495.

[85] Rahman I. 2000. Regulation of nuclear factor-κB, activator protein-1, and glutathione levels by tumor necrosis factor-α and dexamethasone in alveolar epithelial cells. *Biochem. Pharmacol.* 60:1041–1049.

[86] Shi Q, Cao J, Fang L, Zhao H, Liu Z, Zheng X, LiX, Zhou Y, Ge D, Zhang H, Wang L, Ran Y, Fu J. 2014. Geniposide suppresses LPS-induced nitric oxide, PGE2 and inflammatory cytokine by downregulating NF-κB, MAPK and AP-1 signaling phathway in macrophages. Int Immunopharmacol. 20(2):298-306.

[87] Chen, J.; Y. Liu; L. Hongmei; P. L. Hermonat y J. L. Mehta. 2006. "Lectin-like oxidized low-density lipoprotein receptor-1 (LOX-1) transcriptional regulation by Oct-1 in human endothelial cells: implications for atheroscle- rosis", *Biochem.* J. 393.

[88] Yang C.-Y., S.-H. Chen, and S. H. Gianturco. 1986. "Sequence, structure, receptor-binding domains and internal repeats of human apolipoprotein B-100," *Nature.* 323(6090)738–742.

[89] Li, D., Chen, H., Romeo, F., Sawamura, T. and Mehta, J. L. 2002. Statins modulate oxidized low-density lipoprotein-mediated adhesion molecule expression in human coronary artery endothelial cells: role of LOX-1. *J. Pharmacol. Exp. Ther.* 302:601–605.

[90] Mehta, J. L., Li, D. Y., Chen, H. J., Joseph, J. and Romeo, F. 2001. Inhibition of LOX-1 by statins may relate to upregulation of eNOS. *Biochem. Biophys. Res. Commun.* 289:857–861.

[91] Chen, J., Mehta, J. L., Haider, N. and Li, D. 2004. Role of caspases in ox-LDL-induced apoptotic cascade in human coronary artery endothelial cells. *Circ. Res.* 94:370–376.

[92] N. Kume, T. Kita. 2001. Roles of lectin-like oxidized LDL receptor-1 and its soluble forms in atherogenesis *Curr. Opin. Lipidol.* 12:419–423.

[93] Ford, M. A.; T. G. Allison y A. Lerman. 2008. "New Approaches to the Concept of Primary Prevention of Atherosclerosis", *Current Treatment Options in Cardiovascular Medicine.* Vol. 10.

[94] Ze Y, Zheng L, Zhao X, Gui S, Sang X, Su J, Guan N, Zhu L, Sheng L, Hu R, Cheng J, Cheng Z, Sun Q, Wang L, Hong F. 2013. Molecular mechanism of titanium dioxide nanoparticles-induced oxidative injury in the brain of mice. *Chemosphere.* 92(9): 1183-9.

[95] Sheng L, Wang X, Sang X, Ze Y, Zhao X, Liu D, Gui S, Sun Q, Cheng J, Cheng Z, Hu R, Wang L, Hong F. 2013. Cardiac oxidative damage in mice following exposure to nanoparticulate titanium dioxide. *J. Biomed. Mater Res. A.* 101(11):3238-46.

[96] Wang J, Ma J, Dong L, Hou Y, Jia X, Niu X, Fan Y. 2013. Effect of anatase TiO2 nanoparticles on the growth of RSC-364 rat synovial cell. *J. Nanosci. Nanotechnol.* 13(6): 3874-9.

[97] Wang ZY, Yu XL, Gao DM, Feng WQ, Xing BS, Li FM. 2010. Effect of nano-rutile TiO2 and multi-walled carbon nanotubes on the growth of maize (Zea mays L.) seedlings and the relevant antioxidant response]. *Huan Jing Ke Xue.* 31(2): 480-7.

[98] Montiel-Dávalos A, Ventura-Gallegos JL, Alfaro-Moreno E, Soria-Castro E, García-Latorre E, Cabañas-Moreno JG, del Pilar Ramos-Godinez M, López-Marure R. 2012. TiO_2 nanoparticles induce dysfunction and activation of human endothelial cells. *Chem. Res. Toxicol.* Apr 16:25(4):920-30.

[99] Ramos Godinez M del Pilar, Gonzalez-Gomez BE, Montiel-Davalos, Lopez-Marure R, Alfaro-Moreno E. 2013. TiO2 nanoparticles induce endotelial cell activation in a pneumocyte-endothelial co-culture model. *Toxicol. In vitro*. 27(2):774-81.

[100] Huerta-García E, Montiél-Dávalos A, Alfaro-Moreno E, Gutiérrez-Iglesias G, López-Marure R. 2013. Dehydroepiandrosterone protects endothelial cells against inflammatory events induced by urban particulate matter and titanium dioxide nanoparticles. *Biomed. Res. Int.* 1-7.

[101] Tang M, Zhang T, Xue Y, Wang S, Huang M, Yang Y, Lu M, Lei H, Kong L, Wang Y, Pu Y. 2011. Metabonomic studies of biochemical changes in the serum of rats by intratracheally instilled TiO2 nanoparticles. *J. Nanosci. Nanotechnol.* 11(4):3065-74.

[102] Wang J, Deng X, Zhang F, Chen D, Ding W. 2014. ZnO nanoparticle-induced oxidative stress triggers apoptosis by activating JNK signaling pathway in cultured primary astrocytes. *Nanoscale Res. Lett.* 19(1):117. 102

[103] Avalos A, Haza AI, Mateo D, Morales P. 2014. Cytotoxicity and ROS production of manufactured silver nanoparticles of different sizes in hepatoma and leukemia cells. *J. Appl. Toxicol.* 34(4):413-23.

[104] Akhtar MJ, Kumar S, Alhadlaq HA, Alrokayan SA, Abu-Salah KM, Ahamed M. 2013 Dose-dependent genotoxicity of copper oxide nanoparticles stimulated by reactive oxygen species in human lung epithelial cells. Toxicol Ind. Health.

[105] Kumari M, Singh SP, Chinde S, Rahman MF, Mahboob M, Grover P. 2014 Toxicity study of cerium oxide nanoparticles in human neuroblastoma cells. *Int. J. Toxicol.* 33(2):86-97.

[106] Karin, M and A Lin. 2002. NF-kappaB at the cross roads of life and dheat. "*Nat. immunol.* 3(3):221-227.

[107] Sun B, Fan H, Honda T, Fujimaki R, Lafond–Walker A, Masui Y, Lowenstein CJ, Becker LC. 2001. Activation of NF kappa B and expression of ICAM–1 in ischemic–reperfused canine myocardium. *J. Mol. Cell Cardiol.* 33(1):109–119.

[108] Morgan E, Boyle E, Yun W, Griscavage–Ennis J, Farr AL, Canty T, Pohlman TH, Verrier ED. 1999. An essential role for NF–kappaB in the cardioadaptive response to ischemia. *Ann. Thorac. Surg.* 68(2):377–382.

[109] Feeley BT, Miniati D, Park AK, Hoyt EG, Robins RC. 2000. Nuclear factor–KappaB transcription factor decoy treatment inhibits graft coronary artery disease after cardiac transplantation in rodents. *Transplant.* 70(11):1560–1568.

[110] Valen G, Yan Z, Hansson G. 2001. Nuclear Factor Kappa–B and the Heart. *J. Am. Coll. Cardiol.* 38(2): 307–314.

[111] Perkins, N.D. 2007."integrating cell-signalling pathways with NF-kappaB and IKK function" *Nat. Rev. Mol. Cell Biol.* 8(1):49-62.

[112] Perkins ND. 2006. Post-translational modifications regulating the activity and function of the nuclear factor kappa B pathway. *Oncogene.* 25(51): 6717-30.

[113] Pye J, Ardeshirpour F, McCain A, Bellinger DA, Merricks E, Adams J, Elliott PJ, Pien C, Fischer TH, Baldwin AS Jr, Nichols TC. 2003. Proteasome inhibition ablates

activation of NF-kappa B in myocardial reperfusion and reduces reperfusion injury. *Am. J. Physiol. Heart Circ. Physiol.* 284(3): 919-26.

[114] Yamamoto Y, Gaynor RB. 2001. Therapeutic potential of inhibition of the NF-kappaB pathway in the treatment of inflammation and cancer. *J. Clin. Invest.* 107(2): 135.

[115] Baldwin AS, Jr. 2001. Series introduction: the transcription factor NF-kappaB and human disease. *J. Clin. Invest.* 107(1): 3-6.

[116] Janssen-Heininger, Y. M., Poynter, M. E., and Baeuerle, P. A. 2000. Recent advances towards understanding redox mechanisms in the activation of nuclear factor kappaB. *Free Radical Biol. Med.* 28: 1317−1327.

[117] Zhou, L., Tan, A., Iasvovskaia, S., Li, J., Lin, A., and Hershenson, M. B. 2003. Ras and mitogen-activated protein kinase kinase kinase-1coregulate activator protein-1- and nuclear factor-kappaB-mediated gene expression in airway epithelial cells. *Am. J. Respir. Cell Mol. Biol.* 28:762−769.

[118] Nakajima, H, Fujiwara, H, Furuichi, Y, and Shimbara, N. 2008. A novel small-molecule inhibitor of NF-κB signaling. *Biochem. Biophys.* 368:1007−1013.

[119] Pham, C. G., Bubici, C., Zazzeroni, F., Papa, S., Jones, J., Alvarez,K., Jayawardena, S., De Smaele, E., Cong, R., Beaumont, C., Torti, F.M., Torti, S. V., and Franzoso, G. 2004. Ferretin heavy chain upregulation by NF-κB inhibits TNF-α induced apoptosis by suppressing reactive oxygen species. *Cell.* 119:529−542.

[120] Churg, A., Xie, C, Wang, X, Vincent, R., and Wang, R. D. 2005. Air pollution particles activate NF-kappaB on contact with airway epithelial cell surfaces. *Toxicol. Appl. Pharmacol.* 208:37−45.

[121] Chen P, Migita S, Kanehira K, Sonezaki S, Taniguchi A. 2011. Development of sensor cells using NF-κB pathway activation for detection of nanoparticle-induced inflammation. *Sensors* (Basel). 11(7):7219-30.

[122] Halbleib J. M., y Nelson W. J. 2006. Cadherins in development: cell adhesion, sorting, and tissue morphogenesis". *Genes Dev.* 20(23):3199-214.

[123] Weiss U. 2002. Nature insight: Inflammation. *Nature*.420:845.

[124] Alberts B, Johnson A, Lewis J, Raff M, Roberts K, Walter P. 2002. *Molecular Biology of the Cell*. 4th ed. New York: Garland Publishing Group, Inc.

[125] Aird WC. 2007. Phenotypic Heterogeneity of the Endothelium: I. Structure, Function and Mechanisms. *Circ. Res.* 100:158-173.

[126] Vogel RA. 2003. Heads and hearts. The endothelial connection. *Circulation.* 107:2766-68.

[127] Bonetti PO, Lerman LO, Lerman A. 2003. Endothelial dysfunction. A marker of atherosclerotic risk. *Art Thromb. Vasc. Biol.* 23:168-75.

[128] Mikkelsen L, Jensen KA, Koponen IK, Saber AT, Wallin H, Loft S, Vogel U, Møller P. 2013. Cytotoxicity, oxidative stress and expression of adhesion molecules in human umbilical vein endothelial cells exposed to dust from paints with or without nanoparticles. *Nanotoxicology.* 7(2):117-34.

[129] Han SG, Newsome B, Hennig B. 2013. Titanium dioxide nanoparticles increase inflammatory responses in vascular endothelial cells. *Toxicology.* 5(306):1-8.

[130] Napierska D, Quarck R, Thomassen LC, Lison D, Martens JA, Delcroix M, Nemery B, Hoet PH. 2013. Amorphous silica nanoparticles promote monocyte adhesion to human endothelial cells: size-dependent effect. *Small.* 9(3): 430-8.

[131] Li CH, Liao PL, Shyu MK, Liu CW, Kao CC, Huang SH, Cheng YW, Kang JJ. 2012. Zinc oxide nanoparticles-induced intercellular adhesion molecule 1 expression requires Rac1/Cdc42, mixed lineage kinase 3, and c-Jun N-terminal kinase activation in endothelial cells. *Toxicol.* 126(1): 162-72.

[132] Sommer C, White F. 2010. *Cytokines, chemokines and Pain*, em: Beaulieu P, Lussier D, Porreca F. Pharmacology of Pain. 1 st Ed, Seatttle, IASP Press. 279-302.

[133] Lin E, Calvano SE, Lowry SF. 2000. *Inflammatory cytokines and cell response in surgery.* 127:117-126.

[134] Gonçalves DM, Chiasson S, Girard D. 2010. Activation of human neutrophils by titanium dioxide nanoparticles. *Toxicol. In Vitro.* 24(3): 1002-8.

[135] Zhang JM, An J. 2007. Cytokines, inflammation, and Pain. *Int. Anesthesiol. Clin.* 45:27-37.

[136] Alfaro-Moreno E, Nawrot TS, Vanaudenaerde BM, Hoylaerts MF, Vanoirbeek JA, Nemery B, Hoet PH. 2008. Co-cultures of multiple cell types mimic pulmonary cell communication in response to urban PM10. *Eur. Respir. J.* 32(5): 1184-94.

[137] Emmerechts J, Alfaro-Moreno E, Vanaudenaerde BM, Nemery B, Hoylaerts MF. 2010. Short-term exposure to particulate matter induces arterial but not venous thrombosis in healthy mice. *J. Thromb. Haemost.* 8(12): 2651-61.

[138] Scherbart AM, Langer J, Bushmelev A, van Berlo D, Haberzettl P, van Schooten FJ, Schmidt AM, Rose CR, Schins RP, Albrecht C. 2011. Contrasting macrophage activation by fine and ultrafine titanium dioxide particles is associated with different uptake mechanisms. *Part. Fibre Toxicol.* 13:8:31.

[139] Ekstrand-Hammarström B, Akfur CM, Andersson PO, Lejon C, Osterlund L, Bucht A. 2012. Human primary bronchial epithelial cells respond differently to titanium dioxide nanoparticles than the lung epithelial cell lines A549 and BEAS-2B. *Nanotoxicology.* 6(6):623-34.

[140] Mendoza A, Torres-Hernandez JA, Ault JG, Pedersen-Lane JH, Gao D, Lawrence DA. 2014. Silica nanoparticles induce oxidative stress and inflammation of human peripheral blood mononuclear cells. *Cell Stress Chaperones.* 5:789-895.

[141] FGF Place LW, Sekyi M, Kipper MJ.2014. Aggrecan-mimetic, glycosaminoglycan-containing nanoparticles for growth factor stabilization and delivery. *Biomacromolecules.* 2:680-9.

[142] Murugesan S, Mousa SA, O'connor LJ, Lincoln DW, Linhardt RJ. 2007. Carbon inhibits vascular endothelial growth factor- and fibroblast growth factor-promoted angiogenesis. *FEBS Lett.* 581(6):1157–1160.

[143] Khatri M, Bello D, Pal AK, Cohen JM, Woskie S, Gassert T, Lan J, Gu AZ, Demokritou P, Gaines P. 2013. Evaluation of cytotoxic, genotoxic and inflammatory responses of nanoparticles from photocopiers in three human cell lines. *Part. Fibre Toxicol.* 10:42.

[144] Ferrara N. 2002. VEGF and the quest for tumor angiogenesis factor. *Nat. Rev. Cancer.* 2:795-803.

[145] Bir SC, Shen X, Kavanagh TJ, Kevil CG, Pattillo CB. 2013. Control of angiogenesis dictated by picomolar superoxide levels. *Free Radic. Biol. Med.* 63:135-42.

[146] Xin Q. Brown, Erzsebet Bartolak-Suki, Corin Williams, Mathew L. Walker, Valerie M. Weaver, and Joyce Y. Wong. 2010. Effect of substrate stiffness and PDGF on the

behavior of vascular smooth muscle cells: implications for atherosclerosis *J. Cell Physiol.* 225(1):115–122.

[147] Raines EW. PDGF and cardiovascular disease. 2004. *Cytokine Growth Factor Rev.* 15(4): 237–254.

[148] Castrogiovanni P1, Scuderi R, Travali S, Failla A, Failla G, Imbesi R. 2013. Paracrine activity of heparin-binding EGF-like growth factor in atherogenesis: an immunohistochemical study. *Minerva Med.* 104(1):85-91.

[149] Jang S, Thirupathi P, Neupane LN, Seong J, Lee H, Lee WI, Lee KH. 2012. *Highly sensitive ratiometric fluorescent chemosensor for silver ion and silver nanoparticles in aqueous solution.* 21:14(18):4746-9.

[150] Karthikeyan B, Kalishwaralal K, Sheikpranbabu S, Deepak V, Haribalaganesh R, Gurunathan S. 2010. Gold nanoparticles downregulate VEGF-and IL-1β-induced cell proliferation through Src kinase in retinal pigment epithelial cells. *Exp. Eye Res.* 91(5):769-78.

[151] Al-Rasheed NM, Faddah LM, Mohamed AM, Abdel Baky NA, Al-Rasheed NM, Mohammad RA. 2013. Potential impact of quercetin and idebenone against immuno-inflammatory and oxidative renal damage induced in rats by titanium dioxide nanoparticles toxicity. *J. Oleo.* 62(11): 961-71.

[152] Nogueira CM, de Azevedo WM, Dagli ML, Toma SH, Leite AZ, Lordello ML, Nishitokukado I, Ortiz-Agostinho CL, Duarte MI, Ferreira MA, Sipahi AM. 2012. Titanium dioxide induced inflammation in the small intestine. *World J. Gastroenterol.* 14:18(34):4729-35.

[153] Armand L, Dagouassat M, Belade E, Simon-Deckers A, Le Gouvello S, Tharabat C, Duprez C, Andujar P, Pairon JC, Boczkowski J, Lanone S. 2013. Titanium dioxide nanoparticles induce matrix metalloprotease 1 in human pulmonary fibroblasts partly via an interleukin-1β-dependent mechanism. *Am. J. Respir. Cell Mol. Biol.* 48(3):354-63.

[154] Glista-Baker EE, Taylor AJ, Sayers BC, Thompson EA, Bonner JC. 2012. Nickel nanoparticles enhance platelet-derived growth factor-induced chemokine expression by mesothelial cells via prolonged mitogen-activated protein kinase activation. *Am. J. Respir. Cell Mol. Biol.* 4:552-61.

[155] Wood JP, Ellery PE, Maroney SA, Mast AE. 2014. Biology of tissue factor pathway inhibitor. *Blood.* Mar 11.

[156] Kinev AV, Roubey RA. 2008. Tissue factor in the antiphospholipid syndrome. *Lupus.* 17(10): 952-8.

[157] Panasiuk A, Zak J, Panasiuk B, Prokopowicz D. 2007. Increase in expression of monocytic tissue factor (CD142) with monocytes and blood platelet activation in liver cirrhosis. *Blood Coagul. Fibrinolysis.* 18(8):739-44.

[158] Verdonk K, Visser W, Van Den Meiracker AH, Danser AH. 2014. The renin-angiotensin-aldosterone system in pre-eclampsia: the delicate balance between good and bad. *Clin. Sci.* (Lond). 126(8):537-44.

[159] Liu X, Sun J. 2013. Time-course effects of intravenously administrated silica nanoparticles on blood coagulation and endothelial function in rats. *J. Nanosci. Nanotechnol.* 13(1):222-8.

[160] Bröker L.E., Kruyt F.A. y Giaccone G. 2005. Cell death independent of caspases: a review. *Clin. Cancer Res.* 11:3155-3162.

[161] Danial N.N. y Korsmeyer S.J. 2004. Cell death critical control points. *Cell* 116: 205-219.

[162] Baehrecke E.H. 2002. How death shapes life during development. *Nat. Rev. Mol. Cell Biol.* 3:779-787.

[163] Mackay I.R., Leskovsek N.V. y Rose N.R. 2008. Cell damage and autoimmunity: a critical appraisal. *J. Autoimmun.* 30:5-11.

[164] Huerta S., Goulet E.J., Huerta-Yepez S. y Livingston E.H. 2007. Screening and detection of apoptosis. *J. Surg. Res.* 139:143-156.

[165] Savill J. y Fadok V. 2000. Corpse clearance defines the meaning of cell death. *Nature.* 407:784-788.

[166] Takaki K, Higuchi Y, Hashii M, Ogino C, Shimizu N. 2014. Induction of apoptosis associated with chromosomal DNA fragmentation and caspase-3 activation in leukemia L1210 cells by TiO2 nanoparticles. *J. Biosci. Bioeng.* 117(1):129-33.

[167] Shi Y, Wang F, He J, Yadav S, Wang H. 2010. Titanium dioxide nanoparticles cause apoptosis in BEAS-2B cells through the caspase 8/t-Bid-independent mitochondrial pathway. *Toxicol. Lett.* 196(1):21-7.

[168] Zhao J, Bowman L, Zhang X, Vallyathan V, Young SH, Castranova V, Ding M. 2009. Titanium dioxide nanoparticles induce JB6 cell apoptosis through activation of the caspase-8/Bid and mitochondrial pathways. *J. Toxicol. Environ. Health A.* 72(19):1141-9.

[169] Hussain S, Thomassen LC, Ferecatu I, Borot MC, Andreau K, Martens JA, Fleury J, Baeza-Squiban A, Marano F, Boland S. 2010. Carbon black and titanium dioxide nanoparticles elicit distinct apoptotic pathways in bronchial epithelial cells. *Part. Fibre Toxicol.* 16:7-10.

[170] Vamanu CI, Cimpan MR, Høl PJ, Sørnes S, Lie SA, Gjerdet NR. Dröge W. Free. 2008. Induction of cell death by TiO2 nanoparticles: studies on a human monoblastoid cell line. *Toxicol. In Vitro.* 22(7):1689-96.

[171] Shukla RK, Kumar A, Pandey AK, Singh SS, Dhawan A. 2011. Titanium dioxide nanoparticles induce oxidative stress-mediated apoptosis in human keratinocyte cells. *J. Biomed. Nanotechnol.* 1:100-1.

[172] Sang X, Fei M, Sheng L, Zhao X, Yu X, Hong J, Ze Y, Gui S, Sun Q, Ze X, Wang L, Hong F. 2013. Immunomodulatory effects in the spleen-injured mice following exposure to titanium dioxide nanoparticles. *J. Biomed. Mater Res. A* [Epub ahead of print].

[173] Gui S, Sang X, Zheng L, Ze Y, Zhao X, Sheng L, Sun Q, Cheng Z, Cheng J, Hu R, Wang L, Hong F, Tang M. 2013 F. Intragastric exposure to titanium dioxide nanoparticles induced nephrotoxicity in mice, assessed by physiological and gene expression modifications. *Part. Fibre Toxicol.* 10:1-16.

In: Titanium Dioxide
Editor: Jerri Brown

ISBN: 978-1-63321-391-3

Chapter 11

TITANIUM DIOXIDE FOR VISIBLE-LIGHT PHOTOCATALYTIC PURIFICATION OF GASEOUS INDOOR AIR POLLUTANTS

Mark Daniel G. de Luna[1] and Ming-Chun Lu[2]
[1]Department of Chemical Engineering, University of the Philippines, Diliman, Quezon City, Philippines
[2]Department of Environmental Resources Management, Chia Nan University of Pharmacy and Science, Tainan, Taiwan

ABSTRACT

Indoor air pollution from toxic gaseous chemicals poses adverse effects on human health, comfort and productivity. This is a serious concern nowadays since people spend more time (over 80%) indoors. In addition, recent studies on indoor air quality show that certain gaseous contaminants have greater concentration levels indoor than in the outdoor environment. Strategies to address this problem include increased ventilation and air cleaning. The former involves the transfer of pollutants from indoor to outdoor while conventional air cleaning methods utilize air filtration and adsorption. Meanwhile, heterogeneous photocatalysis, an advanced oxidation process (AOP), can completely degrade gaseous organics to carbon dioxide and water. In the past two decades, photocatalysis has emerged as a promising technology for the purification of indoor air and exhibited greater effectiveness over conventional air cleaning technologies. A catalyst that has taken center stage in the development of this technology is titanium dioxide (TiO_2), also known as titania, mainly due to these advantages: (1) TiO_2 is relatively inexpensive, non-toxic and chemically stable, (2) it promotes ambient temperature oxidation of the major classes of indoor air pollutants, (3) the complete degradation of a broad class range of pollutants can be achieved under certain operating conditions, and (4) chemical additives are not required for the process to proceed. The photogenerated holes of TiO_2 are highly oxidizing while its photogenerated electrons are reducing enough to produce superoxide from dioxygen. However, TiO_2 has a low solar photoconversion efficiency because of the large band gap of 3.2 and 3.0 eV for the anatase and rutile phases, respectively. Thus, TiO_2 can only be excited by irradiation using ultraviolet (UV) light, which covers only ~5% of the light spectrum. The development of photocatalysts exhibiting high reactivity under visible light (λ > 400 nm) is of great interest because it would allow the main part of the light spectrum to be utilized and photocatalytic activity would be observed even under poor illumination of interior lighting. This requires engineering the band-gap of TiO_2 to less than 3.2 eV. Studies on TiO_2 modification have been carried out to increase its photocatalytic activity under visible light illumination through doping metals and non-metals onto TiO_2. This chapter offers a review of the status of research related to (1) indoor air pollution by gaseous contaminants such as formaldehyde, benzene, toluene and nitrogen oxide; and (2) the synthesis and application of visible-light active TiO_2 by metal and non-metal doping for the treatment of these gaseous pollutants in indoor air.

1. INTRODUCTION

An indoor setting is essentially an enclosed space with gas composition, pressure, or temperature unlike that of the surrounding air in the natural atmosphere. The typical indoor environment is not limited to areas within buildings, but includes those in underground garages, vehicles and aircrafts [1]. Poor air quality in closed environments has been identified as an imminent risk to human health and this condition drives the demand for more efficient and environmentally friendly technologies for air purification [2]. Various organic and inorganic compounds mainly formaldehyde, benzene, toluene and other volatile organic compounds (VOCs), nitrogen oxides (NOx) and sulfur oxides (SOx) contaminate indoor air and pose significant hazards to human health, with symptoms such as skin irritation, chest tightness, headache, fatigue, dullness and dehydration [3]. The sources of these pollutants include emissions from nearby vehicles, from materials such as furniture, carpets, cleaning products, office equipment, furnishings and paints and from activities such as cooking, use of electrical appliances and other electrical devices [3–5]. The concentration of chemical contamination is usually higher indoor than outdoor [1] and is affected by the interaction of several physical and chemical parameters including humidity, temperature, air exchange rate, emission rates of gaseous pollutants from various materials, diffusion and adsorption rates of gaseous contaminants to sinks and generation of secondary pollutants as a result of chemical reactions [6].

The global energy crisis in the 1970s spurred worldwide awareness for energy conservation. Energy-saving strategies including the use of materials for insulation, construction of airtight buildings and reduction of fresh air utilization in air conditioning have been widely implemented. These measures caused a decline in indoor air circulation while the volatile chemical content of various decorative and building materials for the indoor setting continued to increase. Furthermore, indoor air is contaminated by the fumes from various indoor activities such as smoking and the influx of outdoor pollutants via the openings of some doors and windows [7]. When the outdoor air environment pose more risk, air dilution by ventilation (i.e., opening doors and windows) should not be considered an option for improving indoor air quality [8].

In recent years, the higher standards of living dictate the need for a more favorable and beneficial indoor conditions. More and more people remain in closed environments 80-90% of the time and air quality in these settings has an immense impact on an individual's health and productivity. The present state of indoor air quality is especially relevant to more vulnerable individuals such as infants and elderlies [7,9]. Thus, it is imperative to develop and apply effective systems for air purification to maintain the highest possible quality of indoor air [5].

Several air treatment technologies are currently being employed and explored to address indoor air quality issues. Conventional air purification technologies can either be of two main categories: one mainly utilizes filters and sorbents to capture pollutants while the other applies heating at elevated temperatures to decompose contaminants in the presence of catalysts.

Though these systems are capable of eliminating volatile organic compounds (VOCs), they have inevitable limitations [2]. Thermal oxidation processes consume large amounts of energy: catalyst-aided thermal oxidation systems usually operate at temperatures ranging from 200°C – 250°C while those that do not require catalysts operate from 730°C – 850°C. In addition, these processes may also trigger side reactions that generate harmful molecules such as nitrogen oxides and sulfur dioxides [2]. On the other hand, filters are economical and efficient devices widely-used in air-conditioning systems to enhance the quality of indoor air. However, the gradual build-up of organic matter collected and deposited on the filter surface makes the unit vulnerable to the growth of microorganisms which causes reduced filter efficiency, eventual filter deterioration and even health problems among individuals in the closed environment [7]. While adsorbents such as activated carbon, zeolite, activated alumina, silica gel, and various molecular sieves have been proven effective in removing pollutants in indoor air due to notable adsorbent properties such as high adsorption capacity and large specific area, adsorption capacity deteriorates over time even after regeneration. Furthermore, the inherent selective adsorption of these materials and the high cost of activated carbon filters (5-100 times more than granular activated carbon (GAC)) hinder the extensive utilization of this technology [7]. Lastly, the latter air purification methods transform and immobilize pollutants onto a solid material and would require added costs for the handling and disposal of spent filters and adsorbents [10,11].

In the recent decade, there has been a growing interest in the exploration of a promising technology utilizing semiconductors TiO_2, SnO_2, ZnO, CdS, Fe_2O_3, WO_3 and MnO_2 as photocatalyst for the non-selective oxidation of organic contaminants to produce harmless CO_2 and H_2O as end-products [2,7]. This air purification process called photocatalytic oxidation (PCO) is largely dependent on reaction rate, humidity, light source, inlet pollutant concentration, photocatalyst characteristics, and reactor type [7].

Based on extensive literature review, no publication offers a comprehensive analysis of research results on the application of visible light active TiO_2 for the removal of gaseous pollutants in the indoor environment. The photoactivity of visible-light-active TiO_2 is usually evaluated using organic molecules in aqueous media such as phenols [12,13]; chlorophenols [14–16]; salicylic acid [17]; acetic acid [18]; dyes [15,19–23]; antibiotics [24] and even bacteria [25]. This chapter presents a review of studies related to indoor air purification by TiO_2-based visible light photocatalysis.

2. Photocatalytic Indoor Air treatment

Photocatalytic oxidation has been extensively investigated for the decomposition of gaseous contaminants in both indoor and outdoor air environments [26] and the most widely applied photocatalytic material is titanium dioxide (TiO_2) because of its high catalytic activity, long-term stability towards photo-corrosion and chemicals, non-toxic properties and low cost [27–30]. One of the main advantages of photocatalytic oxidation is its financial viability since this technology can be operated at room temperature with minimal energy requirements [2]. Air purification by photocatalytic oxidation depends on a number of parameters including: light intensity, substrate concentration, O_2 partial pressure, humidity and substrate type [31]. There are two main categories in the application of TiO_2 photocatalysis to indoor air purification: (1) passive and (2) active treatment.

2.1. Passive Air Purification

Passive air purification is a cost-effective strategy to mitigate, if not eliminate, gaseous contaminants in indoor air. This is done by incorporating TiO_2 onto available surfaces in the indoor environment such as painted walls or transparently coated windows [1,32]. Photocatalysis using TiO_2-loaded cementitious materials has also been applied for air treatment applications [33]. In addition, deodorizing sheets embedded with TiO_2 photocatalysts and active carbon adsorbents have been used to eliminate foul-smelling substances in kitchens, refrigerators, vehicles and shoes. The malodorous compounds are removed by the sheets through adsorption and the adsorbed pollutants are readily oxidized by the action of the TiO_2 photocatalysts when the used sheet is placed under solar irradiation [31]. The effectiveness of the passive air purification method can be maximized when contiguous TiO_2-containing surfaces, whether horizontal or vertical, are placed close to the source of gaseous contaminants [10].

A schematic view of a typical passive air purification system is shown in Fig. 1. When light is irradiated onto the TiO_2 surface, photons having energy at least equal to that of the semiconductor band gap energy will cause electrons (e^-) in the TiO_2 to jump from the valence band (vb) to the conduction band (cb), leaving holes (h^+) or "electron vacancy" in the valence band and forming electron-hole pairs ($e^- - h^+$) as shown in Eq. 1 [34–36].

$$TiO_2 + hv \rightarrow e^-_{cb} + h^+_{vb} \tag{1}$$

When the band gap energy is known, the corresponding wavelength of light that can cause the electron excitation can be derived from the Planck–Einstein equation given by Eq. 2 [2].

$$E = \frac{hc}{\lambda} \tag{2}$$

where E is band gap energy, h is Plank's constant, c is speed of light and λ is threshold wavelength in nm.

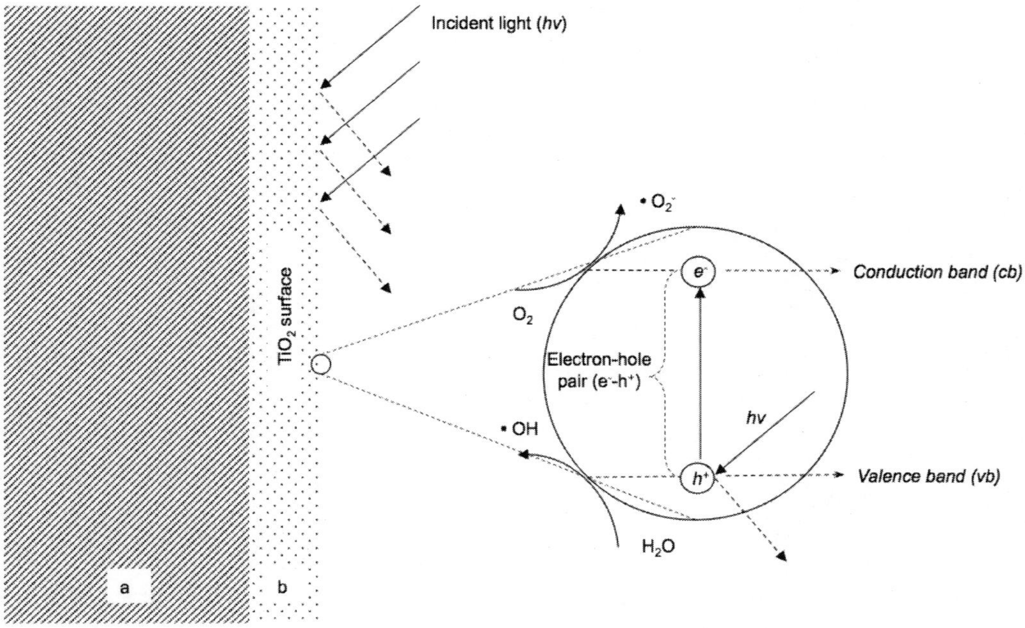

Figure 1. A typical passive photocatalytic indoor air purification system consisting of (a) substrate and (b) TiO_2 coating.

Depending on the photocatalyst molecular structure and the prevailing conditions of the reaction, the electron-hole pairs may recombine in a short time following Eq. 3.

$$e^-_{cb} + h^+_{vb} \rightarrow TiO_2 \quad (3)$$

In most cases, electron-hole pairs, h^+ and e^-, quickly migrate to the surface of the TiO_2 catalyst. The strong oxidation power of h^+ enables it to react with adsorbed water to form hydroxyl radicals ($\bullet OH$) as in Eq. 4 and Eq. 5.

$$h^+_{vb} + H_2O \rightarrow \bullet OH + H^+ \quad (4)$$

$$h^+_{vb} + OH^- \rightarrow \bullet OH_{ad} \quad (5)$$

Hydroxyl radicals are highly energetic and reactive species which disappear as quickly as they are formed [37]. These powerful non-selective oxidants have an oxidizing potential that exceeds that of ozone by 40 percent [38]. The concentration of hydroxyl radicals formed and the mass transfer of these species to the target organic molecules determine the extent of pollutant degradation [39,40].

Hydroxyl radicals destroy organic substances by converting them into highly reactive organic radicals (Eq. 6) which undergo further break down by hydroxyl radical attack until complete mineralization (Eq. 7) into harmless final products [41–43].

$$RH + \bullet OH \rightarrow \bullet R + H_2O \quad (6)$$

$$\bullet R + \bullet OH \rightarrow CO_2 + H_2O + salts \quad (7)$$

The holes (h^+) can even directly oxidize the pollutant R as in Eq. 8

$$h_{vb}^{+} + R_{ads} \rightarrow R^{+} \quad (8)$$

Meanwhile, the electrons (e^-) can reduce molecular oxygen (O_2) to superoxide radical anion ($O_2^- \bullet$) as in Eq. 9.

$$e_{cb}^{-} + O_2 \rightarrow O_2^{-} \bullet \quad (9)$$

The superoxide radical anion may undergo a series of reactions following Eqs. 10 to 13.

$$O_2^{-} \bullet + H^{+} \rightarrow HO_2 \bullet \quad (10)$$

$$HO_2 \bullet + H^{+} + e_{cb}^{-} \rightarrow H_2O_2 \quad (11)$$

$$H_2O_2 + e_{cb}^{-} \rightarrow \bullet OH + OH^{-} \quad (12)$$

$$H_2O_2 + hv \rightarrow 2\bullet OH \quad (13)$$

In addition, the ability of superoxides to decompose organic molecules has been found comparable to those of the holes and hydroxyl radicals [34]. Similar reaction mechanisms are involved during active air treatment by TiO_2 photocatalysis.

2.2. Active Air Purification

Active air treatment utilizes various air purification devices. In 2007 alone, almost 180 patents that dealt with such devices were granted [44]. In general, these indoor air purification devices rely on forced airflow systems such as fans in order to convey polluted air close to the surface of the TiO_2 photocatalysts [1]. A graphical representation of a typical device for indoor air treatment is shown in Fig. 2. The most important feature of this device is the TiO_2-coated honeycomb filter that provides a very large surface area and minimum pressure drop. In some cases, activated carbon and zeolites are also incorporated on the filter surface to provide additional adsorbent capacity for further VOC removal. In this way, gaseous intermediates during photocatalytic VOC oxidation are not readily released to the indoor environment. Black-light lamps have been used to provide UV irradiation on the filter surface at the level of several mW cm^{-2}. Recently, light-emitting diodes (LED) have been employed as the light source for active air purification devices since LEDs provide more benefits including lower energy consumption, longer lifetime, smaller size, and faster switching [45].

Since indoor VOC concentrations only range from several ppb to 1 ppm, complete VOC mineralization is usually achieved. The regeneration of photocatalytic filters poisoned by

HNO_3 or H_2SO_4 during ammonia or hydrogen sulfide removal can be accomplished by washing the spent filters with water [31].

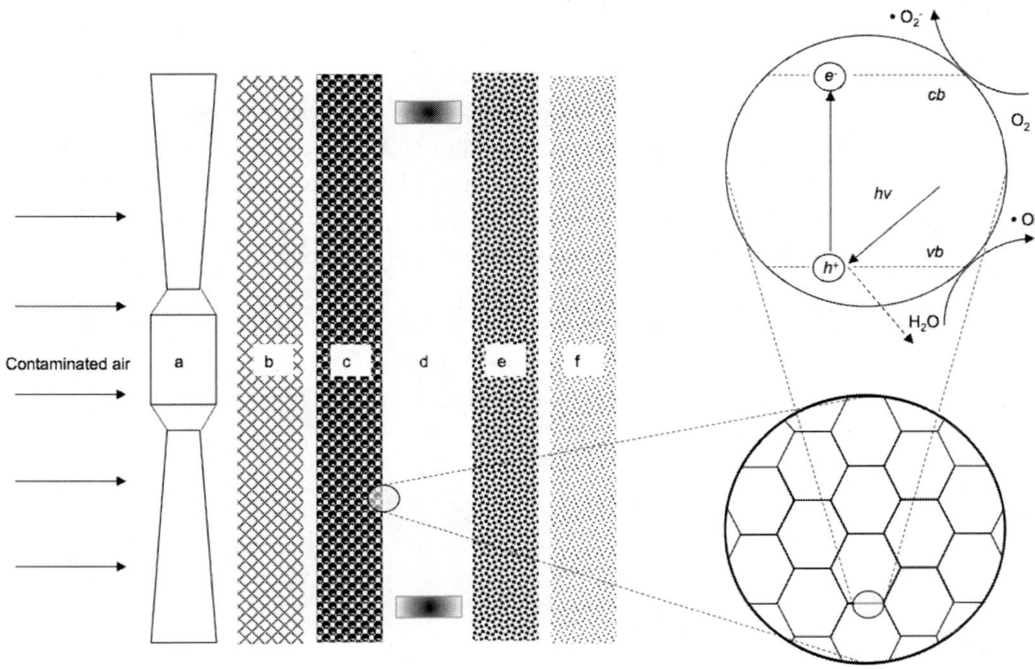

Figure 2. A typical active photocatalytic indoor air purification device consisting of (a) exhaust fan, (b) filter for particulate removal, (c) TiO_2-coated honeycomb-type filter for gaseous contaminant oxidation (d) light source, (e) activated carbon filter for the removal of gaseous intermediates (optional), and (f) ionizer generator (optional)

Various air purification devices utilizing photocatalytic filtration systems are available in the market. In Japan, more than 30 different enterprises fabricate photocatalyst-based air cleaners and their combined output exceeds 100 different models. These air cleaners are widely used in places where indoor air purification is needed such as hospitals, institutions for the elderly, pet stores, offices, smoking rooms, sports clubs, hotels, schools, restaurants and food manufacturers [31].

3. STRATEGIES FOR TiO_2 MODIFICATION

A major drawback in the application of pristine TiO_2 photocatalyst for the oxidation of various indoor air pollutants is the catalyst's inability to utilize the full potential of visible light irradiation. The high band energies of rutile (3.0 eV) and anatase (3.2 eV) phases in pure TiO_2 crystals limit TiO_2 photo-excitation in the near ultraviolet region ($\lambda < 400$ nm) [46,47]. Hence, passive photocatalytic purification of indoor air by pure TiO_2 surfaces will not likely happen since indoor lighting only allows the emission of visible light photons [31]. Since more than a decade ago, numerous ways have been attempted to modify TiO_2 (mainly anatase phase) to make it photoactive under visible light illumination [31]. Anpo et al. [27] found that bombarding TiO_2 photocatalyst with various transition metal ions in the presence of a strong electric field shifts the catalyst's absorption band towards the visible light region and the magnitude of the shift depends on the type of metal ion (V > Cr > Mn > Fe > Ni) implanted on the catalyst. Since then, various strategies have been attempted to modify TiO_2 and make it visible-light active.

In general, TiO_2 modification for visible-light photocatalysis can be classified into: (a) metal-doping, (b) non-metal doping and (c) co-doping. Doping metal or non-metal ions on TiO_2 shifts the valence and conduction bands within the TiO_2 material as shown in Fig. 3. For metal doping such as Fe-doping, the Fe impurity state can happen in two situations: (a) one above the *vb* (Fig. 3a) [48] since the t_{2g} level of *3d* orbital of Fe^{3+} ion lies above the TiO_2 *vb*; and (b) another below the *cb* (Fig. 3b) because the energy level of Fe *3d* electrons is lower than that of Ti *3d* at the bottom of *cb* [49]. Co-doping may shift both the *vb* and *cb* within the TiO_2 structure (Fig. 3c). Another advantage of doping TiO_2 with transition metals is that the metal nanoparticles dispersed in the TiO_2 matrix enhance electron-trapping. As a result, electron-hole recombination during irradiation is prevented thereby enhancing photo-excitation [21,50].

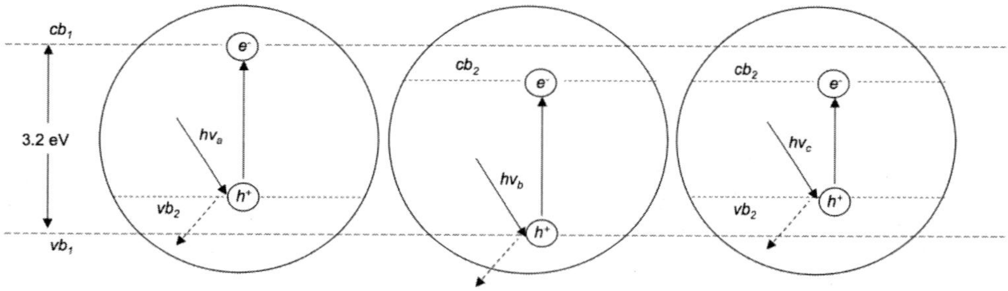

Figure 3. Shifting of the valence and conduction bands within the TiO_2 photocatalyst as a result of metal and non-metal doping.

Over the years, various visible-light active TiO_2 photocatalysts have been synthesized including N-TiO_2 [15], C-TiO_2 [24], I-TiO_2 [46]; Fe-TiO_2 [48]; W-TiO_2 [51]; N:Ni-TiO_2 [52]; Ce:Si-TiO_2 [22]; Zn:Fe-TiO_2 [13] and K:Al:S-TiO_2 [53]. Table 1 lists various visible-light active TiO_2 catalysts having band gap energies of less than 3.2 eV.

Several methods have been used to incorporate metal, non-metal and their various combinations into the TiO_2 matrix including ion-assisted sputtering, plasma, ion-implantation, chemical vapor deposition and sol-gel among others. Of these doping techniques, the sol-gel process remains the simplest and most versatile since it does not require complex instrumentation yet it can achieve high purity nano-sized doped TiO_2 for visible light photocatalysis applications [50]. The sol-gel process (Fig. 4) has been widely-used for the synthesis of visible light-active photocatalysts [13,15,21,51,53,57,58]. A typical sol-gel procedure in preparing N:Fe co-doped TiO_2 is discussed in detail by Shen et al. [21] as follows: the titanium precursor solution is prepared by dissolving titanium isopropoxide in anhydrous ethanol. The dopant precursor solution is then produced by dissolving ammonium nitrate and ferric nitrate in deionized water at room temperature and into this solution, glacial acetic acid is added. Thereafter, the titanium precursor is combined with the dopant precursor in a drop-wise manner. The sol produced is stirred for 2 h, aged for 48 h at room temperature and oven-dried for 12 h at 80°C. The desiccated solids are pulverized and the resulting powder are annealed at predetermined temperatures for 3 h to obtain N:Fe co-doped TiO_2 photocatalysts. The photocatalysts produced via the sol-gel process are greatly affected by the dopant concentration, calcination temperature, and calcination time.

Table 1. Visible-light-active TiO_2 photocatalysts with their respective band gap energies

Doping type	Doped element	Band gap energy (eV)	Reference
metal doping	Fe	3.15	[21]
metal doping	Co	2.14	[23]
non-metal doping	N	2.98	[16]
non-metal doping	N	2.24	[21]
non-metal doping	C	2.90	[54]
non-metal doping	I	2.73	[55]
co-doping	N:Fe	2.22	[21]
co-doping	Bi:N	2.79	[16]
co-doping	Zr:I	2.52-2.71	[55]
co-doping	N:B	2.13	[18]
co-doping	C:N:S	1.88-2.95	[56]

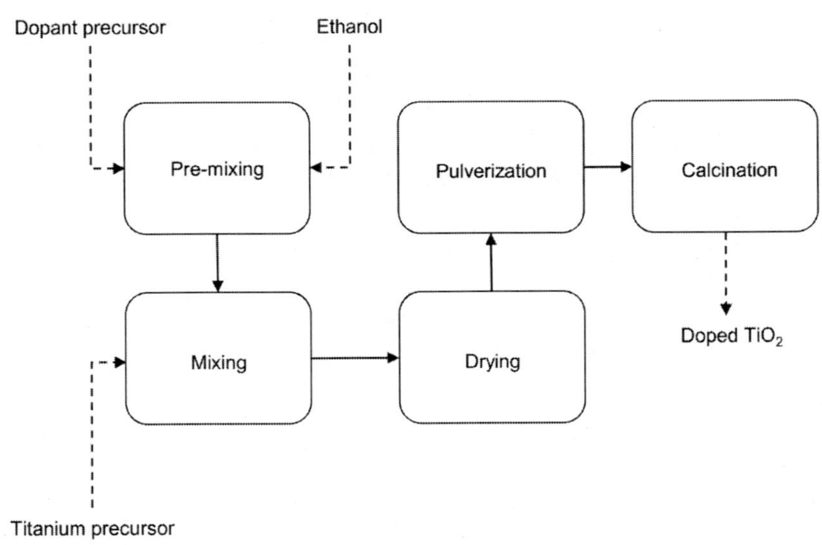

Figure 4. Sol-gel process for the synthesis of visible light active TiO_2 photocatalysts.

Titanium alkoxides ($Ti(OR)_n$) such as titanium isopropoxide [15,21] are the typical TiO_2 precursors in the sol-gel process. Titanium oxo-polymers, which are converted to TiO_2 afterwards, are produced by the hydrolysis and condensation reactions of $Ti(OR)_n$. Selective enrichment of the oxo-bridges of the oxo-polymers occurs during the addition of H_2O to $Ti(OR)_n$. The reactions involved in the process are given by Eqs. 14 to 16 [29].

Hydrolysis: $$M(OR)_n + H_2O \rightarrow M(OR)_{n-1}(OH) + ROH \tag{14}$$

Condensation:

(1) *Dehydration* $M(OR)_n + M(OR)_{n-1}(OH) \rightarrow M_2O(OR)_{2n-2} + ROH$ (15)

(2) *Dealcoholation* $2M(OR)_{n-1}(OH) \rightarrow M_2O(OR)_{2n-2} + H_2O$ (16)

Overall reaction: $M(OR)_n + \frac{n}{2}H_2O \rightarrow MO_{\frac{n}{2}} + nROH$ (17)

4. Applications of Doped-TiO_2

4.1. Volatile Organic Compounds

Volatile organic compounds (VOCs) comprise a class of toxic, carcinogenic and persistent gaseous contaminants having high total concentrations in indoor air (500 ppb to 1 ppm) [59–61]. Any exposure to even low dosage VOCs may already pose considerable risks to human health [62]. Headache, malaise and other respiratory-related and skin-related illnesses can be attributed to elevated concentrations of aromatic and aliphatic hydrocarbons, aldehydes among other VOC compounds [63].

4.1.1. Formaldehyde (CH_2O)

Formaldehyde (CH_2O), a main component of various building materials such as pressed wood, wallpaper, paint, furnishings and other decorating materials, is considered among the key pollutants in indoor air because of its carcinogenic nature and its potential to cause a variety of illnesses [4,64]. Formaldehyde is alarming due to its adverse effects to the human respiratory system and toxicity towards the central nervous system, liver and kidneys. There is even evidence showing that formaldehyde causes harmful effects to human reproductive and developmental health. Formaldehyde's high toxicity can be attributed to its high solubility and high reactivity to nucleophilic units of protein, DNA, and RNA molecules [65]. Indoor concentrations of aldehydes (formaldehyde, acetaldehyde, propioaldehyde, and hexaldehyde) are appreciably greater compared to outdoor concentrations. Typical formaldehyde concentrations in public classrooms reach up to 183.42 ± 12.45 $\mu g\ m^{-3}$ [66]. The average formaldehyde concentration in private reading rooms (119.3 $\mu g\ m^{-3}$) were found to be significantly higher than in public libraries (29.2 $\mu g\ m^{-3}$), children's libraries (29.3 $\mu g\ m^{-3}$), and public reading rooms (40.8 $\mu g\ m^{-3}$). In private reading rooms, the formaldehyde emission rates from desks (255 ± 214.8 $\mu g\ h^{-1}$) and walls (231.7 ± 192.3 $\mu g\ h^{-1}$) were significantly greater compared to those from chairs (79.6 ± 88.5 $\mu g\ h^{-1}$) [67]. In many cases, average indoor formaldehyde concentrations exceed the maximum allowable formaldehyde concentration (0.1 mg/m^3) set by the World Health Organization (WHO) and the U.S. Environmental Protection Agency (USEPA) for indoor air [3]. Gaseous formaldehyde removals by visible light photocatalysis using doped-TiO_2 are shown in Table 2.

Table 2. Preparation, property and photoactivity of modified TiO_2 catalysts under visible light irradiation

Catalyst precursor	Synthesis method	Calcination temp/time (°C/h)	Crystal	Dopant/level (dopa-nt : Ti ratio)	Band gap energy (eV)	Target pollutant/initial pollutant concentration	Light source/light intensity	Pollutant removal (%)	Reference
n.d.	n.d.	n.d.	A	C/n.d.	2.47	CH_2O/8.6 ppm	Vis (30-W fluorescent lamp)/10,000 lux	62	[68]
Titanium isopropoxide ($Ti(OC_4H_9)_4$)/hexamethyltetramine	n.d.	300/8	A	CN/n.d.	2.84	CH_2O/8.6 ppm	Vis (30-W fluorescent lamp)/10,000 lux	45	[68,69]
$Ti(OC_4H_9)_4$	sol-gel	600/2	A, R	C/n.d.	3.17	CH_2O/12.28 mg/m^3	Vis (500-W xenon lamp)/n.d.	4	[52]
$Ti(OC_4H_9)_4$/ammonium carbonate	sol-gel	600/2	A	N/0.010	2.30	CH_2O/12.28 mg/m^3	Vis (500-W xenon lamp)/n.d.	29	[52]
$Ti(OC_4H_9)_4$/nickelous nitrate	sol-gel	600/2	A	Ni/0.015	3.15	CH_2O/12.28 mg/m^3	Vis (500-W xenon lamp)/n.d.	5	[52]
$Ti(OC_4H_9)_4$/ammonium carbonate nickelous nitrate	sol-gel	600/2	A	Ni:N/0.010:0.015	2.29	CH_2O/12.28 mg/m^3	Vis (50-W xenon lamp)/n.d.	82	[52]
Titanium sulfate (TS)/n.d.	sol-gel	300/2	A	N/n.d.	n.d.	CH_2O/2.18 $\mu g\ ml^{-1}$	Vis (blue LED: λ = 458 nm)/1.39 $mW\ cm^{-2}$	58	[45]
Titanium sulfate (TS)/n.d., NH_4F	sol-gel	300/2	A	F, N/1:6 NH_4F to N-TiO_2	n.d.	CH_2O/2.18 $\mu g\ ml^{-1}$	Vis (blue LED: λ = 458 nm)/1.39 $mW\ cm^{-2}$	86	[45]
$TiCl_4$/urea, guanidine	spray pyrolysis	900/n.d.	n.d.	N /	n.d.	CH_3-CHO/930 ppm	Vis (150-W xenon/lamp)/20 $mW\ cm^{-2}$	n.d.	[70]

Table 2. (Continued)

Catalyst precursor	Synthesis method	Calcination temp/time (°C/h)	Crystal	Dopant/level (dopant : Ti ratio)	Band gap energy (eV)	Target pollutant/initial pollutant concentration	Light source/light intensity	Pollutant removal (%)	Reference
$TiCl_4$/urea, guanidine or ammonium fluoride	spray pyrolysis	900/n.d.	n.d.	NF /	n.d.	CH_3-CHO/ 930 ppm	Vis (150-W xenon/lamp)/20 mW cm^{-2}	n.d.	[70]
$TiCl_4$/urea,	spray pyrolysis	900/n.d.	n.d.	N /	n.d.	CHCl=CCl_2/943 ppm	Vis (150-W xenon lamp)/20 mW cm^{-2}	n.d.	[70]
$TiCl_4$/urea, guanidine or ammonium fluoride	spray pyrolysis	900/n.d.	n.d.	NF /	n.d.	CHCl=CCl_2/943 ppm	Vis (150-W xenon lamp)/20 mW cm^{-2}	n.d.	[70]
$Ti(OC_4H_9)_4$/silver nitrate	sol-gel	450/3	A	Ag/2.5	n.d.	toluene/200 ppm	Vis (25-LEDs; λ_{max} =415 nm)	n.d.	[71]
$Ti(OC_4H_9)_4$/potassium tetrachloroplatinate (II) and chloroplatinic acid hexahydrate	sol-gel	450/3	A	Pt/0.5	n.d.	toluene/200 ppm	Vis (25-LEDs; λ_{max} =415 nm)	n.d.	[71]
$Ti(OC_4H_9)_4$/silver nitrate, potassium tetrachloroplatinate (II) and chloroplatinic acid hexahydrate	sol-gel	450/3	A	Ag:Pt/2.5:0.5	n.d.	toluene/200 ppm	Vis (25-LEDs; λ_{max} =415 nm)	99	[71]
n.d.	n.d.	n.d.	n.d.	Mn /	n.d.	NO/200 ppb	Vis (18-W Philips TL-D Super80)/3 Wm^{-2}	95	[72]
$TiCl_4$/urea, guanidine or ammonium fluoride	spray pyrolysis	900/n.d.	n.d.	NF /	n.d.	CHCl=CCl_2/943 ppm	Vis (150-W xenon lamp)/20 mW cm^{-2}	n.d.	[70]
$Ti(OC_4H_9)_4$/silver nitrate	sol-gel	450/3	A	Ag/2.5	n.d.	toluene/200 ppm	Vis (25-LEDs; λ_{max} =415 nm)	n.d.	[71]

Catalyst precursor	Synthesis method	Calcination temp/time (°C/h)	Crystal	Dopant/level (dopant : Ti ratio)	Band gap energy (eV)	Target pollutant/initial pollutant concentration	Light source/light intensity	Pollutant removal (%)	Reference
$TiCl_4$/urea, guanidine or ammonium fluoride	spray pyrolysis	900/n.d.	n.d.	NF /	n.d.	$CHCl=CCl_2$/943 ppm	Vis (150-W xenon lamp)/20 mW cm^{-2}	n.d.	[70]
$Ti(OC_4H_9)_4$/silver nitrate	sol-gel	450/3	A	Ag/2.5	n.d.	toluene/200 ppm	Vis (25-LEDs; λ_{max} =415 nm)	n.d.	[71]
$Ti(OC_4H_9)_4$/potassium tetrachloroplatinate (II) and chloroplatinic acid hexahydrate	sol-gel	450/3	A	Pt/0.5	n.d.	toluene/200 ppm	Vis (25-LEDs; λ_{max} =415 nm)	n.d.	[71]
$Ti(OC_4H_9)_4$/silver nitrate, potassium tetrachloroplatinate (II) and chloroplatinic acid hexahydrate	sol-gel	450/3	A	Ag:Pt/2.5:0.5	n.d.	toluene/200 ppm	Vis (25-LEDs; λ_{max} =415 nm)	99	[71]
n.d.	n.d.	n.d.	n.d.	Mn /	n.d.	NO/200 ppb	Vis (18-W Philips TL-D Super80)/3 Wm^{-2}	95	[72]
$TiCl_4$/urea, guanidine or ammonium fluoride	spray pyrolysis	900/n.d.	n.d.	NF /	n.d.	$CHCl=CCl_2$/943 ppm	Vis (150-W xenon lamp)/20 mW cm^{-2}	n.d.	[70]

*n.d. – no data.

4.1.2. Benzene, toluene, ethyl benzene, xylene (BTEX)

Benzene, toluene, ethyl benzene and xylene (BTEX) are VOCs of significant concern [73]. Unfavorable effects of BTEX to human health were reported by various health organizations, citing xylenes as the most reactive. Mere inhalation of BTEX vapors may cause throat irritation and respiratory failure, among other illnesses [74]. Appropriate air circulation in houses with garages must be ensured since benzene and toluene are most likely present in garages [75].

Benzene is extensively used in different chemical industries as raw material for the production of polymers, resins, and synthetic fibers [76]. Symptoms such as headaches, restlessness, sleeplessness and fatigue have all been associated to benzene inhalation and chronic exposure to benzene even at low concentrations can result in cancer of human blood component, central nervous system depression, fertility and respiratory problems and endocrine related complications leading to hypertension [74,76]. Typical residential indoor benzene concentration is 1.7 – 10.1 $\mu g\ m^{-3}$ [9]. Toluene, on the other hand, most commonly come from paint emissions and tobacco smoke and its indoor air concentration is dependent on the season and ambient temperature [44]. Toluene causes dizziness and drowsiness once inhaled and may also lead to central nervous system depression [74]. Gaseous toluene removals by visible light photocatalysis using doped-TiO_2 are shown in Table 2.

4.2. Nitrogen Oxide

Highly urbanized environments are characterized by elevated levels of nitrogenous oxides (NOx), a mixture of nitric oxide (NO) and nitrogen dioxide (NO_2), coming from vehicle exhausts and other industrialized activities. NOx contamination of indoor air most commonly come from combustion devices used for cooking and from the permeation of heavily polluted outdoor air into the indoor environment [77]. Almost half a billion people, predominantly women, set aside approximately 3 – 5 h daily to cook meals for their families on open fires or stoves in the indoor environment and the exposure to gaseous pollutions from biomass combustion is an important public health hazard especially in developing countries [78]. NO_2 concentrations were found to be highest in kitchens (242 $\mu g/m^3$), followed by living rooms (22 $\mu g/m^3$) and courtyards (12 $\mu g/m^3$) for the rural setting while for the urban setting, NO_2 concentrations were 218 $\mu g/m^3$, 136 $\mu g/m^3$ and 64 $\mu g/m^3$ for kitchens, living rooms, and courtyards, respectively [78]. The World Health Organization and the USEPA, limit NO_2 concentrations in indoor air to 0.2 and 0.05 mg/m^3, respectively [3]. NOx removals by visible light photocatalysis using doped-TiO_2 are shown in Table 2.

CONCLUSION

Gaseous indoor air contaminants that especially pose human health hazards include formaldehyde, benzene, toluene and nitrogen oxides among others. The removal of these gaseous compounds can be achieved by either passive or active TiO_2 photocatalysis. The use of light-emitting diodes (LED) as light source for active air purification devices proved to be more economical. The utilization of visible-light for TiO_2 photo-excitation can be achieved by doping TiO_2 with metal or non-metal elements in order to lower the band gap energy of pure TiO_2 to less than 3.2 eV. Among the various methods of doping TiO_2 for visible-light photocatalysis, the sol-gel process remains the simplest and most versatile. The challenge for future research on visible-light photocatalysis for indoor air purification involves incorporating new dopants into the TiO_2 matrix to enhance visible-light photo-activity, improvement in the doping methods to make the process more simple and cost-effective, and the application of these novel photocatalysts to a mixture of gaseous indoor air pollutants.

ACKNOWLEDGMENTS

The authors acknowledge the financial support from the National Science Council, Taiwan (NSC 101-2923-E-041-001-MY2) and the Philippine Council for Industry, Energy and Emerging Technology Research and Development (PCIEERD), Department of Science and Technology, Philippines.

REFERENCES

[1] Y. Paz, "Application of TiO_2 photocatalysis for air treatment: Patents' overview," *Appl. Catal. B Environ.*, vol. 99, no. 3–4, pp. 448–460, 2010.

[2] J. L. Coutts, L. H. Levine, J. T. Richards, and D. W. Mazyck, "The effect of photon source on heterogeneous photocatalytic oxidation of ethanol by a silica – titania composite," *J. Photochem. Photobiol. A Chem.*, vol. 225, no. 1, pp. 58–64, 2011.

[3] G. Xiao, A. Huang, H. Su, and T. Tan, "The activity of acrylic-silicon/nano-TiO_2 films for the visible-light degradation of formaldehyde and NO_2," *Build. Environ.*, vol. 65, pp. 215–221, 2013.

[4] T. Liu, F. Li, and X. Li, "TiO_2 hydrosols with high activity for photocatalytic degradation of formaldehyde in a gaseous phase," *J. Hazard. Mater.*, vol. 152, no. 1, pp. 347–55, 2008.

[5] B. Sánchez, M. Sánchez-Muñoz, M. Muñoz-Vicente, G. Cobas, R. Portela, S. Suárez, A. E. González, N. Rodríguez, and R. Amils, "Photocatalytic elimination of indoor air biological and chemical pollution in realistic conditions," *Chemosphere*, vol. 87, no. 6, pp. 625–30, 2012.

[6] J. Gunschera, S. Mentese, T. Salthammer, and J. R. Andersen, "Impact of building materials on indoor formaldehyde levels: Effect of ceiling tiles, mineral fiber insulation and gypsum board," *Build. Environ.*, vol. 64, pp. 138–145, 2013.

[7] B. F. Yu, Z. B. Hu, M. Liu, H. L. Yang, Q. X. Kong, and Y. H. Liu, "Review of research on air-conditioning systems and indoor air quality control for human health," *Int. J. Refrig.*, vol. 32, no. 1, pp. 3–20, 2009.

[8] L. Zhong, C.-S. Lee, and F. Haghighat, "Adsorption performance of titanium dioxide (TiO_2) coated air filters for volatile organic compounds," *J. Hazard. Mater.*, vol. 243, pp. 340–9, 2012.

[9] H. K. Lai, M. J. Jantunen, N. Künzli, E. Kulinskaya, R. Colvile, and M. J. Nieuwenhuijsen, "Determinants of indoor benzene in Europe," *Atmos. Environ.*, vol. 41, no. 39, pp. 9128–9135, 2007.

[10] M. Hunger, G. Hüsken, and H. J. H. Brouwers, "Photocatalytic degradation of air pollutants — From modeling to large scale application," *Cem. Concr. Res.*, vol. 40, no. 2, pp. 313–320, 2010.

[11] M. Ouzzine, A. J. Romero-Anaya, M. A. Lillo-Ródenas, and A. Linares-Solano, "Spherical activated carbon as an enhanced support for TiO_2/AC photocatalysts," *Carbon N. Y.*, vol. 67, pp. 104–118, 2014.

[12] Q. Ling, J. Sun, Q. Zhou, H. Ren, and Q. Zhao, "Visible-light-driven titania/silica photocatalyst co-doped with boron and ferrum," *Appl. Surf. Sci.*, vol. 254, pp. 6731–6735, 2008.

[13] S. S. Srinivasan, J. Wade, E. K. Stefanakos, and Y. Goswami, "Synergistic effects of sulfation and co-doping on the visible light photocatalysis of TiO_2," *J. Alloys Compd.*, vol. 424, pp. 322–326, 2006.

[14] Y. Huang, Z. Xuxu, Y. Zhongyi, T. Feng, F. Beibei, and H. Keshan, "Preparation of nitrogen-doped TiO_2 nanoparticle catalyst and its catalytic activity under visible light," *Chinese J. Chem. Eng.*, vol. 15, no. 6, pp. 802–807, 2007.

[15] N. T. Nolan, D. W. Synnott, M. K. Seery, S. J. Hinder, A. Van Wassenhoven, and S. C. Pillai, "Effect of N-doping on the photocatalytic activity of sol – gel TiO_2," *J. Hazard. Mater.*, vol. 211–212, pp. 88–94, 2012.

[16] S. Bagwasi, Y. Niu, M. Nasir, B. Tian, and J. Zhang, "The study of visible light active bismuth modified nitrogen doped titanium dioxide photocatalysts: Role of bismuth," *Appl. Surf. Sci.*, vol. 264, pp. 139–147, 2013.

[17] Y. Ma, J. Zhang, B. Tian, F. Chen, and L. Wang, "Synthesis and characterization of thermally stable Sm, N co-doped TiO_2 with highly visible light activity," *J. Hazard. Mater.*, vol. 182, no. 1–3, pp. 386–393, 2010.

[18] G. Wu, J. Wen, J. Wang, D. F. Thomas, and A. Chen, "A facile approach to synthesize N and B co-doped TiO_2 nanomaterials with superior visible-light response," *Mater. Lett.*, vol. 64, no. 15, pp. 1728–1731, 2010.

[19] K. Rajeshwar, M. E. Osugi, W. Chanmanee, C. R. Chenthamarakshan, M. V. B. Zanoni, P. Kajitvichyanukul, and R. Krishnan-Ayer, "Heterogeneous photocatalytic treatment of organic dyes in air and aqueous media," *J. Photochem. Photobiol. C Photochem. Rev.*, vol. 9, pp. 171–192, 2008.

[20] J. Yang, H. Bai, Q. Jiang, and J. Lian, "Visible-light photocatalysis in nitrogen–carbon-doped TiO_2 films obtained by heating TiO_2 gel–film in an ionized N_2 gas," *Thin Solid Films*, vol. 516, pp. 1736–1742, 2008.

[21] X.-Z. Shen, J. Guo, Z.-C. Liu, and S.-M. Xie, "Visible-light-driven titania photocatalyst co-doped with nitrogen and ferrum," *Appl. Surf. Sci.*, vol. 254, pp. 4726–4731, 2008.

[22] Q. Chen, D. Jiang, W. Shi, D. Wu, and Y. Xu, "Visible-light-activated Ce-Si co-doped TiO_2 photocatalyst," *Appl. Surf. Sci.*, vol. 255, pp. 7918–7924, 2009.

[23] C. Hsieh, W. Fan, W. Chen, and J. Lin, "Adsorption and visible-light-derived photocatalytic kinetics of organic dye on Co-doped titania nanotubes prepared by hydrothermal synthesis," vol. 67, pp. 312–318, 2009.

[24] M. Chen and W. Chu, "Degradation of antibiotic norfloxacin in aqueous solution by visible-light-mediated C-TiO_2 photocatalysis," *J. Hazard. Mater.*, 2012.

[25] J. A. Rengifo-Herrera, E. Mielczarski, J. Mielczarski, N. C. Castillo, J. Kiwi, and C. Pulgarin, "Escherichia coli inactivation by N, S co-doped commercial TiO_2 powders under UV and visible light," *Appl. Catal. B Environ.*, vol. 84, pp. 448–456, 2008.

[26] M. N. Peiris Weerasinghe and K. J. Klabunde, "Chromium oxide loaded silica aerogels: Novel visible light photocatalytic materials for environmental remediation," *J. Photochem. Photobiol. A Chem.*, vol. 254, pp. 62–70, 2013.

[27] M. Anpo, M. Takeuchi, K. Ikeue, and S. Dohshi, "Design and development of titanium oxide photocatalysts operating under visible and UV light irradiation. The applications of metal ion-implantation techniques to semiconducting TiO_2 and Ti/zeolite catalysts," *Curr. Opin. Solid State Mater. Sci.*, vol. 6, pp. 381–388, 2002.

[28] Y. D. Hou, X. C. Wang, L. Wu, X. F. Chen, Z. X. Ding, X. X. Wang, and X. Z. Fu, "N-Doped SiO_2/TiO_2 mesoporous nanoparticles with enhanced photocatalytic activity under visible-light irradiation," *Chemosphere*, vol. 72, pp. 414–421, 2008.

[29] H. Sun, S. Wang, H. M. Ang, M. O. Tadé, and Q. Li, "Halogen element modified titanium dioxide for visible light photocatalysis," *Chem. Eng. J.*, vol. 162, no. 2, pp. 437–447, 2010.

[30] M. Tasbihi, U. Lavrenčič Štangar, U. Černigoj, J. Jirkovsky, S. Bakardjieva, and N. Novak Tušar, "Photocatalytic oxidation of gaseous toluene on titania/mesoporous silica powders in a fluidized-bed reactor," *Catal. Today*, vol. 161, no. 1, pp. 181–188, 2011.

[31] A. Fujishima, X. Zhang, and D. A. Tryk, "TiO_2 photocatalysis and related surface phenomena," *Surf. Sci. Rep.*, vol. 63, pp. 515–582, 2008.

[32] P. Krishnan, M.-H. Zhang, L. Yu, and H. Feng, "Photocatalytic degradation of particulate pollutants and self-cleaning performance of TiO_2-containing silicate coating and mortar," *Constr. Build. Mater.*, vol. 44, pp. 309–316, 2013.

[33] A. M. Ramirez, K. Demeestere, N. De Belie, T. Mäntylä, and E. Levänen, "Titanium dioxide coated cementitious materials for air purifying purposes: Preparation, characterization and toluene removal potential," *Build. Environ.*, vol. 45, no. 4, pp. 832–838, 2010.

[34] J. Chen and C. Poon, "Photocatalytic construction and building materials: From fundamentals to applications," *Build. Environ.*, vol. 44, no. 9, pp. 1899–1906, 2009.

[35] L. Yang, Z. Liu, J. Shi, Y. Zhang, H. Hu, and W. Shangguan, "Degradation of indoor gaseous formaldehyde by hybrid VUV and TiO_2/UV processes," *Sep. Purif. Technol.*, vol. 54, no. 2, pp. 204–211, 2007.

[36] J. Mo, Y. Zhang, Q. Xu, J. J. Lamson, and R. Zhao, "Photocatalytic purification of volatile organic compounds in indoor air: A literature review," *Atmos. Environ.*, vol. 43, no. 14, pp. 2229–2246, 2009.

[37] I. Arslan, I. A. Balcioglu, and D. W. Bahnemann, "Advanced chemical oxidation of reactive dyes in simulated dyehouse effluents by ferrioxalate-Fenton/UV-A and TiO_2/UV-A processes," *Dye. Pigment.*, vol. 47, no. 3, pp. 207–218, 2000.

[38] W. Li, Q. Zhou, and T. Hua, "Removal of organic matter from landfill leachate by advanced oxidation processes: A review," *Int. J. Chem. Eng.*, pp. 1–10, 2010.

[39] N. N. Mahamuni and Y. G. Adewuyi, "Advanced oxidation processes (AOPs) involving ultrasound for waste water treatment: a review with emphasis on cost estimation," *Ultrason. Sonochem.*, vol. 17, no. 6, pp. 990–1003, 2010.

[40] M. D. G. de Luna, R. M. Briones, C.-C. Su, and M.-C. Lu, "Kinetics of acetaminophen degradation by Fenton oxidation in a fluidized-bed reactor," *Chemosphere*, vol. 90, no. 4, pp. 1444–8, 2013.

[41] M. Umar, H. A. Aziz, and M. S. Yusoff, "Trends in the use of Fenton, electro-Fenton and photo-Fenton for the treatment of landfill leachate," *Waste Manag.*, vol. 30, no. 11, pp. 2113–2121, 2010.

[42] M. D. G. de Luna, J. I. Colades, C.-C. Su, and M.-C. Lu, "Comparison of dimethyl sulfoxide degradation by different Fenton processes," *Chem. Eng. J.*, vol. 232, pp. 418–424, 2013.

[43] M. D. G. de Luna, M. L. Veciana, C.-C. Su, and M.-C. Lu, "Acetaminophen degradation by electro-Fenton and photoelectro-Fenton using a double cathode electrochemical cell," *J. Hazard. Mater.*, vol. 217–218, pp. 200–7, 2012.

[44] O. Debono, F. Thevenet, P. Gravejat, V. Hequet, C. Raillard, L. Lecoq, and N. Locoge, "Toluene photocatalytic oxidation at ppbv levels: Kinetic investigation and carbon balance determination," *Appl. Catal. B Environ.*, vol. 106, no. 3–4, pp. 600–608, 2011.

[45] Y. Li, Y. Jiang, S. Peng, and F. Jiang, "Nitrogen-doped TiO_2 modified with NH_4F for efficient photocatalytic degradation of formaldehyde under blue light-emitting diodes," *J. Hazard. Mater.*, vol. 182, no. 1–3, pp. 90–96, 2010.

[46] Z. He, T. Hong, J. Chen, and S. Song, "A magnetic TiO_2 photocatalyst doped with iodine for organic pollutant degradation," *Sep. Purif. Technol.*, vol. 96, pp. 50–57, 2012.

[47] H. S. M. Tabaei, M. Kazemeini, and M. Fattahi, "Preparation and characterization of visible light sensitive nano titanium dioxide photocatalyst," *Sci. Iran.*, vol. 19, no. 6, pp. 1626–1631, 2012.

[48] L. Cui, Y. Wang, M. Niu, G. Chen, and Y. Cheng, "Synthesis and visible light photocatalysis of Fe-doped TiO_2 mesoporous layers deposited on hollow glass microbeads," *J. Solid State Chem.*, vol. 182, no. 10, pp. 2785–2790, 2009.

[49] X. Wang, Y. Tang, M.-Y. Leiw, and T.-T. Lim, "Solvothermal synthesis of Fe–C codoped TiO_2 nanoparticles for visible-light photocatalytic removal of emerging organic contaminants in water," *Appl. Catal. A Gen.*, vol. 409–410, pp. 257–266, 2011.

[50] A. Zaleska, "Doped-TiO_2 : A review," *Recent Patents Eng.*, vol. 2, no. 1, pp. 157–164, 2008.

[51] T. Putta, M.-C. Lu, and J. Anotai, "Photocatalytic activity of tungsten-doped TiO_2 with hydrothermal treatment under blue light irradiation," *J. Environ. Manage.*, vol. 92, no. 9, pp. 2272–2276, 2011.

[52] X. Zhang and Q. Liu, "Visible-light-induced degradation of formaldehyde over titania photocatalyst co-doped with nitrogen and nickel," *Appl. Surf. Sci.*, vol. 254, pp. 4780–4785, 2008.

[53] N. C. Tolosa, M.-C. Lu, H. D. Mendoza, and A. P. Rollon, "The effect of the composition of tri-elemental doping (K, Al, S) on the photocatalytic performance of synthesized TiO_2 nanoparticles in oxidizing 2-chlorophenol over visible light illumination," *Appl. Catal. A Gen.*, vol. 401, no. 1–2, pp. 233–238, 2011.

[54] M. Shen, Z. Wu, H. Huang, Y. Du, Z. Zou, and P. Yang, "Carbon-doped anatase TiO_2 obtained from TiC for photocatalysis under visible light irradiation," *Mater. Lett.*, vol. 60, pp. 693–697, 2006.

[55] S. Song, F. Hong, Z. He, H. Wang, X. Xu, and J. Chen, "Influence of zirconium doping on the activities of zirconium and iodine co-doped titanium dioxide in the decolorization of methyl orange under visible light irradiation," *Appl. Surf. Sci.*, vol. 257, no. 23, pp. 10101–10108, 2011.
[56] P. Wang, P.-S. Yap, and T.-T. Lim, "C–N–S tridoped TiO_2 for photocatalytic degradation of tetracycline under visible-light irradiation," *Appl. Catal. A Gen.*, vol. 399, no. 1–2, pp. 252–261, 2011.
[57] J. Anotai, A. Jevprasesphant, Y.-M. Lin, and M.-C. Lu, "Oxidation of aniline by titanium dioxide activated with visible light," *Sep. Purif. Technol.*, vol. 84, pp. 132–137, 2012.
[58] N. Quici, M. L. Vera, H. Choi, G. L. Puma, D. D. Dionysiou, M. I. Litter, and H. Destaillats, "Effect of key parameters on the photocatalytic oxidation of toluene at low concentrations in air under 254 + 185 nm UV irradiation," *Appl. Catal. B Environ.*, vol. 95, pp. 312–319, 2010.
[59] O. Debono, F. Thévenet, P. Gravejat, V. Héquet, C. Raillard, L. Le Coq, and N. Locoge, "Gas phase photocatalytic oxidation of decane at ppb levels: Removal kinetics, reaction intermediates and carbon mass balance," *J. Photochem. Photobiol. A Chem.*, vol. 258, pp. 17–29, 2013.
[60] Z.-W. Cheng, L. Feng, J.-M. Chen, J.-M. Yu, and Y.-F. Jiang, "Photocatalytic conversion of gaseous ethylbenzene on lanthanum-doped titanium dioxide nanotubes," *J. Hazard. Mater.*, vol. 254–255, pp. 354–63, 2013.
[61] J. B. Zhong, Y. Lu, W. D. Jiang, Q. M. Meng, X. Y. He, J. Z. Li, and Y. Q. Chen, "Characterization and photocatalytic property of Pd/TiO_2 with the oxidation of gaseous benzene," *J. Hazard. Mater.*, vol. 168, no. 2–3, pp. 1632–5, 2009.
[62] M. Hussain, N. Russo, and G. Saracco, "Photocatalytic abatement of VOCs by novel optimized TiO_2 nanoparticles," *Chem. Eng. J.*, vol. 166, no. 1, pp. 138–149, 2011.
[63] T. Baran and W. Macyk, "Photocatalytic oxidation of volatile pollutants of air driven by visible light," *J. Photochem. Photobiol. A Chem.*, vol. 241, pp. 8–12, 2012.
[64] A. Rezaee, H. Rangkooy, A. Jonidi-Jafari, and A. Khavanin, "Surface modification of bone char for removal of formaldehyde from air," *Appl. Surf. Sci.*, vol. 286, pp. 235–239, 2013.
[65] B. A. Katsnelson, T. D. Degtyareva, L. I. Privalova, I. A. Minigaliyeva, T. V. Slyshkina, V. V. Ryzhov, and O. Y. Beresneva, "Attenuation of subchronic formaldehyde inhalation toxicity with oral administration of glutamate, glycine and methionine," *Toxicol. Lett.*, vol. 220, no. 2, pp. 181–186, 2013.
[66] S. Banerjee and I. Annesi-Maesano, "Spatial variability of indoor air pollutants in schools. A multilevel approach," *Atmos. Environ.*, vol. 61, pp. 558–561, Dec. 2012.
[67] J. Kim, S. Kim, K. Lee, D. Yoon, J. Lee, and D. Ju, "Indoor aldehydes concentration and emission rate of formaldehyde in libraries and private reading rooms," *Atmos. Environ.*, vol. 71, pp. 1–6, 2013.
[68] Z. Han, V. W. Chang, X. Wang, T. T. Lim, and L. Hildemann, "Experimental study on visible-light induced photocatalytic oxidation of gaseous formaldehyde by polyester fiber supported photocatalysts," *Chem. Eng. J.*, vol. 218, pp. 9–18, 2013.
[69] X. Wang and T.-T. Lim, "Effect of hexamethylenetetramine on the visible-light photocatalytic activity of C–N codoped TiO_2 for bisphenol A degradation: Evaluation

of photocatalytic mechanism and solution toxicity," *Appl. Catal. A Gen.*, vol. 399, no. 1–2, pp. 233–241, 2011.

[70] D. Li, H. Haneda, S. Hishita, and N. Ohashi, "Visible-light-driven nitrogen-doped TiO_2 photocatalysts: effect of nitrogen precursors on their photocatalysis for decomposition of gas-phase organic pollutants," *Mater. Sci. Eng. B*, vol. 117, pp. 67–75, 2005.

[71] A. Zielińska-Jurek and A. Zaleska, "Ag/Pt-modified TiO_2 nanoparticles for toluene photooxidation in the gas phase," *Catal. Today*, vol. 230, pp. 104–111, 2014.

[72] C. Cacho, O. Geiss, J. Barrero-Moreno, V. D. Binas, G. Kiriakidis, L. Botalico, and D. Kotzias, "Studies on photo-induced NO removal by Mn-doped TiO_2 under indoor-like illumination conditions," *J. Photochem. Photobiol. A Chem.*, vol. 222, no. 1, pp. 304–306, 2011.

[73] C. A. Korologos, C. J. Philippopoulos, and S. G. Poulopoulos, "The effect of water presence on the photocatalytic oxidation of benzene, toluene, ethylbenzene and m-xylene in the gas-phase," *Atmos. Environ.*, vol. 45, no. 39, pp. 7089–7095, 2011.

[74] J. Faber, K. Brodzik, A. Gołda-Kopek, and D. Łomankiewicz, "Benzene, toluene and xylenes levels in new and used vehicles of the same model," *J. Environ. Sci.*, vol. 25, no. 11, pp. 2324–2330, 2013.

[75] M. A. Isbell, R. J. Stolzberg, and L. K. Duffy, "Indoor climate in interior Alaska: simultaneous measurement of ventilation, benzene and toluene in residential indoor air of two homes," *Sci. Total Environ.*, vol. 345, no. 1–3, pp. 31–40, 2005.

[76] H. Bahadar, S. Mostafalou, and M. Abdollahi, "Current understandings and perspectives on non-cancer health effects of benzene: A global concern," *Toxicol. Appl. Pharmacol.*, vol. 276, no. 2, pp. 83–94, 2014.

[77] T. Martinez, A. Bertron, E. Ringot, and G. Escadeillas, "Degradation of NO using photocatalytic coatings applied to different substrates," *Build. Environ.*, vol. 46, no. 9, pp. 1808–1816, 2011.

[78] I. Colbeck, Z. A. Nasir, Z. Ali, and S. Ahmad, "Nitrogen dioxide and household fuel use in the Pakistan," *Sci. Total Environ.*, vol. 409, no. 2, pp. 357–363, 2010.

In: Titanium Dioxide
Editor: Jerri Brown

ISBN: 978-1-63321-391-3

Chapter 12

A SHORT REVIEW: SYNTHESIS OF ANATASE TiO_2 AND ITS COMPOSITE BY IONOTHERMAL SYSTEM UNDER MICROWAVE HEATING AND THEIR WATER TREATMENT APPLICATIONS

Po-I Liu[1,2], Li-Ching Chung[1], Hsin Shao[1], Teh-Ming Liang[1], Ren-Yang Horng[1], M. M. Chen-Chi[1,2] and Min-Chao Chang[1*]

[1]Material and Chemical Research Laboratories, Industrial Technology Research Institute (ITRI), Kuang Fu Road, Hsinchu, Taiwan

[2]Department of Chemical Engineering, National Tsing Hua University, Hsinchu, Taiwan

ABSTRACT

Synthesis of nanostructured anatase TiO_2 and its composite by ionothermal synthesis (sol–gel method containing ionic liquid (IL) under microwave (MW) irradiation and their water treatment applications were reported in this short review.

In ionothermal synthesis, the ionic liquid plays a hydrothermal role in driving surface crystallization of amorphous TiO_2 to the anatase phase by retaining a suitable amount of water through a dissolution-crystallization mechanism. MW heating can lead to rapid temperature increase of chemical reactions. The effects of MW heating on anatase crystallization are diverse and dependent on the initial water content, MW irradiation mode as well as IL type and its content.

The preparation of two kinds of TiO_2 composite is also reported. Firstly, an innovative one-step immobilization of TiO_2 nano-particles on organic polymer (PMMA) substrate under IL/MW condition at ambient condition is described. In this technology, a sol–gel reaction and anatase crystallization are simultaneously occurred at the specific sites of the polymer surface. These sites are the tiny cavities of the rough surface resulting from the softening and swelling effect of an alcohol, such as isopropyl alcohol, on the polymer surface under microwave irradiation. The experimental data show that the anatase TiO_2 prepared by this novel method have high immobilization stability and

* Corresponding author: Material and Chemical Research Laboratories, Industrial Technology Research Institute (ITRI), Bldg. 17, 321 sec. 2, Kuang Fu Road, Hsinchu, 30011, Taiwan. Tel.: +886 357 32633. Fax: +886 357 32349. E-mail address: minchaochang@itri.org.tw; vincenthorng@itri.org.tw; ccma@che.nthu.edu.tw.

photocatalytic water treatment performance. Secondly, the nanostructured anatase TiO_2/activated carbon (AC) composite material for capacitive deionization (CDI) electrode is prepared by a two-step sol-gel method under IL/MW condition In comparison to the pristine AC electrode, higher specific capacitance was observed for the nanostructured anatase TiO_2/AC composite electrode, especially when the composite was prepared with a molar ratio of titanium tetraisopropoxide/H_2O equal to 1:15. The accompanying capacitive deionization result indicates that the decay of electrosorption capacity of this composite electrode is insignificant after five cycle tests. It means that the ion electrosorption–desorption becomes a reversible process.

TiO_2 can be synthesized from hydrolysis and condensation of titanium precursor by sol-gel method. The initial reaction product of this sol-gel reaction is generally amorphous. For achieving crystalline anatase TiO_2 from amorphous product, high temperature calcination or hydrothermal treatment is required. In hydrothermal treatment, the reaction is carried out with aqueous or organo-aqueous media in a closed system. The reaction temperature can be in the range of 100°C to 300°C and the reaction pressure is generally above 1 atmosphere[1]. The adsorbed water molecule around starting material is considered as the key role for the crystallization of the amorphous material. In recent years, it was found that anatase TiO_2 nanocrystals can also be synthesized by sol-gel method with reaction of titanium precursors and water in the presence of ionic liquid (called as ionothermal synthesis) [2,3]. The factors, especially regarding the type of ionic liquid and the role of water, affecting anatase crystallization behavior in ionothermal synthesis of TiO_2 were studied by many authors. However their conclusions are diverse and undecided.

The initial reaction product of sol gel reaction is the mixture of hydrolysis product (hydroxylated titanium) and condensation product with Ti–O–Ti structure. The content of hydrolysis product is dependent upon the degree of being condensated [4]. Consequently, in investigating the factors affecting anatase crystallization behavior in ionothermal synthesis, two things must be clearly identified: 1) the structural chemistry of each sol gel reaction stage (hydrolysis, condensation and thermal annealing) 2) what is the role of ionic liquid in each stage. In recent years, many new findings about the specific physicochemical characteristics of water/ionic liquid mixture system had been demonstrated in the literature [5]. Based upon these findings, a new insight into the anatase crystallization behavior in ionothermal synthesis of nanostructured TiO_2 is reported in this short review.

THE SPECIFIC COEXISTENCE OF WATER AND IONIC LIQUID IS NECESSARY

In a preliminary study, it was found that ionic liquid plays a role in driving a post-condensation crystallization of amorphous TiO_2 to anatase phase during ensuing drying at 100°C and 1 atmosphere [6]. In that study, TiO_2 powder was still amorphous and did not crystallize to anatase by annealing treatment with H_2O and 1-butyl-3-methylimidazolium hexafluorophosphate ($[BuMIm]^+[PF_6]$) respectively at 100°C for 2 hours. However an anatase peak was observed by annealing treatment with mixture of $[BuMIm]^+[PF_6]^-$ and water. It means that that the specific coexistence of water and ionic liquid is an essential factor for achieving anatase crystallization of amorphous TiO_2. Hence an idea that the role of water in ionothermal synthesis system can be similar to that in hydrothermal treatment system is proposed. In this proposition, the role of ionic liquid is considered to retain water molecule around amorphous TiO_2 through hydrogen bonding interaction with it and then inducing surface crystallization of the formed hydroxylated titanium compound from amorphous TiO_2. The self-assembly ability of ionic liquid would lead to kinetically dehydration and crystallization of hydroxylated titanium compound [7]. This proposition is proved by the X-ray photoelectron spectroscopy (XPS)[8]. In the XPS investigation, the chemical bonding types of the TiO_2 powder synthesized without or with 20% ionic liquids (hydrophobic $[BuMIm]^+[PF_6]^-$and hydrophilic 1-butyl-3-methylimidazolium tetrafluoroborate ($[BuMIm]^+[BF_4]^-$) and 300% water were determined by XPS. Figure 1 demonstrated that the time profile of hydroxyl group content and anatase crystallinity during thermal annealing of sample in the presence of ionic liquid. It can be seen that the hydroxyl group content of hydroxylated titanium compound became to be decreasing after crystallization. The appearance of maximum peak was resulted from initially surface water dissolution of the condensation product with Ti–O–Ti structure to form hydroxylated titanium compound. The anatase crystallinity for TiO_2 samples with 20% ionic liquid and 300 % water at different annealing time were shown in Table1. It was found that the higher anatase crystallinity was observed for longer annealing time.

Table 1. The found anatase crystallinity of TiO_2 sample with 20% $[BuMIm]^+[PF_6]^-$ and 300% water at different annealing time

Annealing time (min)	Crystallinity (%)
15	0
60	17
120	26

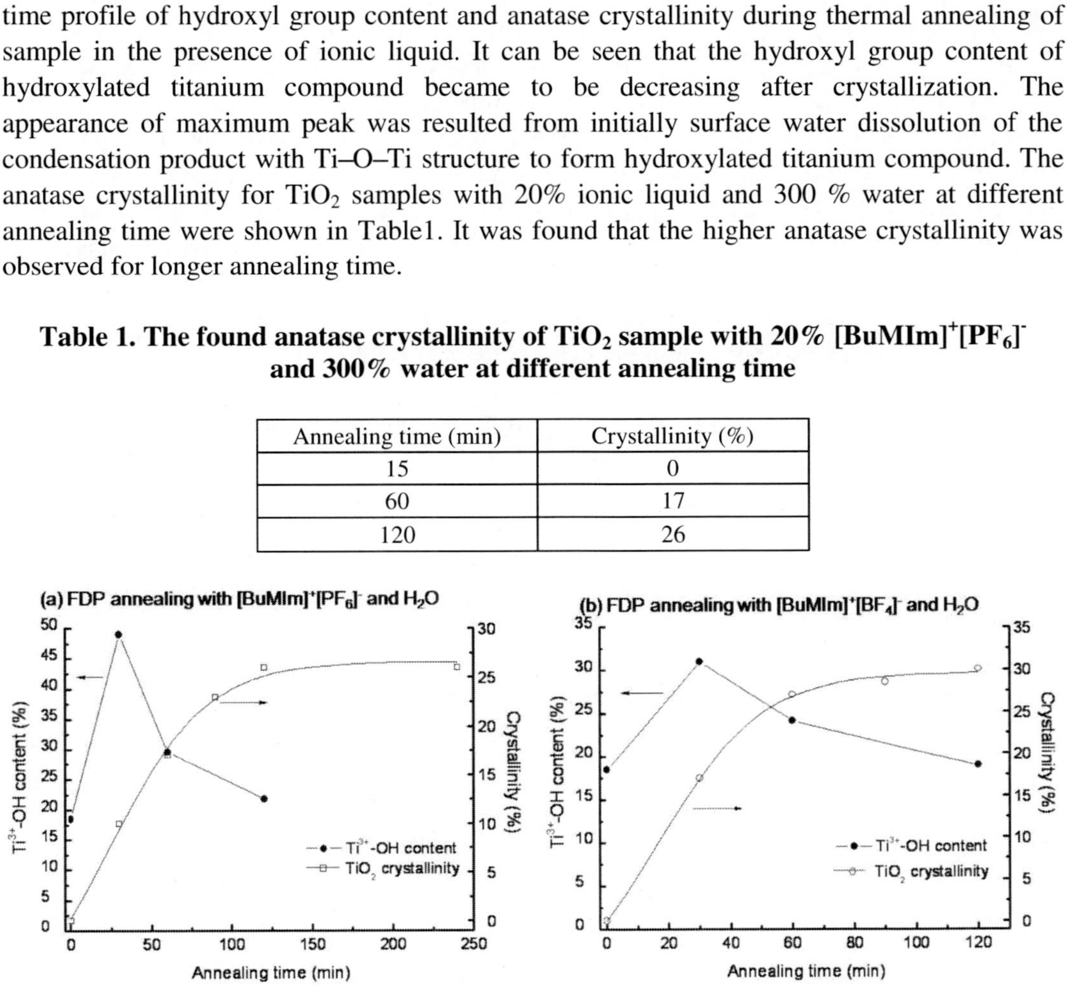

Figure 1. The time profile of hydroloxyl group content and anatase crystallinity during thermal annealing of freeze-dried powder (FDP) sample (a) FDP annealed with 20% $[BuMIm]^+[PF_6]^-$ and 300% water at 100°C for 2hrs, and b) FDP annealed with 20% $[BuMIm]^+[BF_4]^-$ and 300% water at 100°C for 2hrs.

TEMPERATURE, ION LIQUID TYPE AND CONCENTRATION EFFECT

The XRD measurements for the annealed TiO_2 powder prepared with 20% ionic liquid and 300 % water for two ionic liquids ($[BuMIm]^+[PF_6]^-$and $[BuMIm]^+[BF_4]^-$) are conducted [8]. The anatase crystallinity of TiO_2 samples calculated from XRD data are listed in Table 2. The crystallite size is around 3~4 nm. The results indicated that different temperature effect on anatase crystallization of amorphous TiO_2 powder for these two ionic liquids. Higher anatase crystallinity was obtained for $[BuMIm]^+[BF_4]^-$case, especially annealing temperature above 100°C. Obviously, there exists a remarkable different temperature effects for these two ionic liquids. When the temperature was above 100°C, anatase crystallinity was increased with increasing temperature for $[BuMIm]^+[BF_4]^-$case except at160°C and decreased with increasing temperature for $[BuMIm]^+[PF_6]^-$ case. This is due to the difference in temperature stability of ionic liquid/water hydrogen bonded complex. In studying the interactions between the water molecules and ionic liquids, Wang et al. found out that the BF_4^-anion in $[BuMIm]^+[BF_4]^-$ could form stable complex with water molecules while the PF_6^-anion complex is relatively unstable in $[BuMIm]^+[PF_6]^-$ case [9]. Hence when annealing temperature above100°C, some of the adsorbed water molecules could be easily heated away and then a lower anatase crystallinity was observed for the $[BuMIm]^+[PF_6]^-$case. Surely, the hydrogen bonded complex could be broken when temperature is too high. This is why we could observe lower anatase crystallinity at 160°C.

Table 2. The anatase crystallinity of TiO_2 sample calculated from XRD curves

Ionic liquid	Crystallinity (%)				
	80°C	100 °C	120 °C	140 °C	160 °C
$[BuMIm]^+[PF_6]^-$	15	26	18	11	8.5
$[BuMIm]^+[BF_4]^-$	24	27	30	39	29

Table 3. The anatase crystallinity of TiO_2 sample calculated from XRD curves (ionic liquid: $[BuMIm]^+[PF_6]^-$)

Ionic liquid content (%)	Crystallinity (%)
0	---
2.5	---
5	---
7.5	14
10	17
20	26
50	24

The XRD measurements for the TiO_2 samples prepared with different $[BuMIm]^+[PF_6]^-$ content and 300% water are also investigated. Table 3 lists the calculated anatase crystallinity from XRD data. It was shown that a minimum ionic liquid content around 7.5% was required for the formation of anatase TiO_2 nanocrystal. The anatase crystallinity increases with ionic liquid content, from 14% at 7.5 % ionic liquid to a maximum of 26% at 20 % ionic liquid, and then decreases slowly to 24% for 50 % ionic liquid. The maximum anatase crystallinity of TiO_2 could be resulted from the limiting water solubility in ionic liquid. That is to say, ionic liquid concentration is a limiting factor in crystallization of amorphous TiO_2.

H_2O/TIP Molar Ratio Effect

The H_2O/TIP molar ratio is also an important role during the anatase crystallization in ionothermal synthesis system. The anatase crystallinity of TiO_2 samples prepared by the sol gel process with and without ionic liquid at different H_2O/TIP molar ratios is listed in Table 4. No significant anatase peak was observed at a very low H_2O/TIP molar ratio. However, the highest anatase crystallinity was obtained at H_2O/TIP molar ratio of 5, especially in the $[BuMIm]^+[BF_4]^-$ case. In comparison with this medium H_2O/TIP molar ratio case, the anatase crystallinity was also relatively lower for the higher H_2O/TIP molar ratio case.

The specific characteristic of a water/ionic liquid system makes this anatase crystallization behavior being dependent on the H_2O/TIP molar ratio. As described above, the hydroxylated titanium compound must be formed before anatase crystallization. At the medium H_2O/TIP molar ratio, most of the hydroxylated titanium compound was directly formed from the hydrolysis reaction [10]. However at high H_2O/TIP molar ratio, higher proportion of condensed product with a polymeric Ti–O–Ti chain structure can be formed [11]. Water dissolution of the condensation product to form the hydroxylated titanium compound was necessary step. This water-inducing dissolution process of the condensation product is a kinetic control process. A sufficient annealing time and temperature are required to obtain suitable water inducing dissolution efficiency. Hence, lower anatase crystallinity was found for the high H_2O/TIP molar ratio case at the same annealing condition. The fact that no anatase peak was observed in the very low H_2O /TIP molar ratio case could be resulted from most of the ionic liquids being self-associated as separated and tight ion pairs. An incomplete hydrolysis reaction can be another reason.

Table 4. Anatase crystallinity of thermal annealed TiO_2 samples prepared with different ionic liquid and H_2O/TIP mole ratio. The molar ratio for IPA/TIP and IL/TIP was 30 and 3 respectively. Thermal annealing condition was 100 °C for 2 hours

Ionic liquid type	H_2O/TIP mole ratio	Crystallinity (%)	Crystal size (nm)
Without IL	2	No anatase peak	---
	5	No anatase peak	---
	10	No anatase peak	---
	100	No anatase peak	---
$[BuMIm]^+[PF_6]^-$	2	No anatase peak	---
	5	42.7	4.17
	10	36.7	4.10
	100	18.6	3.42
[BuMIm]+[BF4] -	2	No anatase peak	---
	5	57.0	4.12
	10	51.8	4.02
	100	21.0	4.06

X-ray photoelectron spectroscopy (XPS) and thermogravimetric analysis (TGA) data indicated that a higher hydroxyl group ratio content of hydroxylated titanium compound was obtained at medium H_2O/TIP molar ratio as well as the system containing hydrophilic $[BuMIm]^+[BF_4]^-$[12]. TEM results also indicated smaller particle size of TiO_2 was obtained at medium H_2O/TIP molar ratio case [12].

MICROWAVE IRRADIATION EFFECTS

Microwave (MW) heating can lead to a rapid temperature increase of chemical reactions [13]. Such rapid temperature increasing phenomena can affect the hydrolysis and condensation reaction as well as the anatase crystallization in ionothermal synthesis. This effect is dependent on ionic liquid concentration, microwave applied mode and initial water concentration.

When the ionothermal reaction was conducted under microwave irradiation, the reaction mixture could be heated up by two different mechanisms: the ion conduction mechanism and the dipolar polarization mechanism. With a much stronger heating effect than the dipolar polarization, the ion conduction mechanism is able to cause a marked increase in heating rate and attain higher reaction temperature [14].

In ionic liquid/water mixture system, the conductivity of the ionic liquid/water mixture system depends on the ionic liquid concentration [15]. In our study system, the conductivity of the ionic liquid solution increases sharply to a steady value at one molar ratio of IL/TIP [16]. Hence, the dipolar polarization mechanism will dominate at the higher molar ratio of IL/TIP and the ion conduction mechanism will operate when the microwave irradiation was conducted below the 1 molar ratio of IL/TIP. The highest reaction temperature was observed at the case of 1 molar ratio of IL/TIP. Highest annealing temperature results in highest anatase crystallinity.

In conventional ionothermal synthesis of TiO_2, high water content resulted in lower anatase crystallinity [12]. However, different results are obtained under microwave irradiation, where low water contents resulted in lower anatase crystallinity, as shown in Table 5. Hence specific microwave heating effects might exist, which could affect the anatase crystallization behavior of TiO_2. Consequently, three modes (F,S,L) with different microwave irradiation cycles were conducted and the their anatase crystallization characteristics were investigated. In the F mode, microwave irradiation was conducted for the full reaction cycle with an irradiation time of 30 minutes. The major difference in S and L irradiation modes is the initial temperature. It was set to a desired temperature for 10 minutes before microwave irradiation. In the S mode, the reactant solution was mixed and stirred for 10 minutes at 4 °C and then microwave-irradiated for 20 minutes. In the L mode, the reactant solution was mixed and stirred for 10 minutes at room temperature and then microwave-irradiated for 20 minutes. In these three modes, the molar ratio of H_2O/TIP was 5 and 100 respectively. The IPA/TIP molar ratio was fixed at 3, and the IL/TIP molar ratio was varied from 0 to 2.

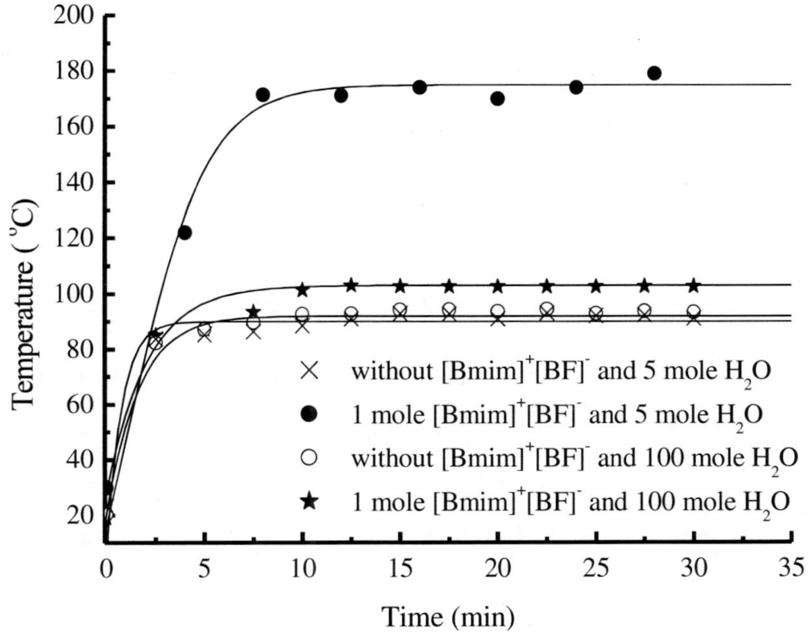

Figure 2. Temperature profiles of TiO_2 samples prepared with and without ionic liquid at two diffrent molar ratios of H_2O/TIP under microwave irradiation in F mode.

Figure 2 demonstrates the temperature profiles at different amounts of ionic liquid and two molar ratios of H_2O/TIP in F mode. The temperature was around or below 100°C at 100 molar ratios of H_2O/TIP irrespective of ionic liquid content. However the temperature increased much faster to 160°C at five molar ratios of H_2O/TIP. The occurrence of lower temperature at 100 molar ratios of H_2O/TIP is because of the fact that it is difficult to heat water to temperatures than 100°C in open vessels [14]. The data of Figure 2 and Table 5 suggests that the initial high temperature increase may have caused the lower anatase crystallinity at five molar ratios of H_2O/TIP.

XRD measurement was conducted for the samples prepared at different concentrations of $[Bmim]^+[BF_4]^-$ at five molar ratios of H_2O/TIP for each mode [16]. Table 6 lists measured anatase crystallinity of TiO_2 from the XRD data. The highest anatase crystallinity of TiO_2 was obtained for the lowest initial reaction temperature (L mode). Hence, the difference in anatase crystallinity of the final TiO_2 products for the three microwave irradiation modes may be due to the reaction temperature effect of the initial 10 minutes on the structure of TiO_2.

Table 5. Anatase crystallinity of TiO_2 samples prepared with different ionic liquid/TIP and H_2O/TIP molar ratios. Microwave irradiation was applied in the F mode

Ionic liquid/ TIP molar ratio	Crystallinity (%)	
	H_2O/TIP molar ratio	
	5	100
0	Amorphous	22
0.5	25	33
1	36	52
2	34	54

Table 6. Anatase crystallinity of samples prepared with different ionic liquid concentrations at five molar ratios of H_2O/TIP for three microwave irradiation modes

$[Bmim]^+[BF4]^-$ (mole)	Anatase crysatllinity (%)		
	L mode	S mode	F mode
No	Amorphous	Amorphous	Amorphous
0.5	72	45	25
1	83	49	36
2	43	33	34

Table 7. Fitting results from XPS curves of TiO_2 samples obtained from the initial 10 minutes for each mode

mode	Water/ TIP molar ratio	Ionic liquid/TIP molar ratio	O1s (Ti–O–Ti) area ratio (%)	O1s (Ti–OH) area ratio (%)
L	5	0	77	23
		1	73	27
		2	76	24
S	5	0	79	21
		1	75	25
		2	77	23
F	5	0	85	15
		1	76.5	23.5
		2	77.5	22.5

Hydroxylated titanium compounds were a key intermediate for forming anatase TiO_2 in ionothermal synthesis of nanostructured TiO_2 [8]. Table 7 demonstrates the XPS data and the calculated amounts of the hydroxylated titanium compound of the reaction products were from the initial 10 minutes for each mode. In S or L mode, a higher proportion of hydroxylated titanium compounds derived from hydrolysis of titanium precursors. High initial reaction temperatures in F mode can enhance the condensation of hydrolysis products. The hydroxylated titanium compound in F mode is resulted from ionic liquids inducing water dissolution of this higher molecular weight condensation product. The F mode obtained a lower amount of hydroxylated titanium. It could explain the lower anatase crystallinity in the F mode.

ONE-STEP IMMOBILIZATION ANATASE TiO_2 ON POLYMER MATERIALS

In photocatalytic water treatment applications, anatase TiO_2 can be directly suspended in aqueous phase or immobilized on a substrate support [17]. In the former applications, it needs an additional separation step which can increase the operation costs. Hence, the issue of immobilized TiO_2 on a substrate support has come to be an active research topic. Due to the high calcination temperature (generally above 400 °C) required to form anatase TiO_2 [18], high temperature-resistant inorganic materials, such as glass, steel, and the like, are required. This is an expensive and energy-intensive process. Accordingly, low cost organic polymer has been considered as an alternative immobilizing support material. Because of the thermal sensitivity property of organic polymer materials, this immobilization is generally carried out by a two-step process [19]. In this two-step process, the anatase TiO_2 nano-particles are first synthesized, and then spread on the surface of the polymer materials by an appropriate method. However, anatase TiO_2 nanoparticles are easily aggregated and difficult to disperse and the TiO_2 nanoparticles can be scratched from the polymer support material.

Based on the multiple ionic liquid characteristics, including structure-ordering, lower temperature anatase crystallization and high microwave absorption ability, a one-step immobilization of anatase TiO_2 nanoparticles on polymer material at ambient condition can be realized by microwave assisted ionothermal synthesis (MAIS) method.

The conceptual diagram of this one-step immobilization of TiO_2 on the polymer material is depicted in Figure 3. The polymer can be thermoplastic polymer such as PMMA and the like. Thermoplastic polymers are commonly considered as microwave transparent materials[20]. Hence only the polymer surface containing polar compounds such as water, ionic liquid, and alcohol, etc., are microwave superheated. The temperature of the polymer material can be kept below the polymer's glass transition temperature (T_g) (The T_g of PMMA is 116 °C). Consequently, the dimensional stability and mechanical properties of the whole polymer material are not affected.

In this immobilization technology, isopropyl alcohol acts as plasticizer or softener of PMMA [21]. The adsorption and diffusion of isopropyl alcohol into polymer producing a softened and swollen polymer surface [22]. As a result, the weak bond between the polymer chains was broken and then the formed tiny cavities leading to rough surface. The sol gel reaction occurs to form the TiO_2 nano-particles and then directly immobilized in the tiny cavities of the polymer surface. When TiO_2 nanoparticles are immobilized, the surface anatase crystallization of the TiO_2 nanoparticle can happen simultaneously.

In one typical experiment, the reaction mixture containing TIP, IPA, IL, H_2O and virgin PMMA chips was stirred and then microwave irradiated for 30 minutes. Atomic Force Microscope (AFM) is used to measure the roughness of material. The surface roughness can be quantitatively analyzed using AFM images [23]. Quantitative AFM pictures and data for the roughness average (Ra) of our softened sample treated by isopropyl alcohol for 30 minutes under microwave irradiation is shown in Figure 4, showing a Ra roughness of 80 nm. The SEM pictures and element distribution by Energy Dispersive Spectrometer (EDS) analysis of one immobilized TiO_2/PMMA chip after ultrasonic treatment indicate that the surface was covered with irregular TiO_2 clusters and contained about 20−33 wt% of Ti. The crystal phase of the immobilized TiO_2 was investigated by Raman spectroscopy. The Raman spectra of virgin PMMA and immobilized TiO_2/PMMA chip indicated that the surface of immobilized TiO_2 shows the anatase characterization [24]. The PMMA chips after immobilization are shown in Figure 5. The immobilization strength between the TiO_2 nanoparticles and the polymer material is still retained after 30 minutes of ultrasonic treatment.

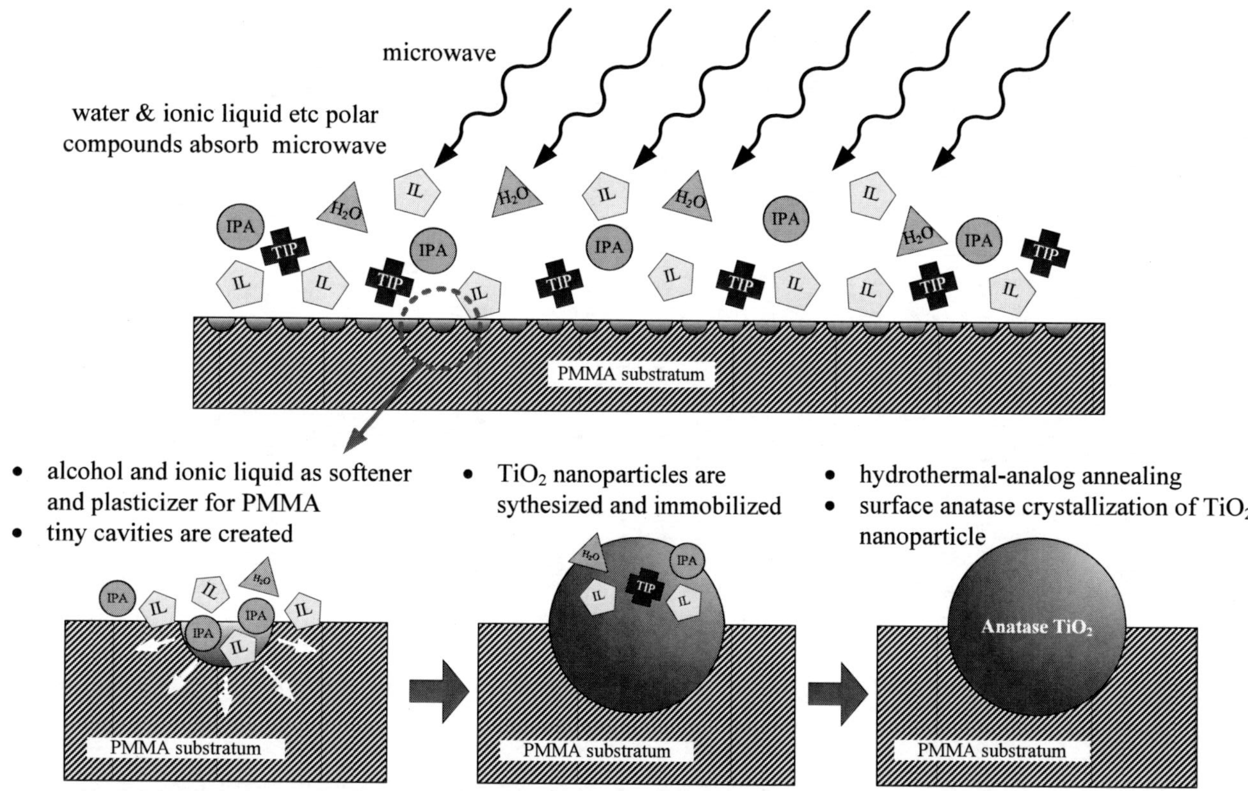

Figure 3. Conceptional diagram of one-step immobilization of TiO_2 nanoparticles on a thermoplastic polymer by ionothermal method under microwave irradiation.

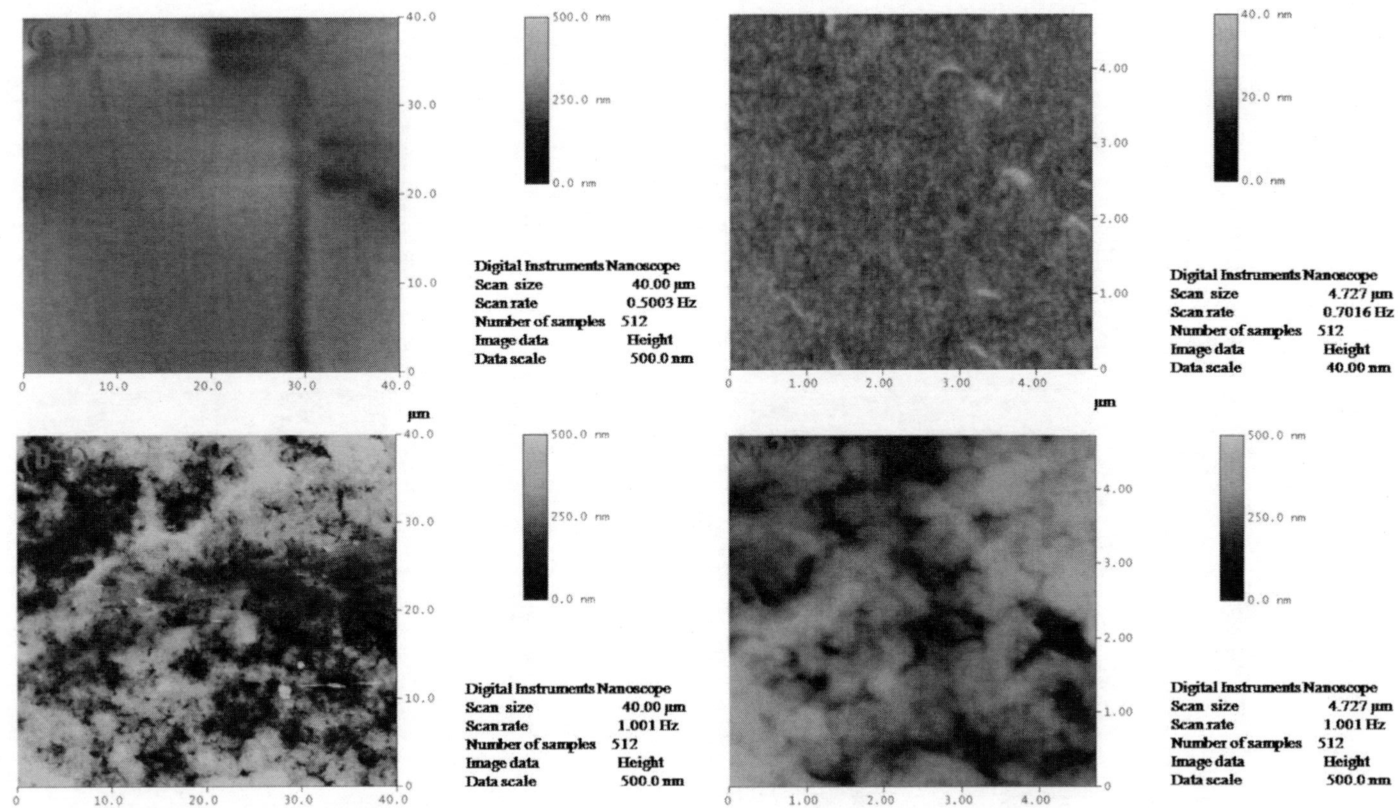

Figure 4. (Continued)

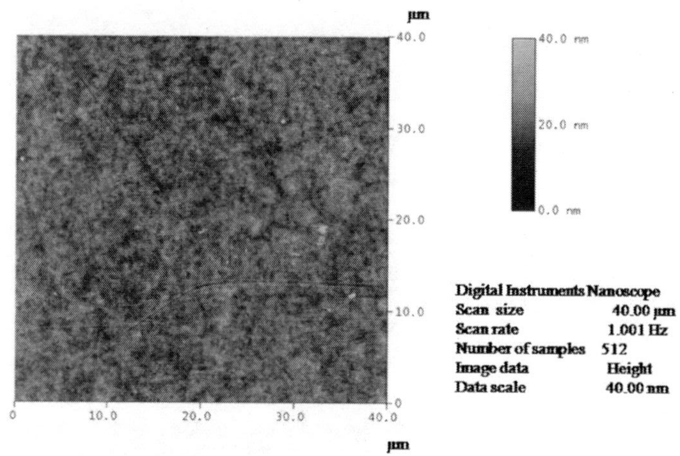

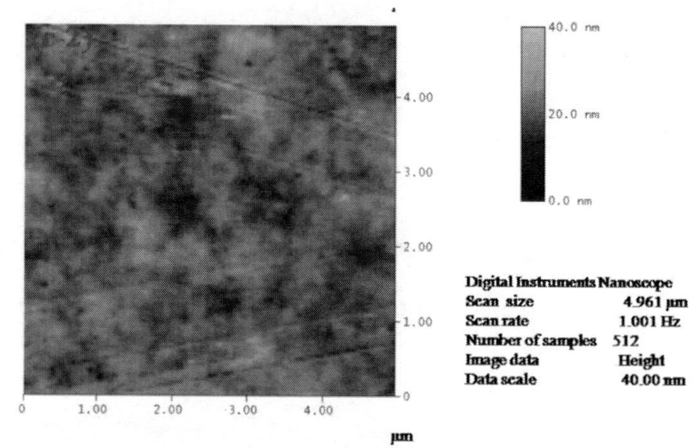

Figure 4. AFM phase images of the PMMA chip mixing with IPA or ionic liquid under microwave heating for 30 minutes, a-1) and a-2): virgin PMMA at different magnification levels, b-1) and b-2): with IPA after microwave heating at different magnification levels, c-1) and c-2): with ionic liquid after microwave heating at different magnification levels.

IONOTHERMAL SYNTHESIS OF NANOSTRUCTURED ANATASE TiO_2/AC COMPOSITE

A nanostructured anatase TiO_2/AC composite can be prepared by a two-step MAIS method. This nanostructured anatase TiO_2/AC composite can be used as electrode material for CDI application.

The CDI has been considered as a green technology for low strength brackish water desalination due to its low energy consumption and no requirement of chemical cleaning process [25]. The principle of CDI is that the cations and anions in a salt solution can be adsorbed in the electrical double layer near the charged electrode surface. The ions of salt solution are then removed from the water stream [26]. AC has been widely studied as a CDI electrode material. However, the major drawback of AC is that an irreversible physical adsorption can be happened to lower the electrical double-layer capacitance of the AC electrode and its CDI efficiency (electrosorption capability). It was found that the electrosorption capability of the AC electrode can be enhanced by incorporating TiO_2 [27]. Among the TiO_2 polymorphs, the scientific research of anatase has increased recently due to its interesting electrochemical properties for energy application [28]. It is also known that the nanostructured material is characterized by a short electron transport path and high surface area. Combination of these two features makes the nanostructured anatase TiO_2 or its composite attractive to numerous electrochemical electrode applications [29].

Nanostructured anatase TiO_2/AC composite can be prepared by two methods. One is the direct blending of nanostructured anatase TiO_2 with AC. The other is the sol-gel method. In the former method, the aggregation of anatase TiO_2 nanoparticles is a severe problem. As to the conventional sol-gel method, a high temperature (above 400 °C) and long post-calcination time (over 12 hours) are required to obtain the anatase crystalline phase. In addition, this energy-consuming process must be conducted under an inert or vacuum condition otherwise the AC could be oxidized and broken down[30, 31]. Therefore, there is a need to develop a low temperature sol-gel method for preparing the nanostructured anatase TiO_2/AC composite. This wish can be achieved by two step MAIS method.

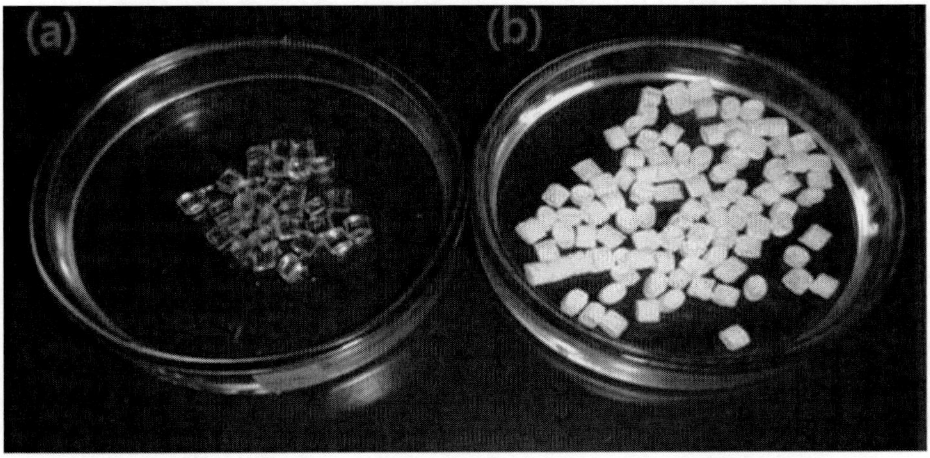

Figure 5. Pictures of virgin (left, a) and immobilized (right, b) PMMA chips, with a TIP: IPA: IL: H_2O=1: 3: 1: 100 molar ratio.

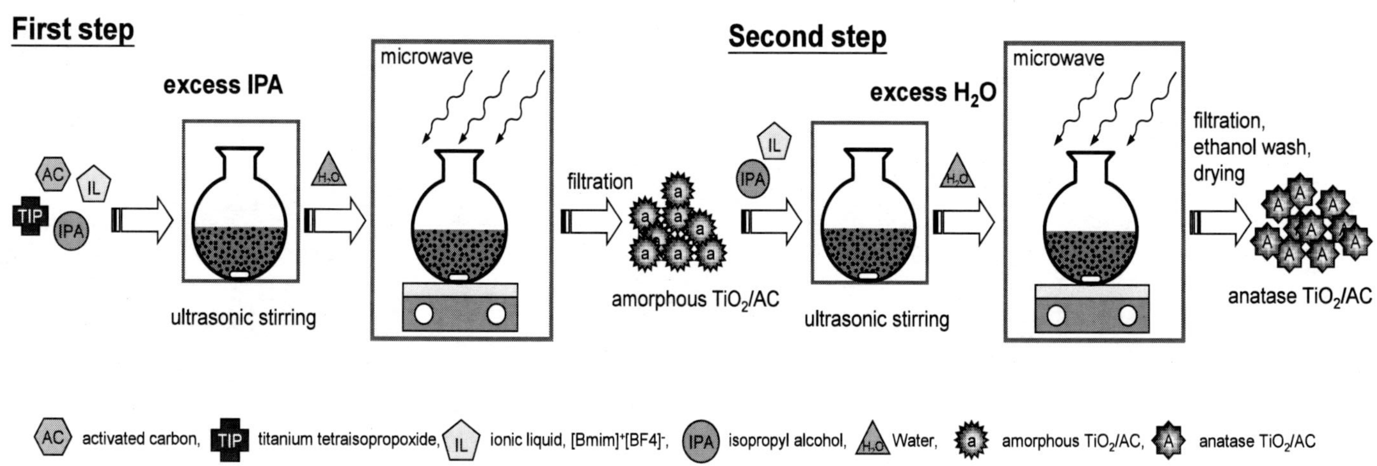

Figure 6. Microwave-assisted ionothermal synthesis of the nanostructured anatase TiO_2/AC composite.

NANOSTRUCTURED ANATASE TiO_2/AC COMPOSITE PREPARED BY THE TWO STEP MAIS METHOD

It includes a controlled sol-gel reaction step and a subsequent IL-inducing crystallization step. The flow chart of this two-step MAIS method is shown in Figure 6 [32]. In a typical experiment, the TIP/H_2O molar ratio was set at 1:30 at a constant TIP/IL molar ratio of 1:1. In the reaction step, the TiO_2 products formed from sol-gel reaction can be loaded in the AC matrix. In the crystallization step, the filtrated solid product from reaction step was treated in an IL/H_2O mixture under microwave irradiation. The molar ratio of IL and H_2O was set at 1:18.

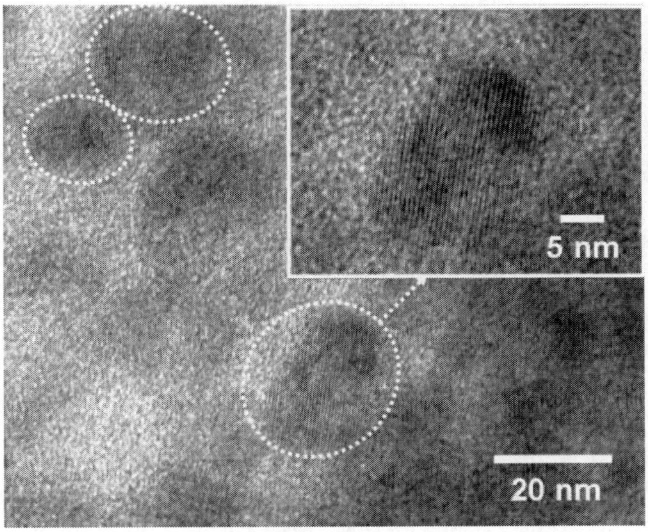

Figure 7. TEM micrograph of the nanostructured anatase TiO_2/AC composite and the inset shows the crystalline size of anatase TiO_2 (at TIP/H_2O molar ratio of 1:30).

The phase of the deposited TiO_2 in TiO_2/AC composite was investigated by Raman spectroscopy. A strong main anatase peak is observed at 147.3 cm^{-1} for the sample from crystallization step [32]. The TEM picture of the anatase TiO_2/AC composite is shown in Figure 7. The magnified image indicates the crystalline size of the anatase TiO_2 particle is around 7-12 nm. In addition, BJH data indicates the average pore diameter AC after modification is 2.23 nm. It implies that the pore blocking of AC by the TiO_2 particles is not significant.

SPECIFIC CAPACITANCE OF THE FABRICATED TiO_2/AC COMPOSITE ELECTRODES

The specific capacitance of the electrode fabricated from the TiO_2/AC composite prepared by MAIS method was measured by performing Cyclic Voltammetry (CV). For comparison, the CV of Degussa-P25-TiO_2/AC composite electrode was also measured. This composite was obtained by the direct blending of Degussa-P25-TiO_2 and AC powder. Figure 8 showed the specific capacitance of these TiO_2/AC composite electrodes in 0.5 M NaCl electrolyte solution at a scan rate of 10 mV s^{-1}. The CVs of these TiO_2/AC composite electrodes are rectangular-like. An ideal double-layer capacitive behavior is observed for these TiO_2/AC composite electrodes. Table 8 lists the measured specific capacitance for each TiO_2/AC composite electrode. The results of Table 8 indicate that the nanostructured anatase TiO_2/AC composite electrode prepared by MAIS method has the highest specific capacitance.

Table 8. The specific capacitance of TiO_2/AC composite electrodes with different sources of TiO_2 in 0.5 M NaCl electrolyte (at TIP/H_2O molar ratio of 1:30)

Electrode material	Specific capacitance/F g^{-1}	Increment /%
AC	44.9	-
Degussa-P25 TiO_2/AC	45.9	2.2
amorphous TiO_2/AC	47.5	5.8
anatase TiO_2/AC	51.5	14.7

The specific capacitance of the three different nanostructured anatase TiO_2/AC composite electrodes was investigated. They were prepared with different molar ratios of TIP/H_2O. Figure 9 shows the CVs of these TiO_2/AC composite electrodes in 0.5 M NaCl electrolyte solution at a scan rate of 10 mV s^{-1}. The incorporated amounts of TiO_2 in composite were measured by TGA. Table 9 lists the incorporated amounts of TiO_2, electrode resistivity and the specific capacitance of the pristine AC, as well as the three TiO_2/AC composite electrodes. The results of Table 9 indicate that the highest specific capacitance and the lowest electrode resistivity were observed for the TIP/H_2O molar ratio of 1:15. However, the highest resistivity was obtained for the TIP/H_2O molar ratio of 1:30. This is owing to the fact that the amount of TiO_2 increases with increasing water concentration, and therefore increases the resistance of electron transfer to the surface of AC [33].

In AC electrode, the surface texture and conductivity of an AC electrode can be changed after the TiO_2 modification. Different structural characteristics of TiO_2 have different modification effects. XPS was used to examine the structural characteristic of the TiO_2 of our nanostructured anatase TiO_2/AC composite. Table 10 lists that the XPS data and the calculated amounts of the Ti-OH and Ti-O of each TiO_2/AC composite. They were varied with the molar ratio of TIP/H_2O and the highest amount of Ti-OH was observed for the molar ratio of 1:15. Higher amount of the Ti-OH can improve the surface wetting ability of the AC electrode, thereby facilitating the access ability of solution ions to the surface of the AC electrode and a higher uptake capacity of the ions on the electrode. Meanwhile, the resistivity measurement also indicates that the lowest resistivity value for the molar ratio of 1:15. These two results lead to that this composite electrode has the highest specific capacitance.

Table 9. The electrode resistivity and specific capacitance of the pristine AC and nanostructured anatase TiO_2/AC composite electrodes prepared with different TIP/H_2O molar ratios

TIP/H_2O (molar ratio)	TiO_2 content /%	Electrode resistivity /μΩ cm	Specific capacitance/ Fg^{-1}
			0.5 mol L^{-1} NaCl
AC	0	80.8	44.9
1:5	4.65	71.3	66.4
1:15	5.41	65.6	84.7
1:30	7.62	91.1	51.5

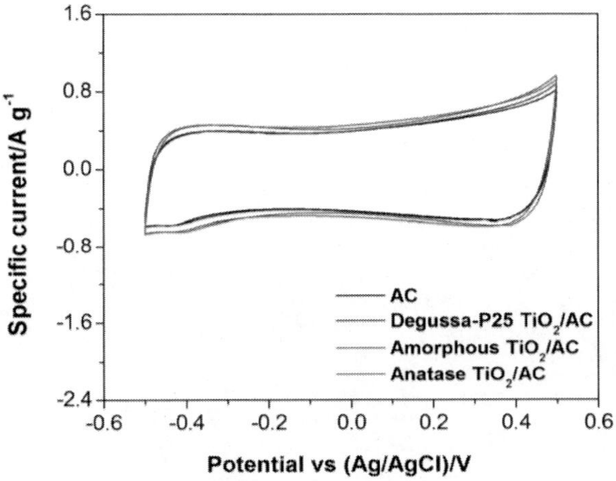

Figure 8. The specific capacitance of the TiO_2/AC composite electrodes with different sources of TiO_2 in 0.5 M NaCl electrolyte (at TIP/H_2O molar ratio of 1:30).

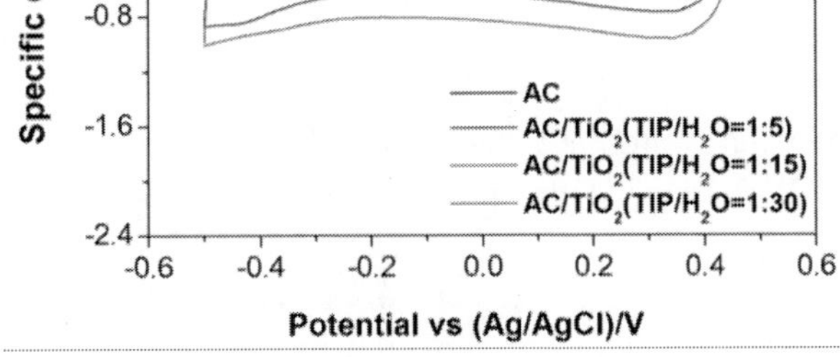

Figure 9. Cyclic voltammograms of the pristine AC electrode and the nanostructured anatase TiO_2/AC composite electrodes at a scan rate 10 mV s-1 in 0.5 M NaCl electrolyte.

The surface morphology of this nanostructured anatase TiO_2/AC composite was characterized by SEM, as shown in Figure 10. It can be seen that the TiO_2 nanoparticles were mounted on the surface of AC. The TEM picture and its EDS map of the TiO_2/AC composite demonstrate that the TiO_2 nanoparticles are evenly deposited within the AC matrix [32]. This is to be expected due to the formation of hydrogen bond between the Ti-OH of the anatase TiO_2 particle and the hydrophilic IL [34]. In such case, better dispersion of the anatase TiO_2particles in the matrix of AC is obtained.

Table 10. XPS analysis of three nanostructured anatase TiO_2/AC composite materials prepared with different TIP/H_2O molar ratios

TIP/H_2O (molar ratio)	O-C=O/%	C-OH/%	Ti-OH/%	Ti-O/%
1:5	17.9	52.4	18.5	11.2
1:15	14.9	49.8	22.3	13.0
1:30	17.8	41.8	19.3	21.1

Figure 10. SEM micrograph of the pristine AC (a) and the nanostructured anatase TiO_2/AC composite (b) (at TIP/H2O molar ratio of 1:15).

WATER TREATMENT APPLICATIONS

Facing to world water shortage and the rising need for clean water, previously unusable water sources such as brackish water, seawater, and other wastewater can be as an available source of water supply. Before using these water sources, some technologies are need to produce high quality water. Two water treatment technologies, including photocatalytic water treatment and water desalination technology are often mentioned. For photocatalytic water treatment, it is mainly utilized for the decomposition of trace organic compounds as well as destruction of viruses and bacteria. Water desalination technology is used to remove dissolved minerals. The water treatment performance of anatase TiO_2 and its composite prepared by our ionothermal synthesis system are reported

PHOTOCATALYTIC PERFORMANCE OF TiO_2 AND THE IMMOBILIZED TiO_2

Figure 11 shows the photodegradation of methylene blue in aqueous suspension containing TiO_2 sample treated without and with ionic liquid under UV radiation. A good photodegradation performance was obtained for sample treated with $[BuMIm]^+[BF_4]^-$case [8].

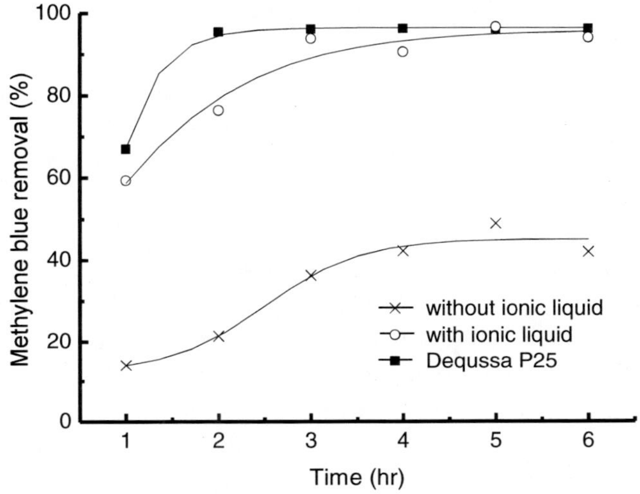

Figure 11. The photodegradation of methylene blue in aqueous suspension containing TiO_2 sample treated without and with ionic liquid under UV irradiation.

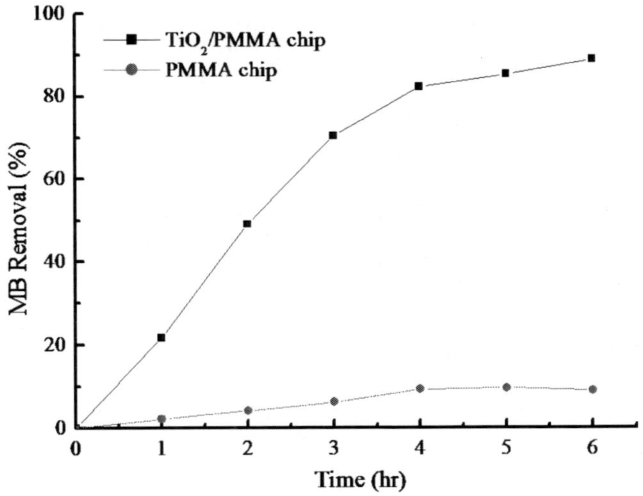

Figure 12. Photodegradation of methylene blue in aqueous solution containing PMMA chips with or witohut immobilized TiO_2, with a TIP: IPA: IL: H2O=1: 3: 1: 100 molar ratio under UV irradiation.

The photodegradation of methylene blue aqueous solution by immobilized TiO_2 was carried out as follows [24]. First, 300 mL of aqueous solution with 5 ppm methylene blue was prepared. Several immobilized TiO_2/PMMA chips were placed in this solution and irradiated by UV. The UV light was applied with a wavelength of 365 nm and a light intensity of 13 W.

The total irradiation time was 6 hours. An aliquot sample was taken for analysis every 1 hour, and the results are shown in Figure 12. The immobilized TiO_2 sample prepared by this novel method has significant removal efficiency of the target compound to over 80% after four hours.

CDI PERFORMANCE OF THE NANOSTRUCTURED ANATASE TIO_2/AC COMPOSITE ELECTRODE

The CDI performance of one nanostructured anatase TiO_2/AC composite electrode was studied. This composite was prepared with the TIP/H_2O molar ratio of 1:15. The CDI performance is evaluated by its desalting efficiency. It is calculated according to the difference between the influent and the effluent conductivity. The electrosorption capacity was equal to the concentration change of NaCl solution divided by the weight of active material. The CDI test was conducted by repeating five charging (1.2 V) and discharging (0.0 V) processes in 100 mg l^{-1} of NaCl solution. Figure 13 shows the ion adsorption/desorption cycles of the pristine AC electrode and the nanostructured anatase TiO_2/AC composite electrode. Table 11 lists the average electrosorption capacity of these two electrodes for the 5 cycle tests. It can be seen that the CDI performance was improved for the nanostructured anatase TiO_2/AC composite electrode. Moreover the decay of its electrosorption capacity is insignificant. This means that the ion electrosorption process of the nanostructured anatase TiO_2/AC composite electrode is a reversible process.

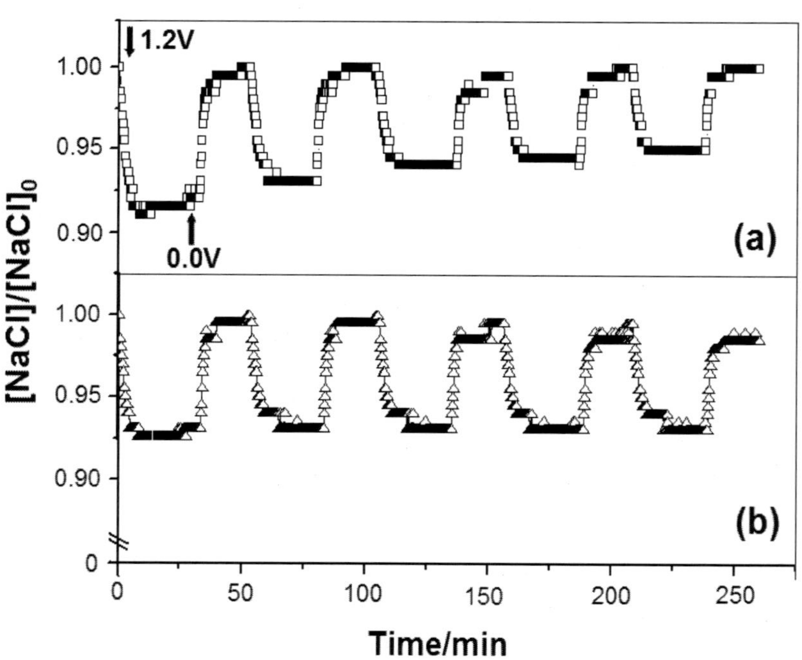

Figure 13. The adsorption-desorption cycles of AC electrode (a) and the nanostructured anatase TiO_2/AC electrode (b) at $[NaCl]_0$ of 100 mg l-1 and applied potential difference of 1.2 V (at TIP/H_2O molar ratio of 1:15).

Table 11. Electrosorption capacity of the pristine AC and nanostructured anatase TiO_2/AC composite electrodes (at TIP/H_2O molar ratio of 1:15)

Electrode material	NaCl electrosorption capacity/mg g^{-1}
AC	5.43±0.91
TiO_2/AC	8.05±0.34

REFERENCES

[1] Burnside, S. D., Shklover, V., Barbě, C., Comte, P., Arendse, F., Brooks, K. & Grãtzel, M. (1998). *Chem. Mater.*, *10*, 2419.

[2] Antonietti, M., Kuang, D., Smarsly, B. & Zhou, Y. (2004). *Angew. Chem. Int. Ed.*, *43*, 4988.

[3] Zhou, Y. (2005). *Curr. Nanosci. 1*, 35.

[4] Shin, H., Jung, H. S., Hong, K. S. & Lee, J. K. (2005). *J. Solid State Chem.*, *178*, 15.

[5] Liu, W., Zhao, T., Zhang, Y., Wang, H. & Yu, M. (2006). *J. Solu. Chem.*, *35*, 1337.

[6] Liu, Y. H., Lin, C. W., Chang, M. C., Shao, H. & Yang, A. C. M. (2007). (Europacat VIII). 26-31 August 2007 Turku/Åbo, Finland.

[7] Cooper, E. R., Andrews, C. D., Wheatley, P. S., Webb, P. B., Wormald, P. & Morris, R. E. (2004). *Nature*, *430*, 1012.

[8] Liu, Y. H., Lin, C. W., Chang, M. C., Shao, H. & Yang, A. C. M. (2008). *J. Mater Sci.*, *43*, 5005.

[9] Wang, Y., Li, H., & Han, S. (2006). *J. Phys. Chem. B*, *110*, 24646.

[10] Bessekhouad, Y., Robert, D. & Weber, J. V. (2003). *J. Photochem. Photobiol. A. Chem.*, *157*, 47.

[11] Livage, J., Henry, M. &, Sanchez, C. (1988). *Prog. Solid State Chem.*, *18*, 259.

[12] Liu, Y. H., Chang, M. C., Shao, H., Huang, M. S. & Yang, A. C. M. (2010). *J. Mater. Sci.*, *45*, 369.

[13] Tompsett, G. A., Conner, W. C. & Yngvesson, K. S. (2006). *ChemPhysChem*, *7*, 296.

[14] Dallinger, D. &, Kappe, C. O. (2007). *Chem. Rev.*, *107*, 2563.

[15] Jarosik, A., Krajewski, S. R., Lewandowski, A. & Radzimski, P. (2006). *J. Mol. Liq.*, *123*, 43.

[16] Liu, Y. H., Liu, P. I., Chung, L. C., Shao, H., Huang, M. S., Horng, R. Y., Yu, S. W., Yang, A. C. M. & Chang, M. C. (2011). *J. Mater. Sci.*, *46*, 4826.

[17] Gaya, U. I., Abdullah, A. H. J. (2008). *Photochem. Photobiol. C. Photochem. Rev.*, *9*, 1.

[18] Chen, Y. & Dionysiou, D. D. (2007). *Appl. Catal. A: Gen.*, *317*,129.

[19] Daoud, W. A. & Xin, J. H. J. (2004). *Sol-Gel Sci. Technol.*, *29*, 25.

[20] Weinheim, Microwaves in Organic Synthesis (Andre Loupy Ed.), Wiley-VCH, 2006

[21] Scott, M. P., Rahman, M. & Brazel, C. S. (2003). *Eur. Polym. J.*, *39*, 1947.

[22] Jeon, J. Y. & Kim, H. Y. (2000). *Eur. Polym. J.*, *36*, 895.

[23] Walther, F., Heckl, W. M. & Stark, R. W. (2008). *Appl. Surf. Sci.*, *254*. 7290.

[24] Liu, Y. H., Chang, M. C., Liu, P. I., Chung, L. C., Shao, H., Huang, M. S., Horng, R. Y. & Yang, A. C. M. (2010). *J. Mater. Sci.*, *45*, 6212.

[25] Oren, Y. (2008). *Desalination*, *228*, 10.

[26] Li, H. B., Pan, L. K., Nie, C. Y. & Sun, Z. (2012). *J. Nanosci. Lett.*, *2*, 1.

[27] Ryoo, M. W., Kim, J. H. & Seo, G. (2003). *J. Colloid Interf. Sci.*, *264*, 414.
[28] Su, X., Wu, Q.L., Zhan, X., Wu, J., Wei, S. & Guo, Z. (2012). *J. Mater. Sci.*, *47*, 2519.
[29] Das, S. K., Darmakolla, S. & Bhattacharyya, A. J. (2010). *J. Mater. Chem.*, *20*, 1600.
[30] Huang, D., Miyamoto, Y., Matsumoto, T., Tojo, T., Fan, T., Ding, J. Guoa, Q. & Zhang, D. (2011). *Sep. Purif. Technol.*, *78*, 9.
[31] Tatsuda, N., Itahara, H., Setoyama, N. & Fukushima, Y. (2005). *Carbon*, *43*, 2358.
[32] Liu, P. I., Chung, L. C., Shao, Liang, T. M., Horng, R. Y. & C. C. M. M, M. C. Chang. (2013). *Electrochim. Acta*, *96*, 173– 179.
[33] Liang, H., Chen, F., Li, R., Wang, L. & Deng, Z. (2004). *Electrochim. Acta*, *49*, 3463.
[34] Wittmar, A. & Ulbricht, M. (2012). *Ind. Eng. Chem. Res.*, *51*, 8425.

INDEX

A

B

C

D

E

F

G

H

I

J

K

L

M

N

O

P

Q

R

S

T

U

V

W

X

Y

Z